AF573824

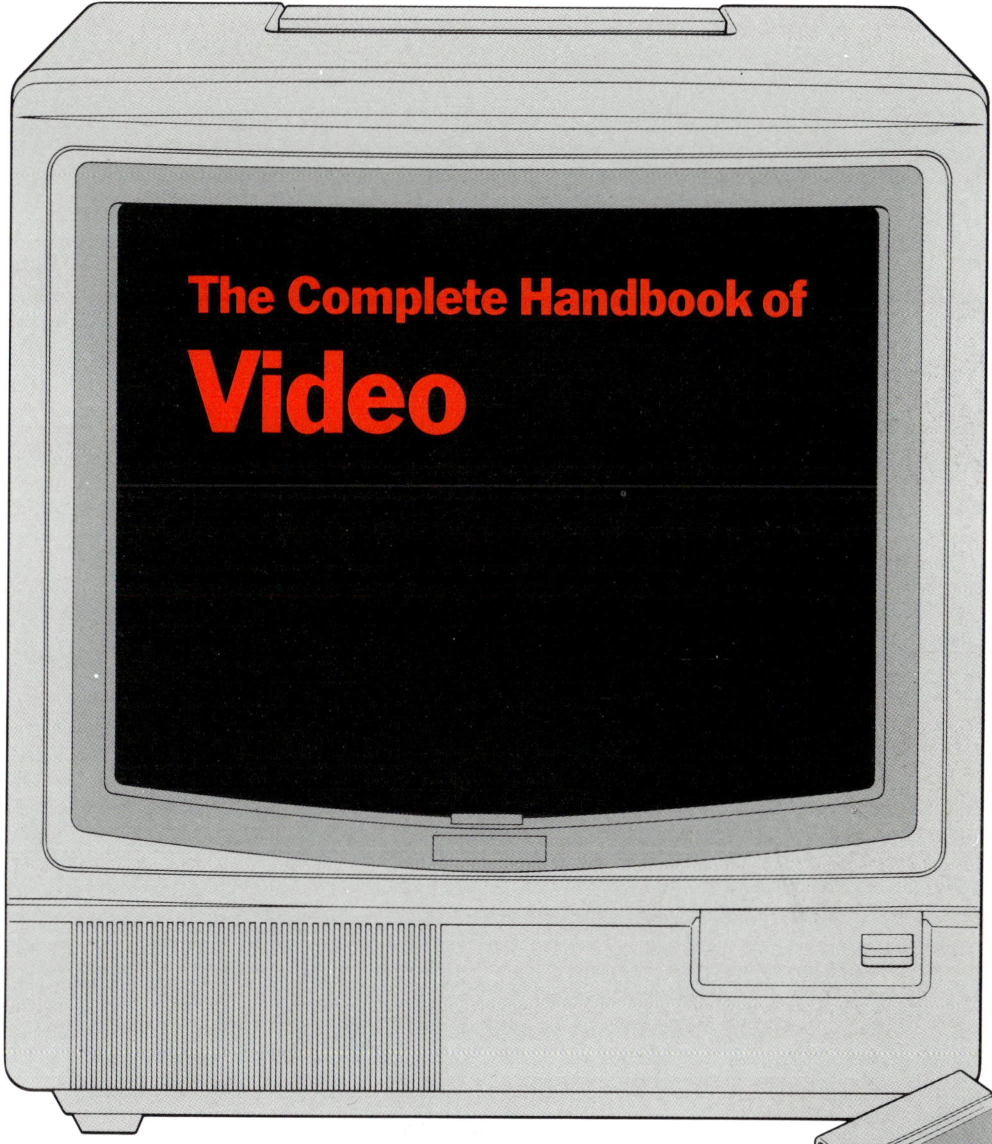
The Complete Handbook of
Video

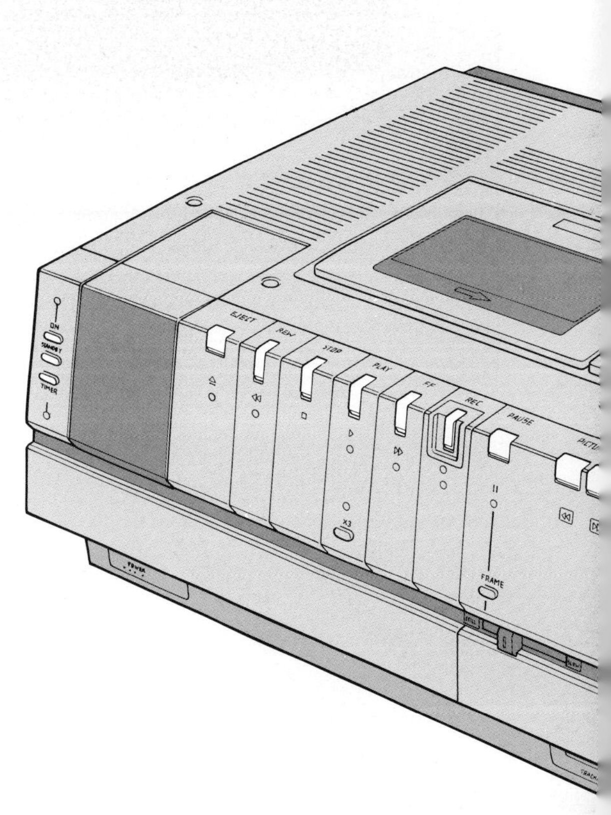
ON
TIMER
EJECT
REW
STOP
PLAY
FF
REC
PAUSE
X3
FRAME

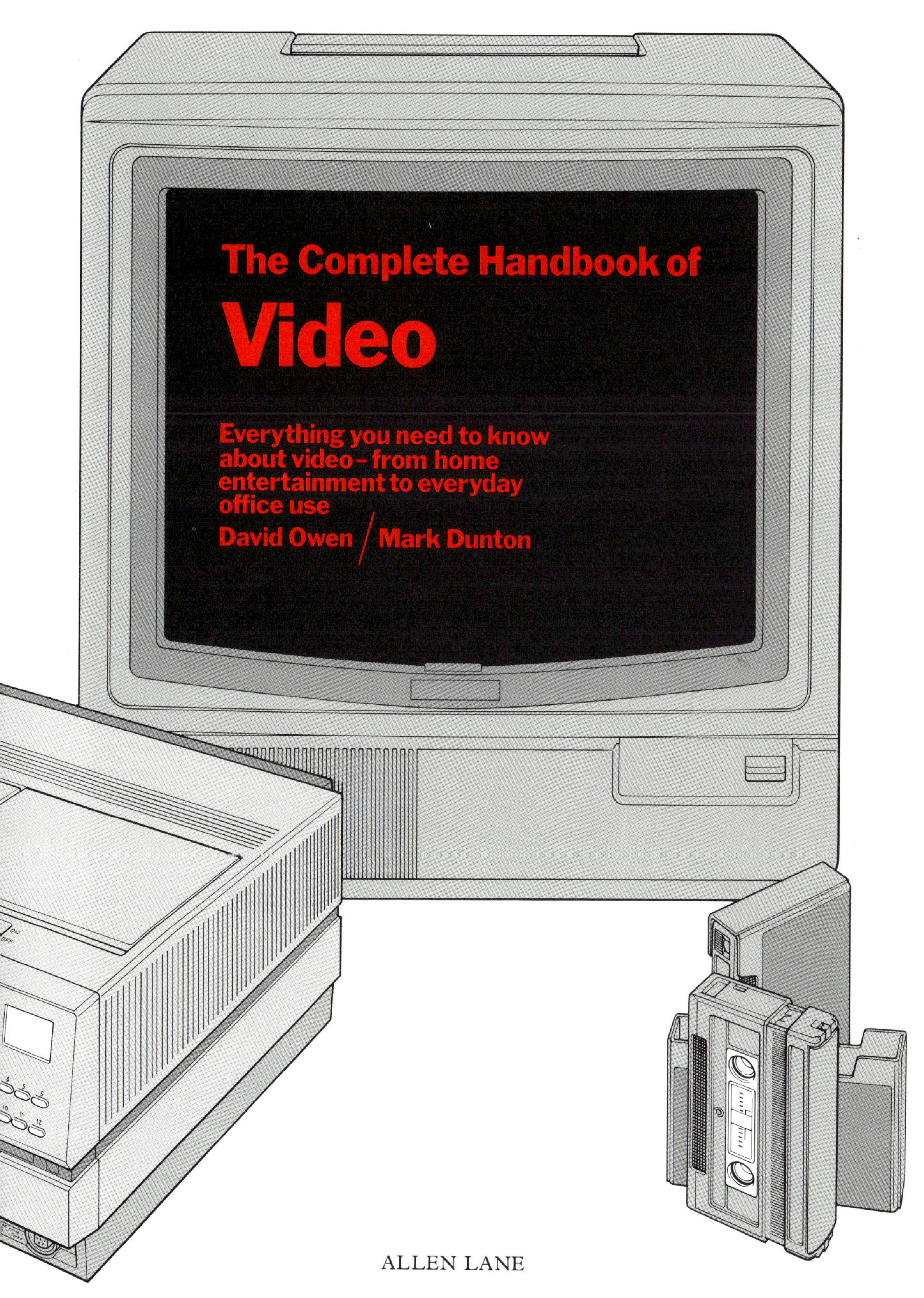

The Complete Handbook of Video

Everything you need to know about video – from home entertainment to everyday office use

David Owen / Mark Dunton

ALLEN LANE

ALLEN LANE
Penguin Books Ltd.,
536 King's Road,
London SW10 0UH
First published 1982
Published simultaneously by Penguin Books

Copyright © Marshall Editions Ltd., 1982

All rights reserved. No part of this publication may be reproduced, stored in a retrieval system, or transmitted in any form or by any means, electronic, mechanical, photocopying, recording or otherwise, without the prior permission of the copyright owner.

ISBN 0 7139 1493 9
Set in Plantin 110
Typeset by Servis Filmsetting Ltd.,
Manchester, U.K.
Reproduced by Gilchrist Bros. Ltd.,
Leeds, U.K.
Printed and bound in Belgium by Brepols S.A.

Edited and designed by
Marshall Editions Ltd
71 Eccleston Square
London SW1V 1PJ

Editor: Helen Varley
Assistant Editor: Helen Armstrong
Art Director: Paul Wilkinson
Artwork: Hayward & Martin
Retouching and Make-up Roy Flooks
Picture researchers: Helen Armstrong;
Mary Corcoran

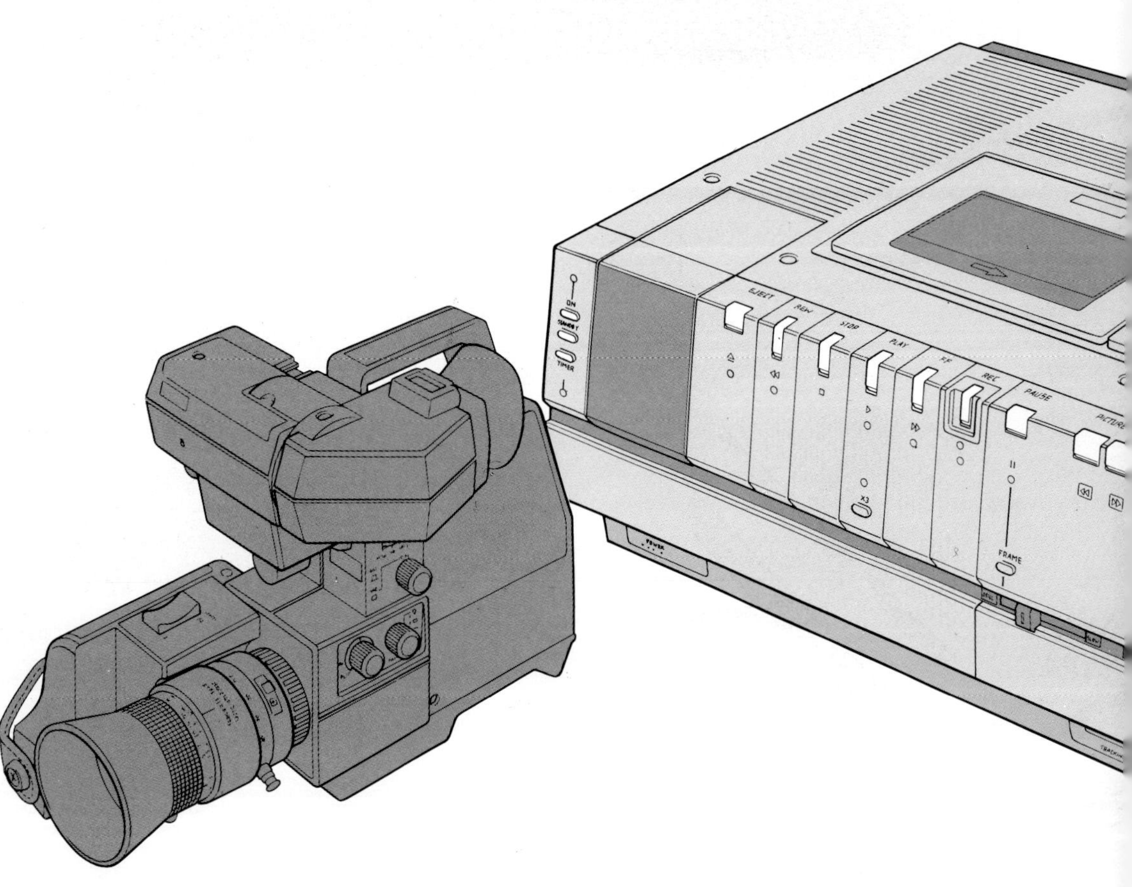

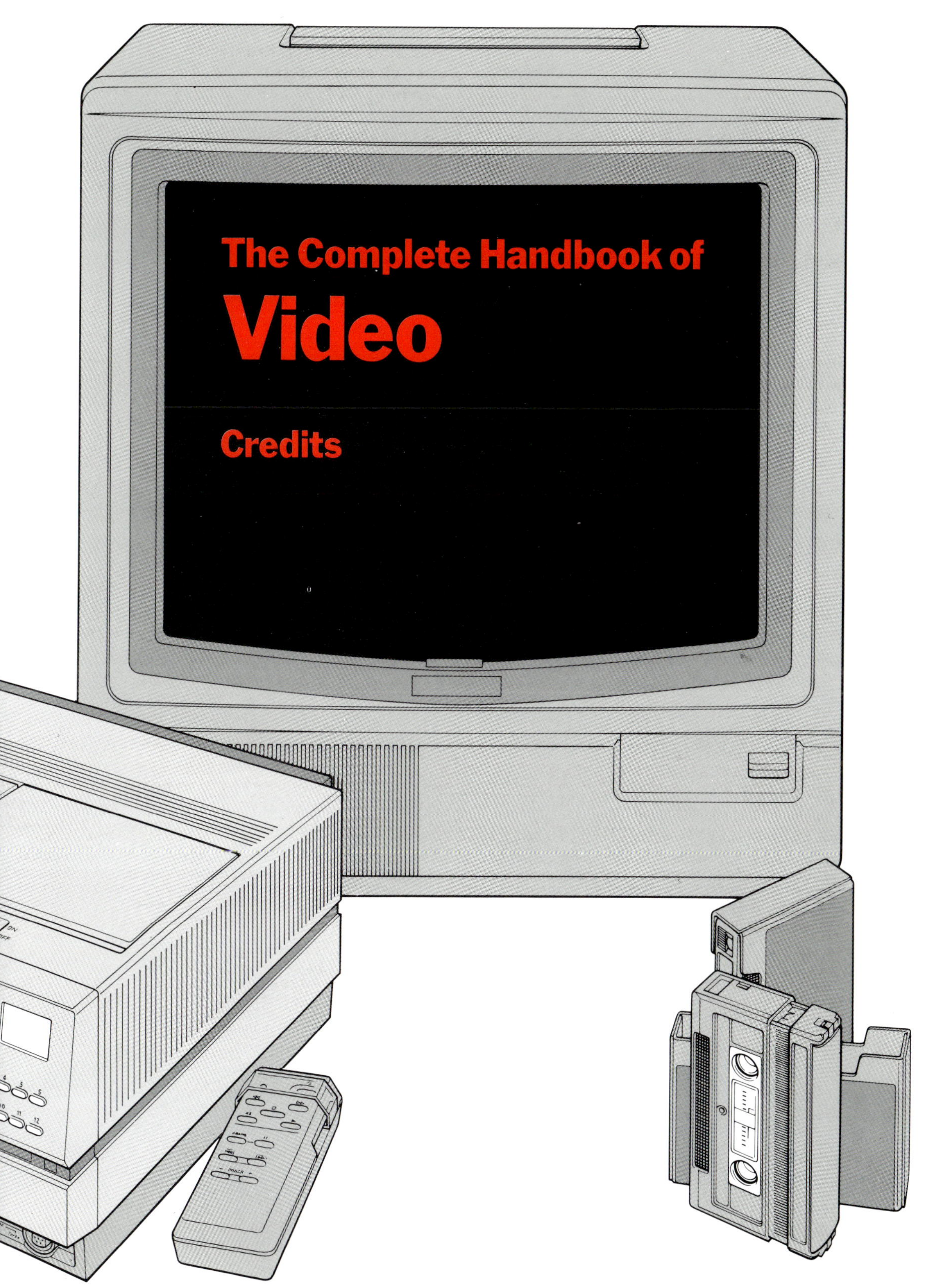
The Complete Handbook of
Video
Credits

continued overleaf ⟶

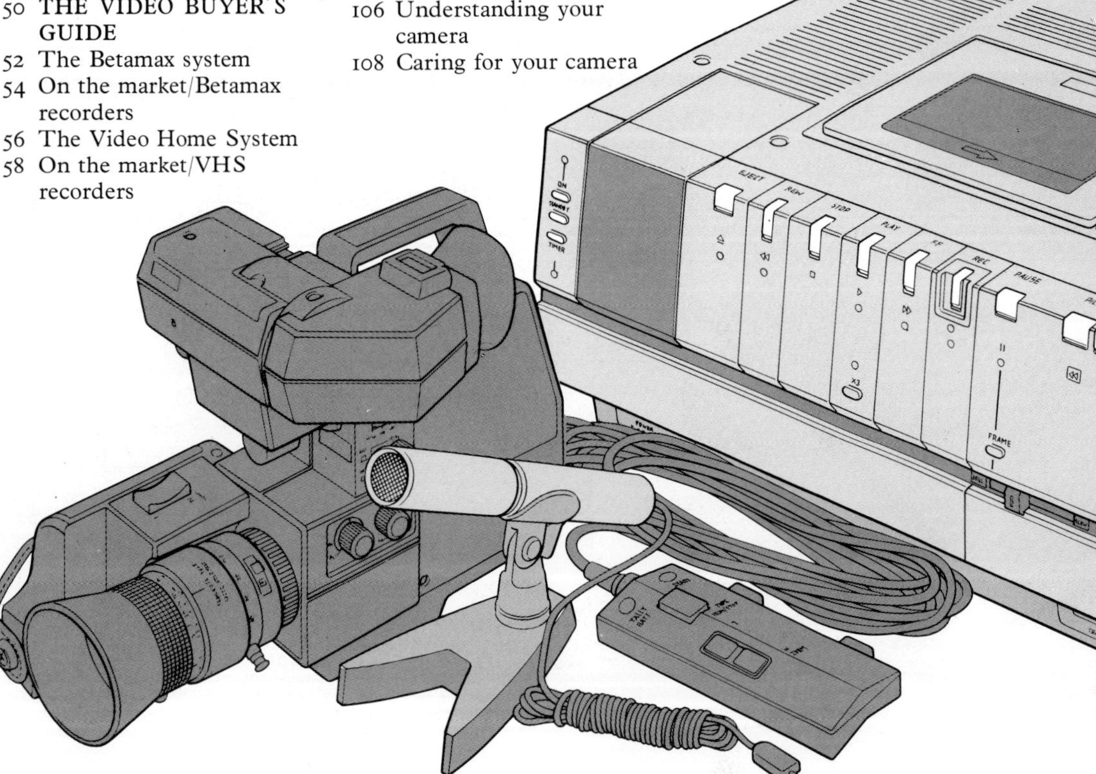

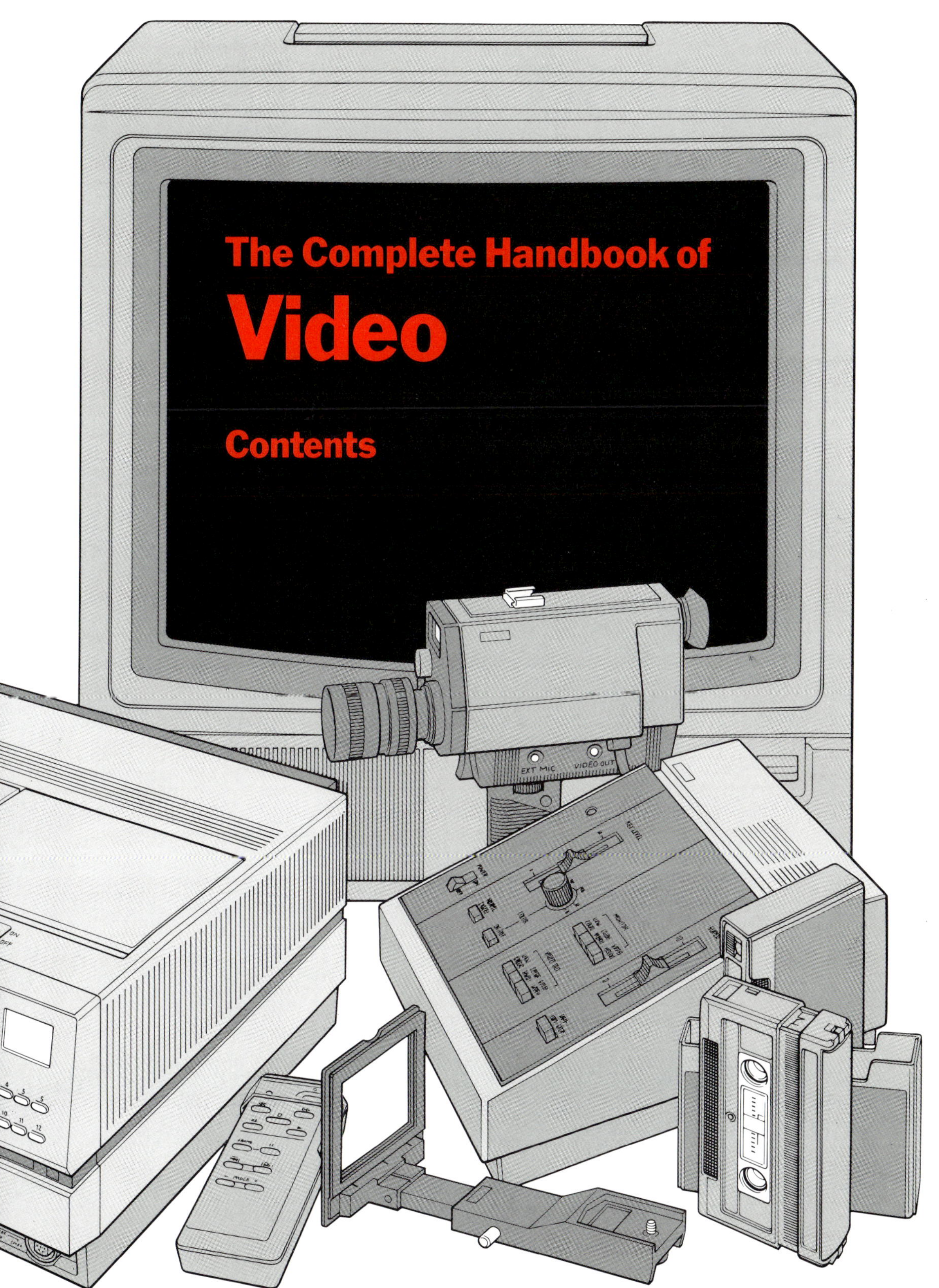
The Complete Handbook of
Video
Contents
EXT MIC
VIDEO OUT
ON
OFF

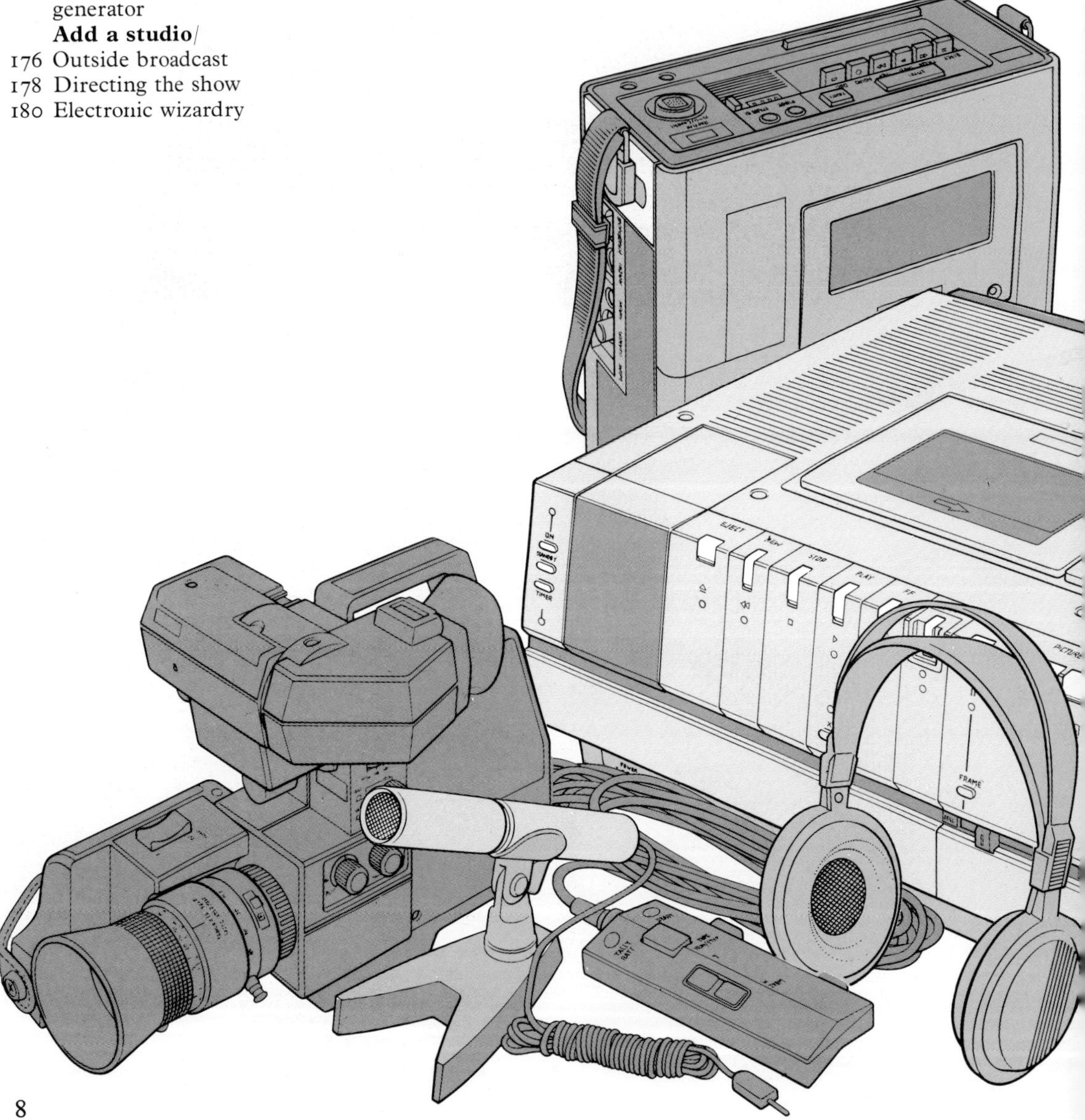

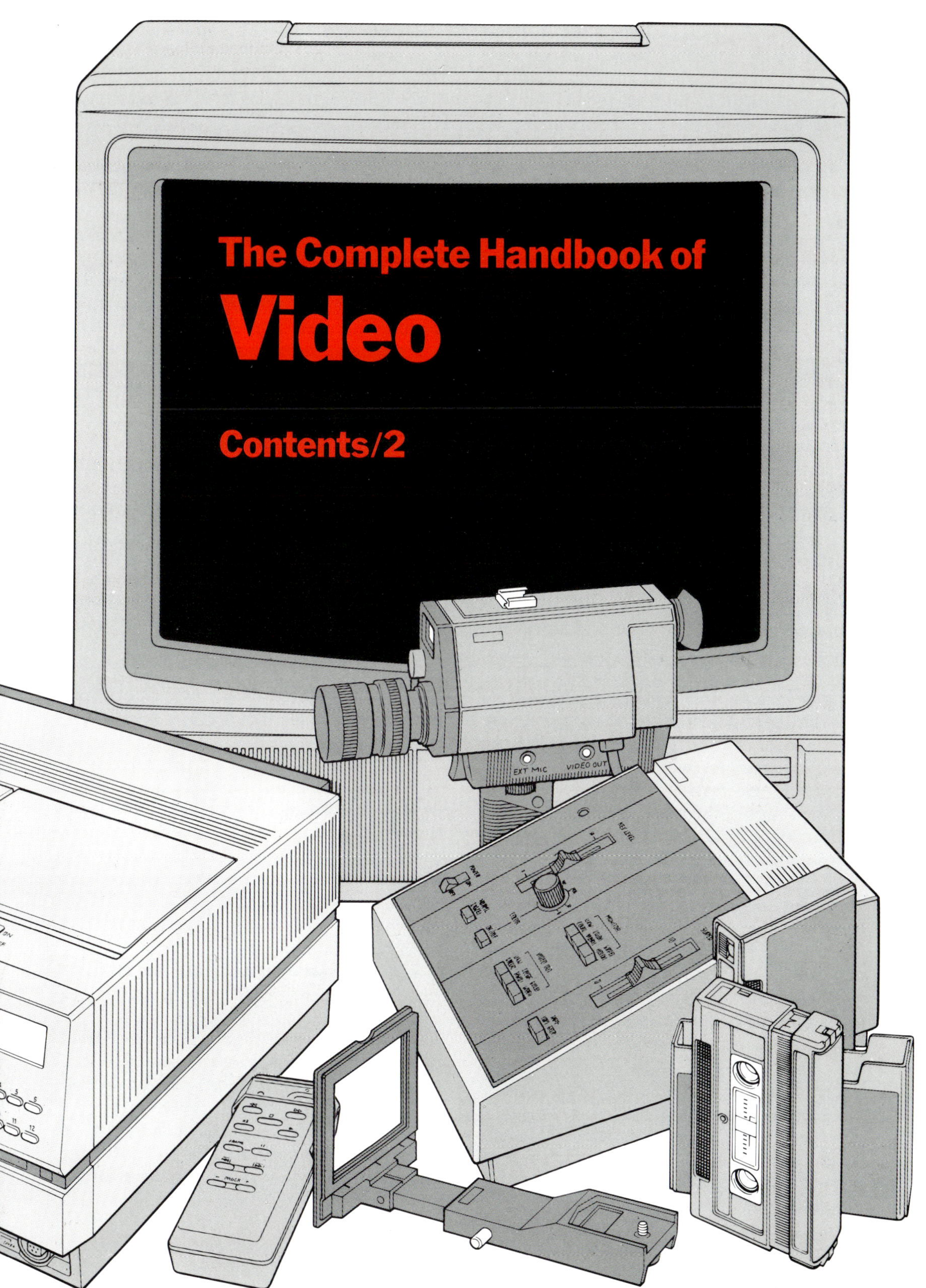
The Complete Handbook of
Video
Contents/2
EXT MIC
VIDEO OUT
ON
OFF

The video user's guide

Video is a word which has become so familiar, and so fashionable, that it is all too easy to forget how recent is its impact on home entertainment. The idea of recording and playing back television pictures is as old as television: a primitive video disk was on sale in Selfridges of London in 1938.

Yet the first post-World War II experiments in recording television pictures and playing them back were not made with the domestic market in mind. The intention was to remove the stranglehold live television broadcasting imposed on the medium. Before video recording, sequences had to be shot on film, which was expensive, or they had to be transmitted live. Since broadcasting was closely tied to the evenings, this overloaded the studios and, moreover, any mistakes or hiccups in the production went out on the air immediately, giving the producer no chance to reshoot or edit them out afterwards. Repeats were impossible unless the production was shot all over again, or transferred to film.

Recording the signals which went out over the air on magnetic tape, just as in audio recording, seemed an obvious solution to the problem, but it proved amazingly difficult to achieve. To preserve a picture, vastly more complex information has to be recorded than with the highest-quality sound reproduction. Simply capturing the finer details of a picture meant a huge increase in the area of tape used to store the information. This meant, in turn, that the speed of the tape over the recording and playback heads had to be much higher than in audio recording.

After years of development, the American company Ampex produced the first working video recorders for broadcast use. These used four recording heads on a rotating drum, and two-inch (five-centimetre) wide magnetic tape. They were first shown to the public in April 1956 and used over the air at CBS's Hollywood studios seven months later.

Ever since then, more and more television has been recorded on tape, so that nowadays it is rare to see live broadcasts. The main exceptions are sports and other topical events, and even these are often recorded as they are transmitted, for later repeats or instant action replays.

Moving into the domestic market called for a much more portable system. Home recorders use narrower tape and a helical scan system. Although the first recorder using this system was patented in Germany in 1953, the prototypes were far too expensive and difficult to use. Friction between the tape and the heads meant that a tape could be used only once, and that every time it was used the machine had to be cleaned of tape particles.

Home video arrived in 1972, when the first cassette systems were produced: the original Philips VCR, still to be found in schools and colleges, and Sony's U-matic format, followed later by Sony's Betamax and the even more commercially successful VHS format, which used narrow half-inch (19 millimetre) tape. As the size of the tape became smaller, the cost of both recording and recorders fell quite dramatically and the stage was set for one of the biggest consumer booms of all time.

In most of Europe, sales of video equipment come second only to television sales, overtaking washing machines, radios, refrigerators, freezers and music centres. In 1980 a quarter of a million video cassette recorders (VCRs) were sold; this is rapidly rising to the million a year mark. The largest markets for video are in America and Japan where, by the end of the seventies, more than a million and a half recorders were sold each year.

Its versatility is one of the reasons for video's rapid success. Not only can a VCR be used for recording shows off air, and playing them back at any time, but ever more sophicated timers, and the ability to record the picture broadcast on one channel while watching a different channel, make a video recorder a powerful aid to getting the best from television.

Video's other trump card is the ability to play back pre-recorded tapes from the fast-expanding choice on the market. Since most pre-recorded tapes cost considerably more to produce than the average audio album, they cost proportionately more to buy, but can be rented or borrowed.

So far and so fast has video come in the last decade that any predictions on future uses and markets are almost bound to be overtaken by reality. New developments, such as cable and satellite television, will add to the choice of material available for recording, and ever more sophisticated cameras, recorders and

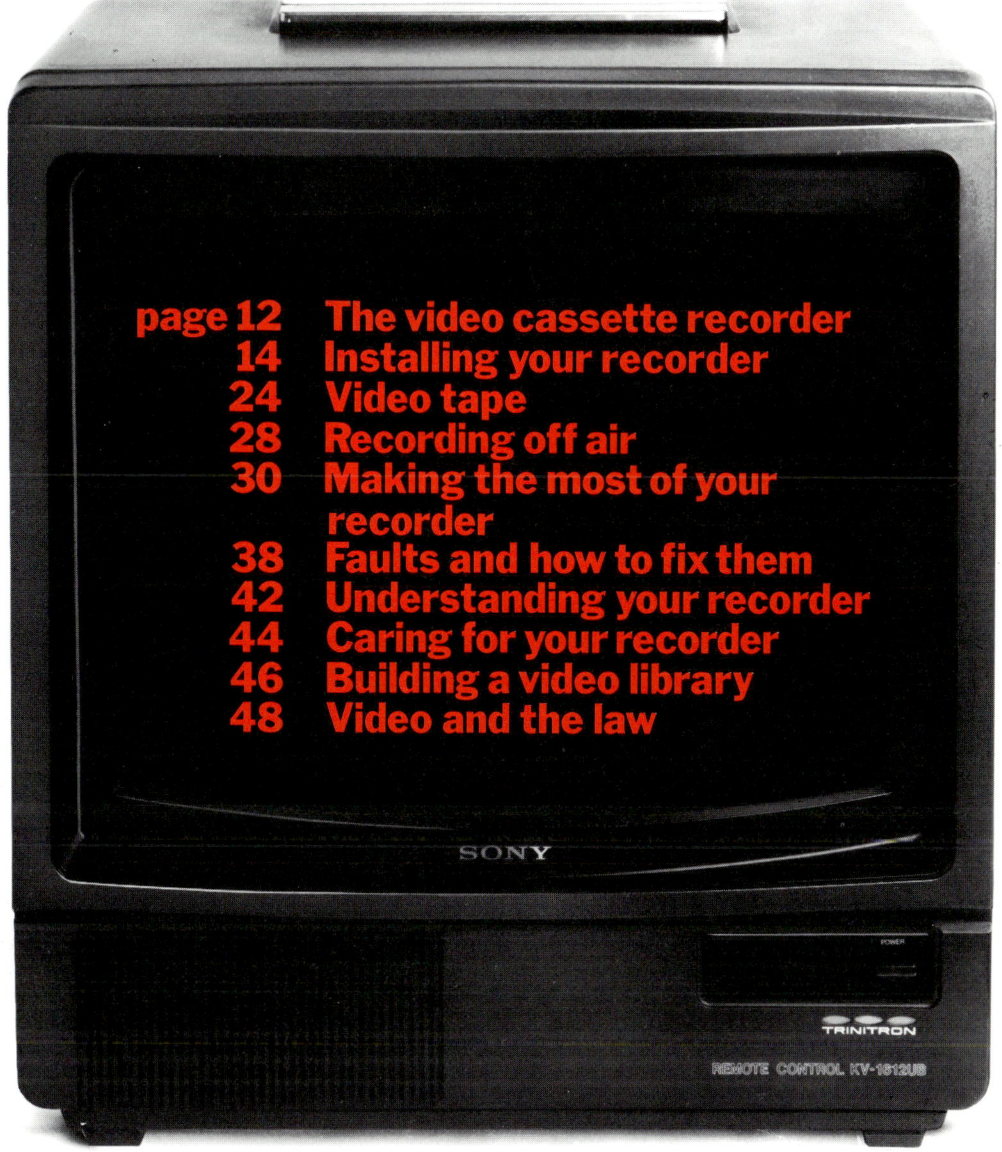

accessories will, in time, begin to rival the technical quality and flexibility of broadcast television.

Yet there is still a technical barrier between many television viewers and the video revolution. In Europe, technical know-how generally falls short of the level needed to operate video hardware with ease. Many a television addict, attracted by the storage capabilities of the VCR, finds it hard to choose from the glut of machines on the market. Those who solve the problem by renting may remain unaware that these expensive machines can do much more than memorize television for replay at a more convenient viewing time.

The next few pages attempt to overcome this technofear by bridging the technical gap. They analyze the basics of a video cassette recorder, and evaluate the trimmings. They explain where and how to install a video system, how it works and how to look after it. The most complex operations, such as tuning and timing, are analyzed.

By the time you reach the end you will know enough to read technical brochures with ease, and you may even begin to read the video press in order to keep up to date.

The video cassette recorder

The video cassette recorder (VCR) is the heart of a home video system. After the television set it is likely to be your first acquisition and it will certainly be more expensive. Whether you rent or buy, and whichever of the home video systems you eventually decide upon, all the machines you will consider having in your home have certain functions and facilities in common.

A VCR is designed to record sound and pictures in both black and white and colour on magnetic tape, and then to replay them when required. There are portable models, and home decks run from the electricity supply. These contain a tuner to enable the machine to record signals received from the antenna. The material recorded can then be played back on an ordinary domestic colour television set. You can also record one show using the VCR's tuner while you watch another station using the television's tuner.

All VCRs have a digital clock, and associated with it, a timer so that a show can be recorded when you are in bed or away from the house.

VCRs will play back tapes recorded on other machines as long as they are compatible. There are three different home video systems, or formats, currently available: VHS, Betamax and Video 2000. Each format will record only on tapes designed for it.

If your VCR and those of your friends are compatible, you can borrow and lend tapes you have recorded, or videograms (pre-recorded tapes). These are expensive, but there are video libraries from which you can borrow cassettes of old and popular movies, educational or sports tapes, usually in VHS and Betamax, if not in all three formats. Pre-recorded tapes have been available in VHS and Betamax for some time, and are now becoming available for the new and expanding Video 2000 format.

Some features, such as a footage counter which shows how much tape has been used, are standard on all VCRs. Others, such as fast picture search – the video equivalent of fast forward on an audio cassette recorder – have been available on the more expensive models, but are now appearing on budget machines. The more sophisticated facilities offered on the luxury models may seem confusing at first; they are hardly essential to the basic functioning. However, it is as well to be fully conversant with all these facilities before you decide on a VCR because some of the apparently gimmicky extras may be just what you need.

For instance, if you always fall asleep during a movie, a warning buzzer to tell you when the the show is over might be a good idea. If you hate commercials, you need a remote control to fade them out. If you go away on business for weeks at a time, you need a sophisticated timer to keep up with your favourite television serials, but if you want a VCR to watch movies with your friends, and record television shows to watch later, a basic machine is all you require.

The Toshiba V-8600B, illustrated here, is a luxury Betamax model that offers a wide selection of the features currently available.

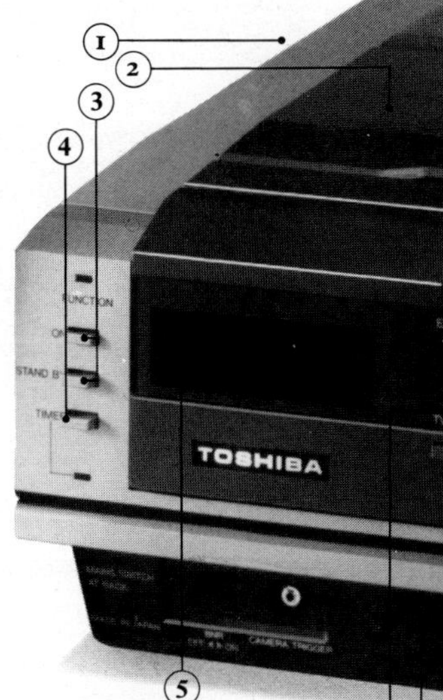

The 'Function' switch, 3, enables you to select manual or automatic (timer) use. When it is switched off, the signal from the antenna passes through to the TV set for normal viewing.

When the 'Function' switch is set to 'On' the light above it illuminates.

When the 'Timer' switch, **4**, is pressed, the lamp below it illuminates and the VCR will begin recording at the time pre-set in the timer compartment, **2**. On this machine you can pre-set start and stop times.

On many machines, the power indicator light (above the 'Function' switch on this machine) will switch off when the 'Timer' switch is on.

All VCRs have a 24-hour clock, **5**, which can display pre-set recording times. This panel can display more information, including time left on the tape, or a warning that you have not yet loaded a cassette.

A recorded tape will never replay perfectly on any machine other than the one on which it was recorded. The 'Tracking' knob, **6**, will minimize interference if you play a borrowed cassette.

When the 'TV/VTR' switch, **7**, is set to the TV mode, you can watch one TV channel while recording another.

Into this panel you can connect a mike, **9**; an audio input, **10**; a camera, **11**; and a remote control, **12**.

Switch on your VCR by pressing the 'Mains' or 'Power' switch. This is usually located on the back panel, **1**, but may be in the position occupied by the 'Function' switch, **3**.

On this VCR, the 'Function' switch must also be pressed before the machine will record.

Only the most advanced home decks have an input socket for a video camera, so consider buying a portable deck (called a portapack) if you intend making home movies. A video camera, unlike a film camera, is not loaded with film. Instead, the picture is transmitted along a cable to the portapack and recorded on a video cassette.

Portapacks are battery powered, and more compact than home decks. The RCA SelectaVision 170, weighs 11 lb (about 5 kg). If you buy a tuner/timer unit it can be used to record broadcast TV and to run from the home electricity supply. It has all the features of a good home deck, including a high-speed picture search, slow-motion playback and remote control.

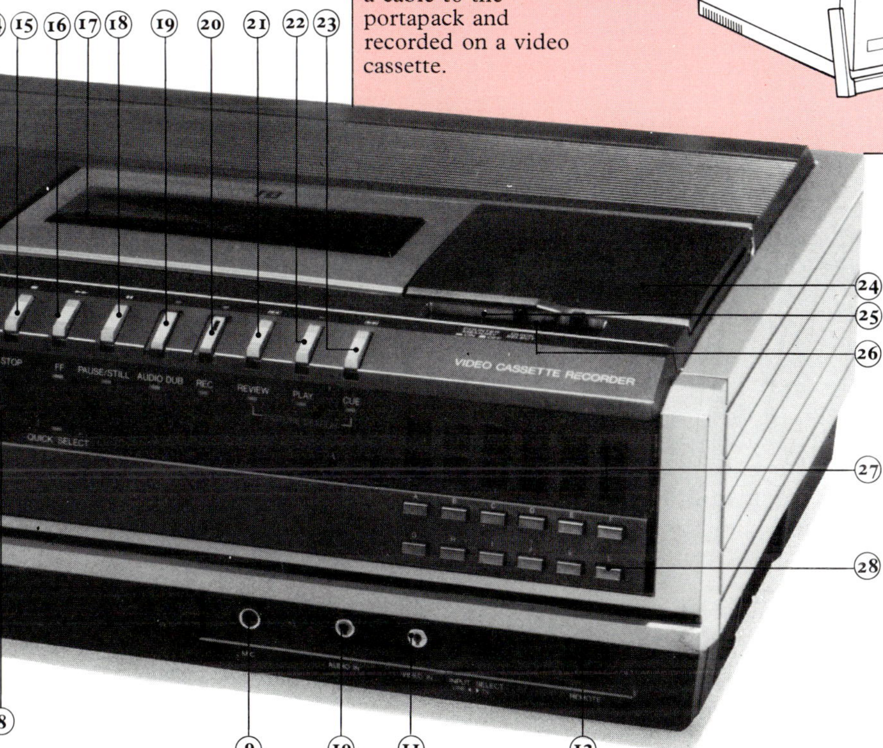

The 'Record' button, 20, is usually operated together with the 'Play' button, **22**, to start recording. On this machine each controls a different function.

In the tuning compartment, 24, are controls for tuning each VCR channel to the locally available stations. Once the VCR is tuned in, the selector buttons, **28**, can be used to select a pre-tuned channel, and the appropriate indicator light, **27**, identifies it.

The 'Dew Indicator Lamp', 8, is a safety device not found on all machines. It warns that all functions have temporarily ceased because moisture has accumulated on the tape heads.

'Rewind', 14; 'Stop', 15; and 'Fast-Forward', 16; function as on an audio recorder. The tape disengages from the heads when 'Stop' is pressed.

The cassette-holder, 17, on a top-loading VCR rises when the 'Eject' button, **13**, is pressed. It may be spring-loaded or hydraulically operated, but has to be pressed down once the tape is in the holder.

Front-loading holders have a dust flap that folds away as the tape is pushed in, then closes over it.

'Pause', 18, stops the tape while keeping it in contact with the heads. This gives a still picture in playback, or temporarily interrupts a recording.

Press the 'Audio dub' button, 19, to record over a pre-recorded cassette using a mike. This automatically erases the previous recording.

Most VCRs now have a picture search facility. This sophisticated machine has two: 'Cue', **23**, accelerates the tape forward and 'Review', **21**, reverses it, while retaining an image on the screen.

The tape counter, 25, can be reset by a button, **26**, to 0000 at the beginning of a tape so that the starts and finishes of different shows can be noted.

Siting the recorder

If you are to get the best results from your VCR, it is worth while giving some consideration to its location in a room. Like all electronic devices, a VCR likes a dry, well-ventilated environment. Excessive heat or dust can damage its delicate circuits and precise moving parts.

Since information is recorded magnetically on video tape, stray magnetic fields can corrupt it. Recordings can suffer if the VCR or cassettes are placed too close to other machines with strong magnetic fields, such as audio recorders or stereo speakers. These have particularly virulent magnetic fields, which increase with the volume of sound. If possible, locate them at least two feet (60 centimetres) from television, tapes and VCR.

Remember that the VCR's magnetic field can affect other components of the video system. Never keep it on top of the television. It is wise to screen one component off from another by means of shelving. As a rule of thumb, leave one foot (30 centimetres) of space between them.

Placing and operating the VCR is largely a matter of common sense and convenience. A VCR is designed to work in a horizontal position, and make sure it is horizontal, because its tolerance for functioning when misaligned is relatively low. Keep the ventilation slots clear of surfaces that might inhibit air circulation through them, and above-all, position the VCR away from sources of heat such as radiators or convector heaters.

Do not stand your VCR on a surface which will transmit vibration from other domestic appliances – such as a suspended floor, for example. Cover the machine with a dust cover when it is switched off.

Damp is the VCR's worst enemy. Keep it in a warm room and do not subject it to extremes of temperature. If it is allowed to go cold and then moved to a warm room, moisture in the air will condense on the metal drum which carries the video heads, making the tape stick to it when it begins to spin. This can damage both tape and recorder.

Most VCRs have a built-in protection against moisture condensing inside the casing. Some have internal heaters, but for these to work, the VCR must be kept connected to the electricity supply.

Never put anything on top of your VCR. You could block the ventilation grilles, and damage the casing.

TV sound quality can be improved by using hi-fi loudspeakers. Never remove the back of a TV receiver to gain access to the sound section. The easiest and safest way is to connect the VCR's audio output to a hi-fi system. For best reproduction of stereo material and mono television, the speakers should be placed at least 6 ft (1.8 m) apart, so that they and the listener form an equilateral triangle.

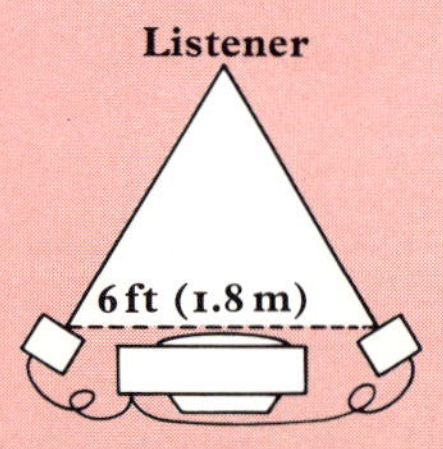

Most VCRs are top loaders. The cassette drawer opens for loading and unloading by rising about 2 in (5 cm) above the top panel. Allow for this when fitting a VCR into shelving.

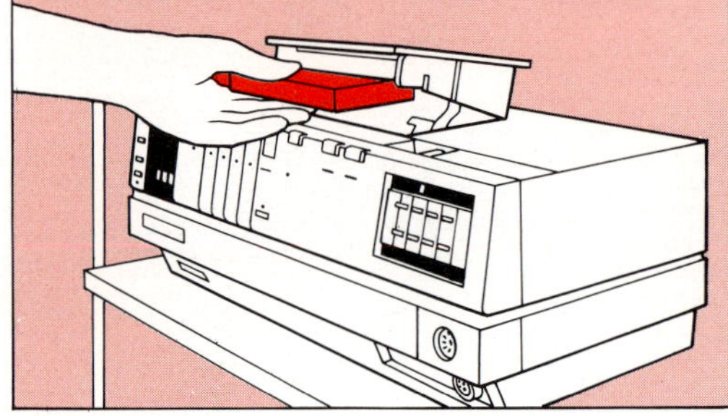

A video recorder should be near a TV receiver and antenna (aerial) input point. If you can, plug the VCR and the TV into different wall sockets to avoid the possibility of picking up an electrical hum. This can cause interference on the picture. If you want audio pick-up, the hi-fi system must also be nearby, but not on top, nor immediately adjacent to the VCR.

Postion the VCR away from doorways and other parts of the room where people walk, or reach across, and out of draughts, which may affect its performance.

Mobilize your video system with a purpose-built video cart, **1**, but beware of tripping over cables, **2**.

Trailing cables are dangerous and also untidy; keep them out of sight. Excess lengths can be wound into 12 to 15-in (30 to 40-cm) loops and secured with tape or a rubber band.

Site your video system well away from windows, **6**, and fires, **3**. Direct heat from a fire, or from sunlight streaming through a window, can damage a VCR. The rapidly changing temperatures and damp found near a window will impair the VCR's tracking ability. If rain falls on the VCR it can cause a short-circuit.

Rack shelving, 4, can be spaced to accommodate video system components of unequal size, and provides the screening necessary to prevent the magnetic field of, say, a stereo speaker, from affecting a colour TV. Make sure the VCR is not on top of or touching the TV – it should be at least 1 ft (30 cm) away.

The casing of a VCR is pierced by ventilation grilles which permit free air circulation. This is essential because parts of the workings heat up with constant use. To maintain an even internal temperature the grilles must be kept clear, so make sure there is at least a half-inch (1.5 cm) space around the casing, **5**.

Never leave your VCR on the floor. Modern carpets, **7**, made from artificial fibres, can build up a static charge which may be transferred to the VCR through its ventilation grilles, with disastrous results. People walking past can raise dust which will damage the video heads.

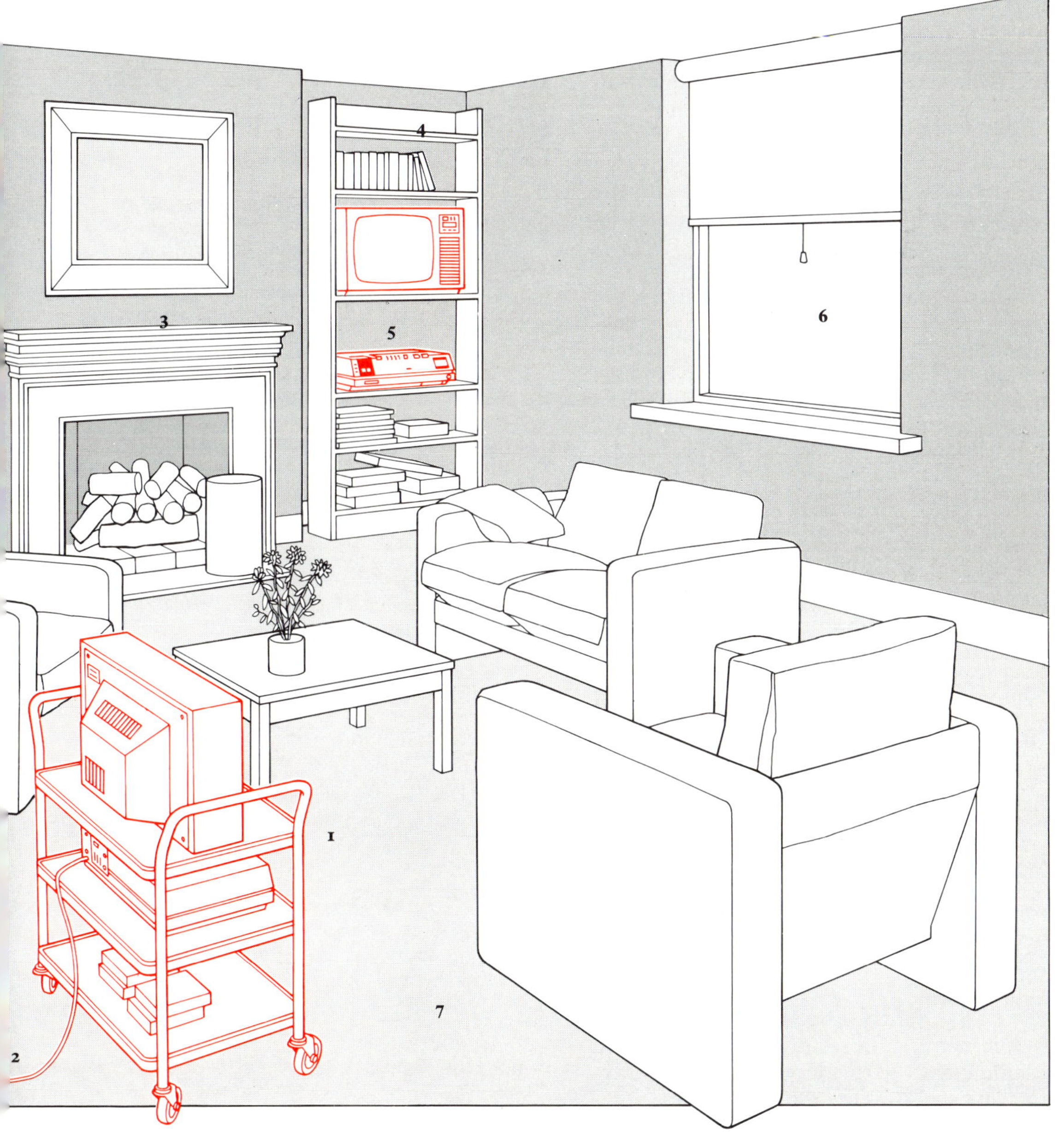

Furnishing with video

Video cabinets come in all shapes and sizes, from a simple rack to hold a VCR and a few tapes to a full console cabinet, holding the television set too. Philips make a complete home Video Centre. The custom-designed case holds a 26-inch (66-centimetre) colour television, a 6-inch (15-centimetre) black and white television monitor and a VCR.

When choosing a cabinet, consider what is to go inside. If it is to take the television set, either built-in behind doors or free-standing on top, the set's size and weight will determine the dimensions and construction. The standard size of $29\frac{1}{2}$ inches (75 centimetres) wide and $18\frac{3}{4}$ inches (45 centimetres) deep and $22\frac{1}{2}$ inches (57 centimetres) high allows for a 26-inch television, with VCR and cassette storage underneath.

Most cabinet designs allow for a reasonable distance between television tube, video motors and tape storage, to prevent the magnetic fields they set up from interacting to cause colour aberration on the tube, or partial erasure of a tape.

When making a cabinet for VCR only, the size depends on the dimensions of the particular model used. VHS and Betamax machines vary in width across the front panel from 17 inches (43 centimetres) for the smaller budget models up to nearly 20 inches (50 centimetres) for the top models. Older machines can be even larger and all Video 2000 models are over 20 inches (50 centimetres) wide. Portables, although smaller, will take up to $23\frac{1}{2}$ inches (60 centimetres) in width when side by side with a tuner. Depth will vary between 10 inches (25 centimetres) for some portables to a more normal 13 to $15\frac{3}{4}$ inches (33 to 40 centimetres) for home decks. Height also varies between 5 and 6 inches (13 and 15 centimetres) for most home decks, more for older machines, and as little as 4 inches (10 centimetres) for the newer portable machines.

Remember to leave enough hand-room at the sides to insert and remove the machine. There needs to be enough room at the back for all the leads to be plugged in without being bent too sharply, and it is important to leave enough air space for ventilation above and behind the recorder. Remember also to leave room for the mechanism to lift.

Alternatively the recorder can be placed on a sliding base, so that it pulls out sufficiently to allow access to the top loader.

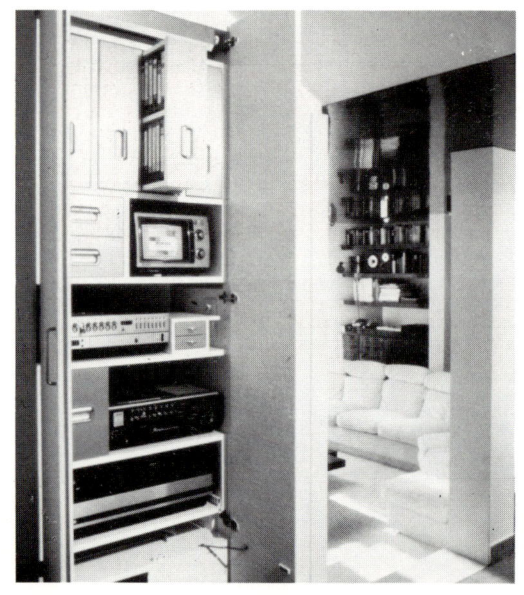

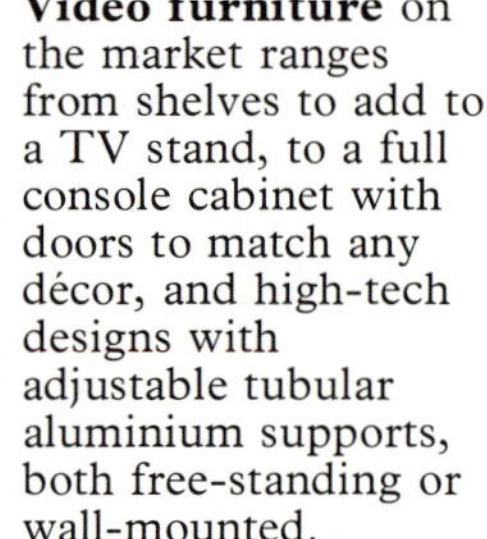

Video furniture on the market ranges from shelves to add to a TV stand, to a full console cabinet with doors to match any décor, and high-tech designs with adjustable tubular aluminium supports, both free-standing or wall-mounted.

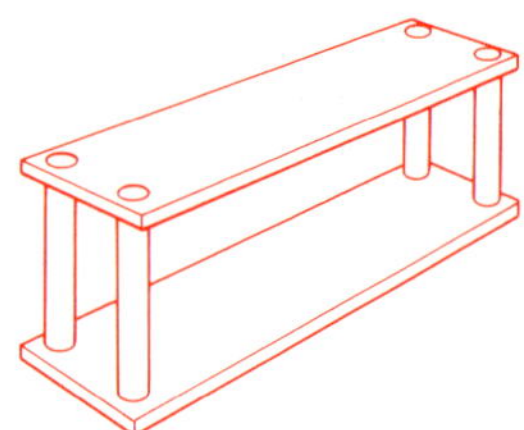

The simplest video stand is a platform, *left above*. The TV sits on top with the VCR underneath: far enough away to prevent interference.

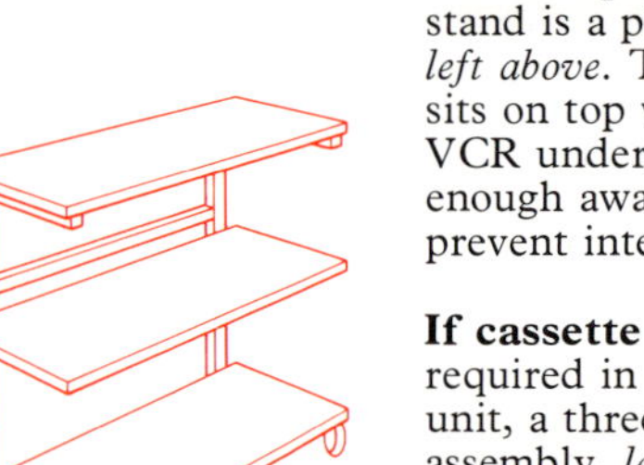

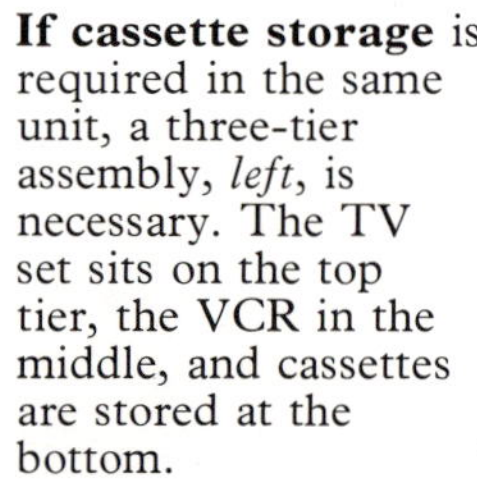

If cassette storage is required in the same unit, a three-tier assembly, *left*, is necessary. The TV set sits on the top tier, the VCR in the middle, and cassettes are stored at the bottom.

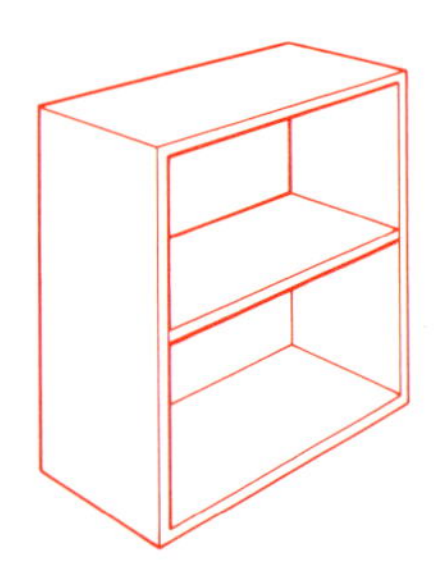

The TV video library case cabinet, *left*, is about 2 ft (60 cm) high, and will accommodate either a VCR and two tiers of cassettes, or a TV set, VCR and one tier of cassettes.

The Kee clamp can be used to build an interlocking pipe framework, in which shelves can be fixed. Different sizes of clamp are used to cope with different diameters of piping.

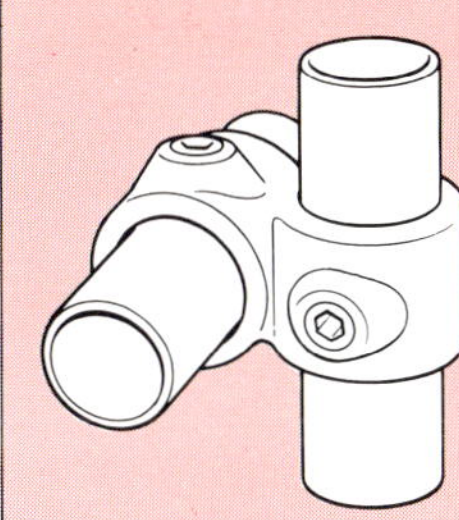

Rota-lock is another pipe-frame system. A chrome 'cage' holds the two pipes loosely together. A packing piece is bolted between them to make the joint rigid and safe for heavy loads.

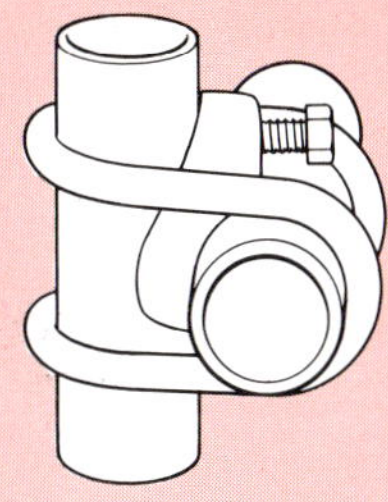

Slotted angle, L-shaped steel sections, with a pattern of horizontal and vertical slot perforations, can easily be cut with a hacksaw, and bolted together to make a video rack.

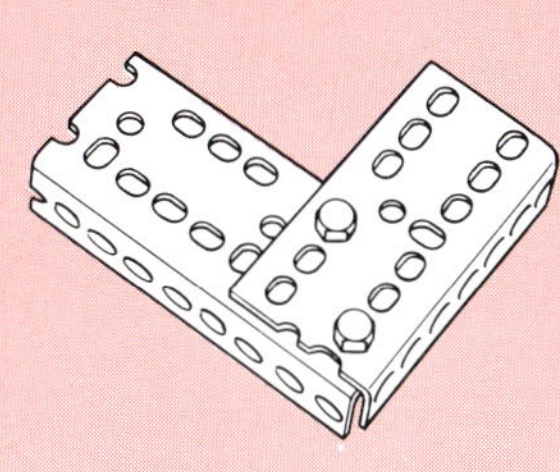

T, X and L-shaped connectors can be screwed into the edges of flat wood panels to link them together in the form of a simple yet sturdy video storage cabinet.

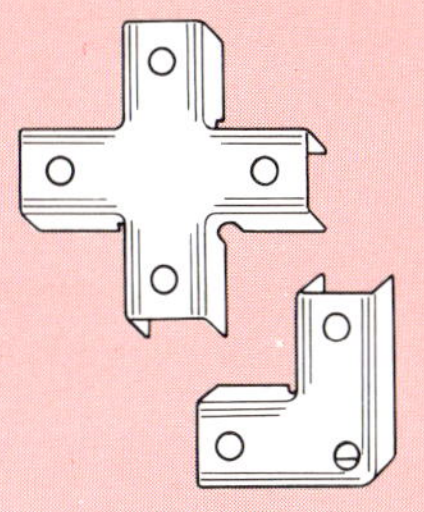

Video stands and carts, *below*, are the most versatile means of supporting video equipment. Carts run on castors and are, therefore, easily moved. The most basic consists of a platform supported by a central column on a four-castor base, *below centre*. The platform is adjustable to take TV sets up to 39 in (100 cm) wide: the left and right halves slide out to the required width.

The supporting column can be altered from 16 in (40 cm) to 39 in (100 cm).

Heavy-duty, two-column designs, *below left*, should be used to support the larger TV sets, and the largest require a cart with the columns set wide apart, *below right*, for improved stability.

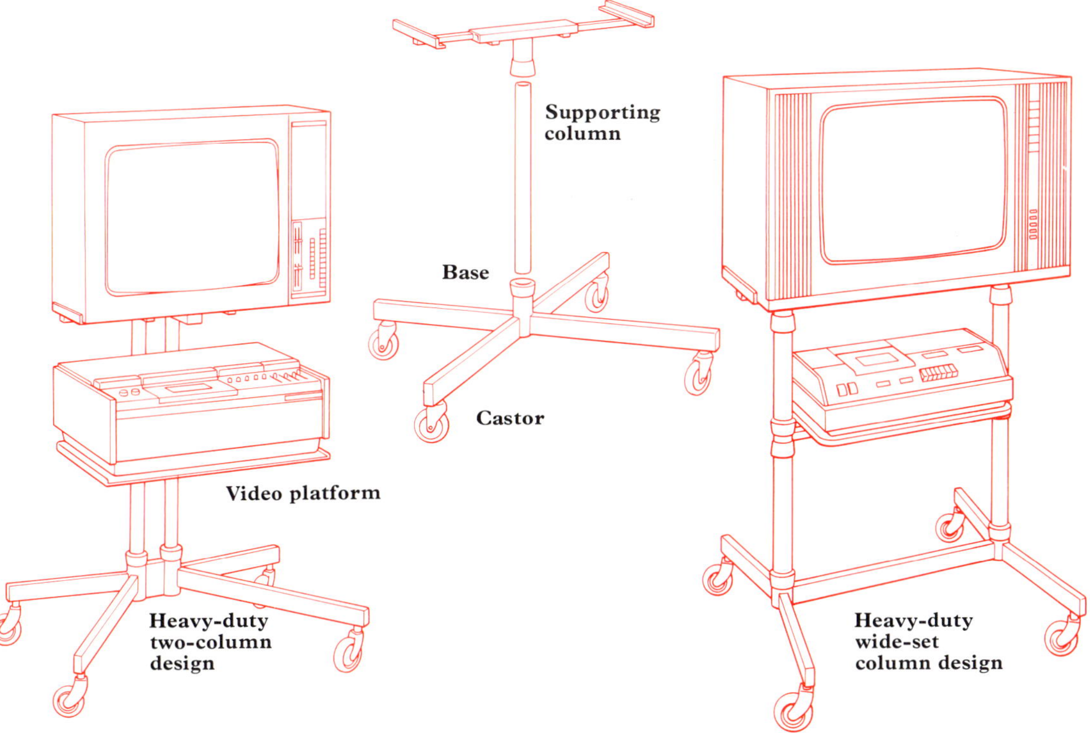

Connecting up

To connect your VCR to your television set you will need no fewer than four cables, and with each component you add to the system, the number multiplies. The AC lead powers the system; each of the other cables contributes to the flexibility, and the quality, of your video system. It is essential, therefore, that each cable is right for the job it has to do.

Why so many different kinds of cables? When the television signals which make up the picture and the accompanying soundtrack are broadcast, in order to make the journey from the transmitter to your receiving antenna, they have to be changed into very high-frequency waves.

These radio frequency (RF) signals cannot recreate the picture inside your television set in the form in which they arrive. They have to be changed back into low-frequency sound and video signals before being fed to the loudspeaker and picture tube. These same low-frequency signals are recorded on the video tape and reproduced when the recording is being played back.

This means that the input cable from the antenna must be coaxial cable (with one lead forming a central core, and the second forming a hollow tube around it) to carry the radio frequency signals from the antenna, and from the VCR, to the television set. Inside the VCR, the RF signal is converted into audio and video signals at a lower frequency for copying tape.

When the tape is played back, those signals must be converted once again into higher RF signals, so that they can be fed into the television set as if they were arriving directly from a transmitter.

This double conversion between VCR and television introduces a slight drop in quality. This is because most sets were designed simply to receive broadcast signals, and VCRs were designed to fit in with this. A television/monitor, however, can take the video and audio signals directly from the VCR and feed them into special sockets on the television set. This involves extra single-core cables used for low-frequency signals.

On semi-professional equipment, where sound quality is important, or where the sound signals have to be carried over long distances – in camera extension leads, for example – twin-core cables, which pick up less electrical interference from outside sources, are used.

To fit a coaxial connector, cut off 1½ in (4 cm) of outer casing, fold back the metallic braiding **1**, and cut away about ½ in (12.5 mm) of core insulation.

Fit the plug collar and the braid clamp, **2**, then fit the pin unit and solder the cable core to the pin. Cut off the surplus wire from the end of the pin and screw on the plug body, **3**.

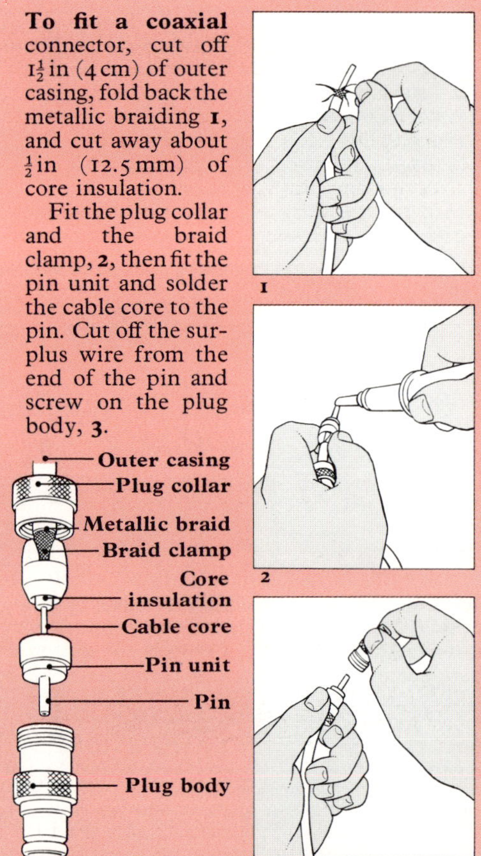

Power

Every VCR, home deck or portable, uses 12-volt (V) direct current (DC) electrical power. Portables have 12V batteries, but models with AC input have to work from the national system. This is usually alternating current (AC); few countries use DC.

AC alternates on a frequency of 50 or 60 Hertz (Hz) per second. This frequency determines the national frame rate: 50Hz systems have a rate of 25 frames per second; 60Hz systems have a rate of 30 frames per second.

Voltages differ nationally. In the USA and Europe the national standard is 110V; in the UK it is 220V. There are adaptors to enable American VCRs to be connected to British power supplies, but most VCRs have a multi-position switch, enabling them to function under either standard. Be careful not to operate it by mistake: if it is set to 110V and plugged into a 220V socket, all the fuses will blow.

While the VCR is recording it consumes only as much electicity as a domestic light bulb.

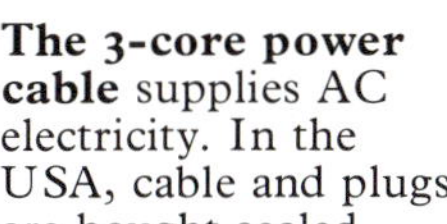

The 3-core power cable supplies AC electricity. In the USA, cable and plugs are bought sealed together; in Europe it is now standard for three-core cables to be colour-coded: brown is live; blue is neutral, and green and white stripes is earth. They are made of multi-strand copper wire, and are invariably separated by a string filling and a layer of braid. Power cables have a thick rubber casing.

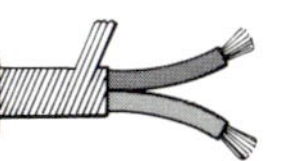

Twin microphone cable is balanced: it allows you to carry two sound signals for stereo recording, and also gives the best sound transmission over long distances.

This is a three-conductor cable; it has two connectors surrounded by braiding, which is earthed to provide a screen against unwanted signals. The whole complex is surrounded by an outer casing of tough plastic.

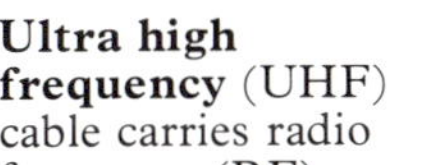

Single-stranded (unbalanced) microphone cable is used for most audio connections. It is used for connections over short distances: to different parts of the video system, for example.

The cable would be subject to considerable interference from outside signals if it were not screened from them by the layer of copper braiding which surrounds it.

Ultra high frequency (UHF) cable carries radio frequency (RF) signals which are of a higher frequency than mike signals. It is occasionally used to carry video signals.

The central copper conductor carries the signal. The inner core is screened by woven metallic braiding, which also makes it more rigid and incapable of being twisted. With an outer plastic casing, it is a tough cable.

Connecting up

1 Disconnect the TV set from the AC electricity supply.
2 Unplug the TV antenna cable from its socket in the back of the set.
3 Plug the antenna cable into the socket in the back of the VCR marked 'Antenna'/'Aerial'/'RF In'.
4 A short lead is provided with the VCR. Plug one end into the socket in the back of the VCR marked 'Antenna'/'Aerial'/'RF Out'; the other into the socket in the back of the TV marked 'Antenna' or 'Aerial'.
5 Connect the VCR and the TV set to the power supply.

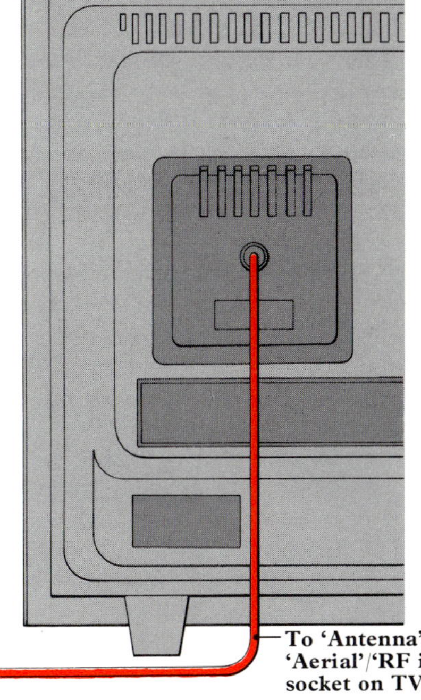

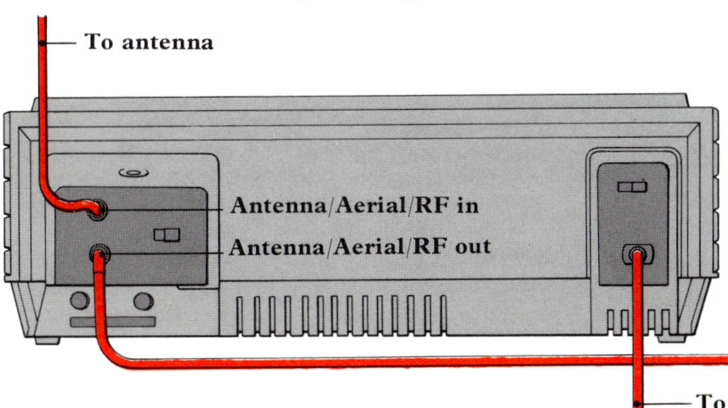

Plugs and sockets simplified

It is almost impossible to feed a cable into the wrong input. Cables which have to carry audio signals are entirely different from those carrying the RF signal from the television antenna. Moreover, by fitting them with different connectors it is possible to eliminate incorrect connections.

Nevertheless, there are pointless variations between the types of connector used in different countries and between one model of VCR or audio system and another. Hitherto, manufacturers have followed their own inclinations regarding cable connections between their products. As a result, as soon as you tried to build up a system of different units from different makers, you needed to have special leads made to link the components. Now there is a degree of uniformity.

Broadly speaking, the connectors, as well as the cables to which they attach, split into three classes: those used for RF signals, for video signals, and for audio signals.

There are, in addition, other more specialized cables and plugs, for example, those needed to supply power from the mains, or a battery pack, or to connect a camera (which involves passing video, audio and other signals down a single cable). These, however, often use specially designed connectors, and cables with ten or more different cores to take all the signals, and have to be made up individually.

Treat your connectors carefully, especially when inserting them and pulling them out. When connecting up multi-pin connectors, check that the pins are in line – there is often a slot or pip inside to guide you; when disconnecting, remember that there may be a release mechanism and do not try and force the connectors. Rough treatment could bend or break a pin.

Keep at hand a spare set of the more important cables and connectors in your system. There are adaptors for some of the connector combinations, enabling a cable intended for one connection to stand in, temporarily, for another.

However daunting it may seem, it is worth while learning about connections. Should you eventually extend your video system to incorporate an audio system, games and a camera, the wiring together of the whole system can add up to a maze of cabling and a whole battery of connectors. It will be useful to know how to simplify it.

Connector care check list

1 Keep connectors and connections clean, using specially formulated fluid solvent cleaners.
2 Do not force connectors into sockets. If they do not fit together easily, check to see if you have aligned them correctly. Forcing a connector may bend a pin.
3 Do not jerk connectors out of sockets. If you cannot draw them gently but firmly toward you, there is probably a ring to twist or unscrew, or a release mechanism to operate.
4 Do not bang connectors by swinging cable ends together, or leave them lying on the floor, where they may be trodden on or kicked, or chewed by animals.
5 Be careful with the pins on connectors. If they break the equipment will not work.
6 Do not pull connectors out of sockets by the cables; it causes faulty connections.
7 Do not bend cables. If you do this frequently, the casing will weaken. Coil them in 12 to 15-in (30 to 40-cm) loops.
8 Colour-code your cables for easy identification.

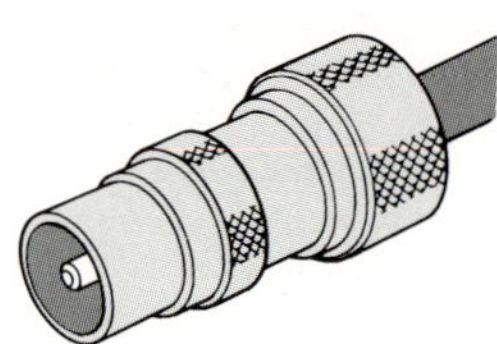

TV Coax Plugs, or Belling & Lee Plugs, *above*, are standard in the UK, Europe, Africa and the Middle East. These push-in connectors transmit RF signals to the VCR and TV set via a coaxial cable.

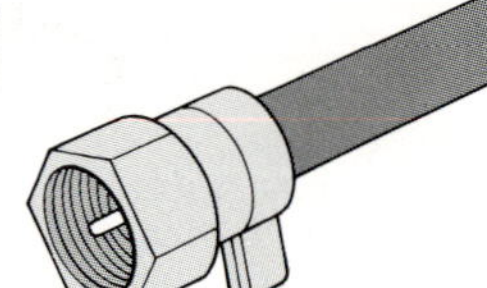

In America, RF cables are either twin-core 300-ohm cables, ending in two spade-type terminals which are screwed on to the TV set, or coaxial 75-ohm cables fitted with an F-connector, *above*.

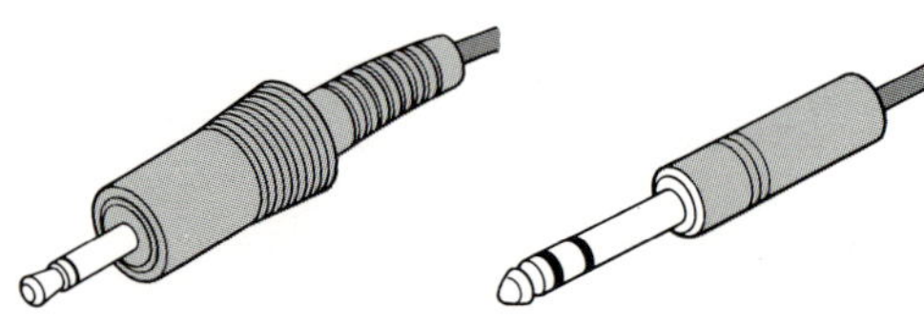

Most domestic video equipment uses unbalanced audio, for which plugs and jacks based on telephone switchboard types are used. The 3.5 mm miniature jack (or mini plug) *above left*, is the most common; there is also a standard (6.3 mm) size. This connector is usually used in audio cassette tape recorders. The ¼-in (6 mm) phone plug *above right*, is used for mike connections by some manufacturers.

One way of keeping the complicated interconnections between different parts of a system as simple and flexible as possible is to wire them through a patch panel, *below*. This may be a panel designed to fit into a rack or a box, and it will have rows of numbered sockets along the front, which can be connected in any combination, using simple jack connectors. By wiring all your video/audio system components into the back, so that each has a set of numbered terminals, you reduce the jumble of connectors, linking them to a neat set of identical cables in one particular spot.

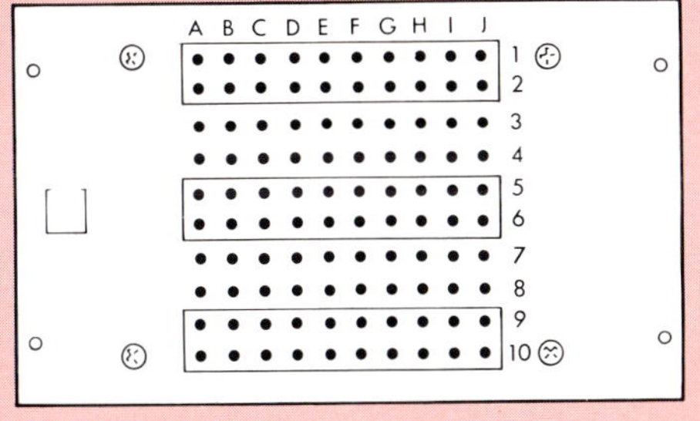

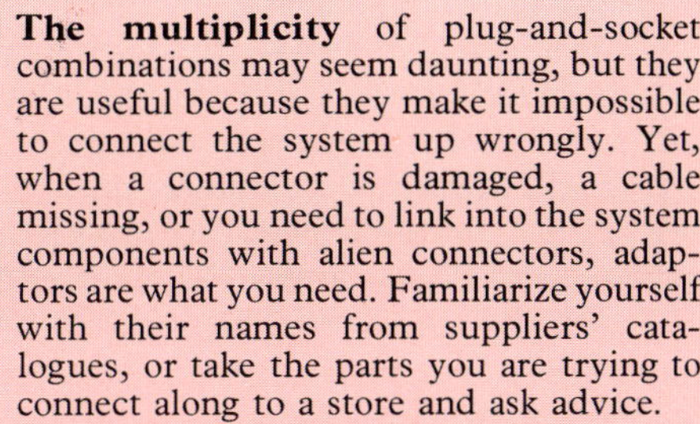

The multiplicity of plug-and-socket combinations may seem daunting, but they are useful because they make it impossible to connect the system up wrongly. Yet, when a connector is damaged, a cable missing, or you need to link into the system components with alien connectors, adaptors are what you need. Familiarize yourself with their names from suppliers' catalogues, or take the parts you are trying to connect along to a store and ask advice.

Adaptors should be temporary. If one seems to be constantly used, have the connector you obviously need made up.

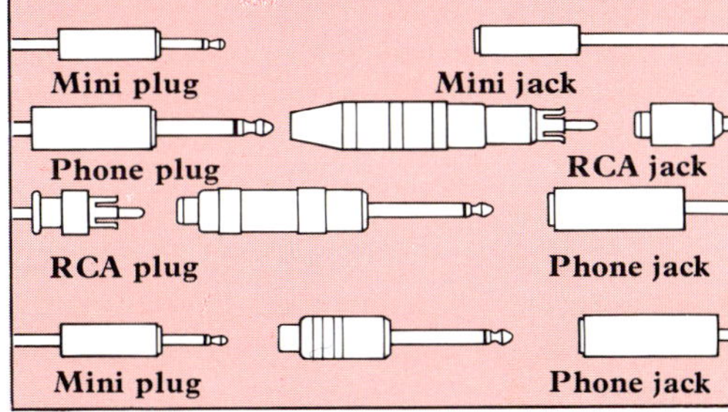

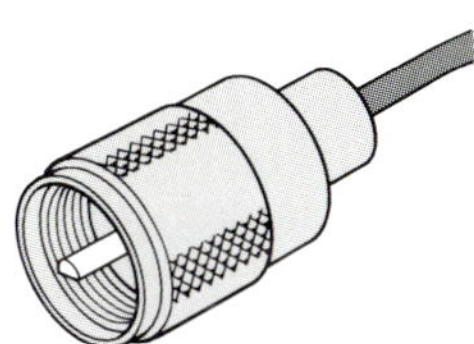

Older models of TV and video equipment usually have UHF connectors, to transmit video signals. They have a central pin and an outer, screw-threaded ring to fasten the plug securely to the socket.

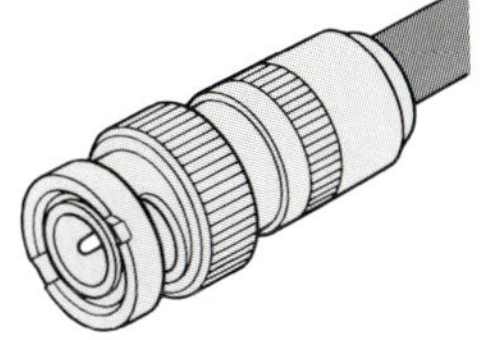

The compact BNC plug, *above*, is gradually replacing the old UHF and F connectors. This is a bayonet-type connector which fastens on a special BNC socket with pins to hold it in place.

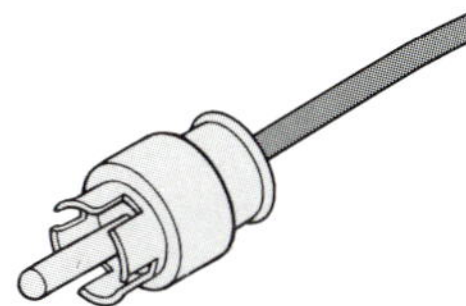

The RCA (phono or cinch) connector is an audio plug for use with unbalanced lines. It loosens easily, and is not really suitable for video use, but it is cheap and, unlike UHF and BNC plugs, is easy to solder.

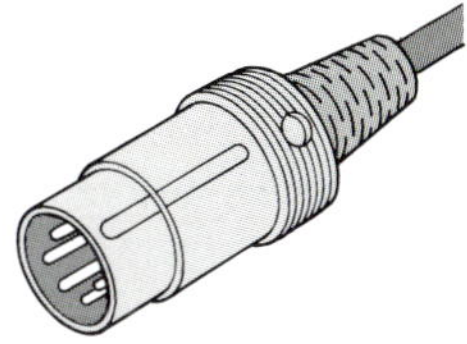

The multi-pin DIN plugs and sockets are standard throughout Europe as connectors for unbalanced audio lines. They combine input and output connections.

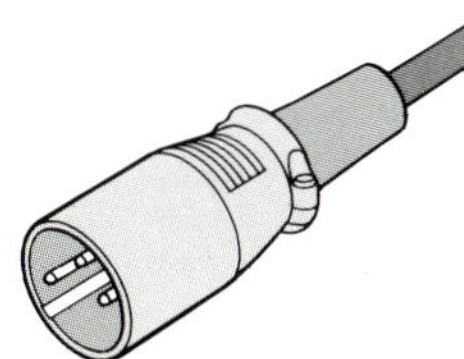

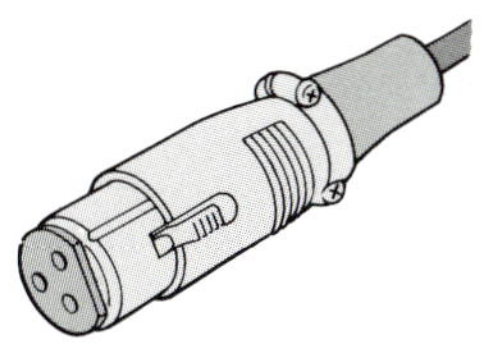

Most good mikes, as well as 1-in (2.5-cm) VTRs and other semi-professional equipment which uses twin-core balanced line cables, will rely on the Cannon (or XLR) three-pin connector, which comes in male *above left*, and female, *above right*, forms. An adaptor is available, with a female Cannon plug on one end and a phone plug on the other, to connect an unbalanced recorder to a balanced mike.

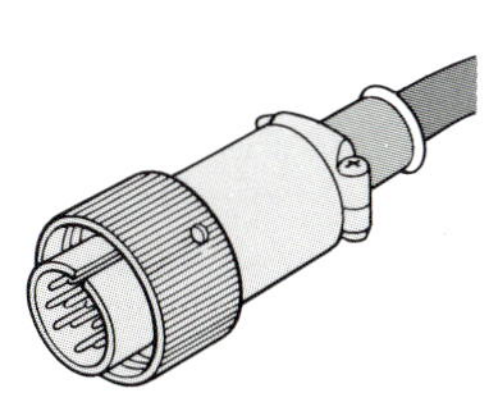

Home video cameras have 10- or 12-pin connectors and professional camera connectors have up to 50 pins. They are usually secured by an O-ring, which must be unscrewed to pull the connector out.

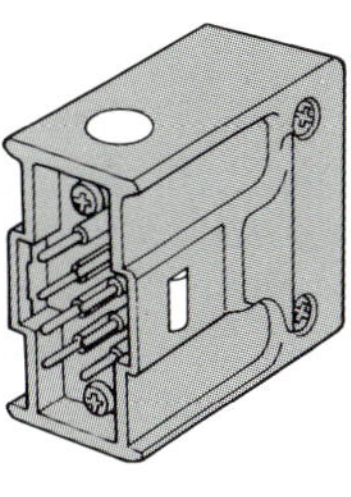

This 8-pin plug transmits both audio and video signals between a VCR and a TV monitor. The 8 pins are unevenly spaced, so check that they are aligned when inserting into the socket.

Tuning in

The VCR has to be supplied with the information which it is to record. If this is a standard broadcast signal from television transmitters it will come via the antenna, which would otherwise be connected to the television. The antenna picks up a mixture of local broadcast signals, but all have different frequencies. By tuning the television to one frequency, it can be made to receive only the signal which has that frequency.

The same is true for VCRs. All non-portable home VCRs have built-in tuners, and portapacks have a tuner/timer as an optional extra. A VCR can be tuned, on its own tuner, to a different station from the one selected on the television's tuner, so that you can watch one channel on the television while recording another on your VCR.

If your television reception is poor, try retuning the set. The tuning knobs and frequency indicators will be underneath a small panel. The fine tuning procedure varies, so if it is unclear, and you have no instruction book, call a service engineer.

Tuning in sounds complicated, but it is quite straightforward. First, choose a spare channel on the television. If this is a new set it may have a button, marked 'AV' or 'VCR', for this purpose. This channel has to be tuned to the channel on your VCR which is pre-set by the manufacturer. This is done by setting the VCR to a special test signal, or by playing a pre-recorded tape. As soon as altering the channel adjustment on the television gives the test pattern or playback picture on the screen, this channel is tuned to the VCR's output channel, which will allow the set to show what the VCR is receiving.

Many tuners on both televisions and VCRs have a button marked 'AFT' (automatic fine tuning), and some have an extra button for making fine adjustments to the tuning. It is important to switch the AFT off while tuning, and to switch it on again when the set is tuned. Once the set is tuned, the AFT will keep the tuning locked into place.

If you get herring-bone interference on your screen the frequency is too close to a broadcast channel. You can alter the VCR's RF channel by a screw control. The manual should tell you how to do this.

The final step is to set the VCR to its normal operating mode and tune in each channel to a different station. Some VCRs have an auto-tuning facility which does this.

Tuning in check list

1 Switch on the TV set and VCR. Select either the 'VCR' or 'AV' channel, or a spare normal channel on the TV.
2 Check with your manual that all the necessary controls are on the right setting for tuning, or you may carry out the procedure with no result.
3 Switch on the VCR's test signal. Most VCRs generate their own test picture, but some have a test signal on a cassette. If you cannot obtain a test signal, substitute any pre-recorded tape, preferably in black and white.
4 Turn the tuning dial on the TV until the test picture is as clear as possible. Then switch the test signal back to its normal setting. If you get herring-bone interference, or if the VCR picks up a TV channel, retune the VCR's RF channel.
5 Tune your VCR channels to the TV as described in the manual. If your VCR has a setting labelled 'AFT' or 'AFC', remember to switch this to 'Off' while tuning and to 'On' again when tuning is completed.
6 Compare the picture you are getting on your VCR channels with the picture you get on the TV on its 'TV' setting. They should not be noticeably inferior. If the picture is poor, first fine-tune the VCR channels again and then, if necessary, the TV. It is worth taking some time and trouble over this, since poor tuning is often the cause of bad video reception.
7 If your picture is persistently poor, consult a service engineer.

Since portable recorders must be as light and compact as possible, they lack the tuning and timing functions of normal VCRs. To enable them to record broadcast television off air, they have separate tuner/timer units. This Panasonic NV-3000 is designed to be used with the NV-V300 tuner/timer, which enables it to be tuned into broadcast RF signals, turning it into the equivalent of a home deck. The most modern tuner/timers have almost as many facilities as other advanced VCRs. To work on the electricity supply, this system needs an AC-DC adaptor/battery charger; some have adaptors built in.

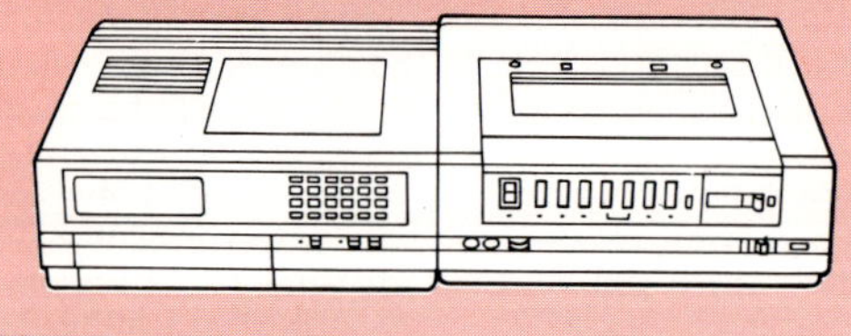

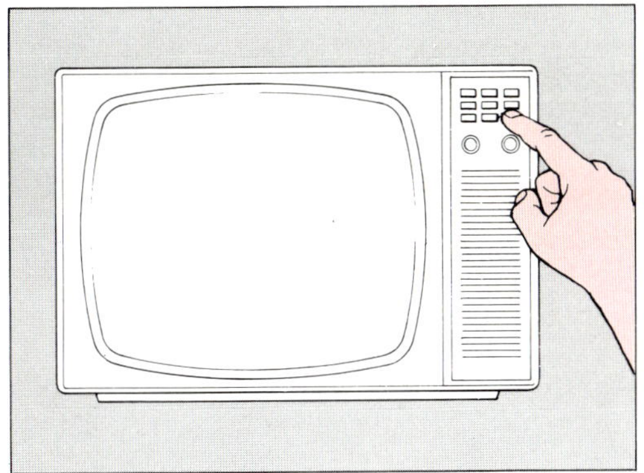

The first stage of tuning in your VCR is to switch on the TV and VCR and select a spare channel on the TV. On new TVs there is a special channel marked 'AV' or 'VCR' for this purpose.

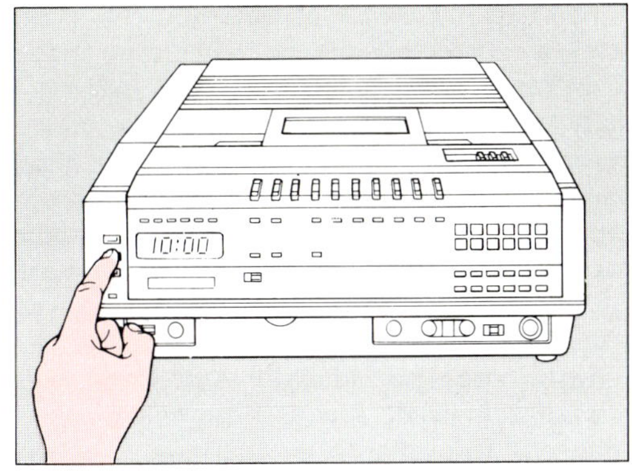

Switch the VCR's 'Function' switch to 'On'. This stage, which bypasses the timer, will vary with the make of VCR, so look through the instructions in your manual again if you are in any doubt.

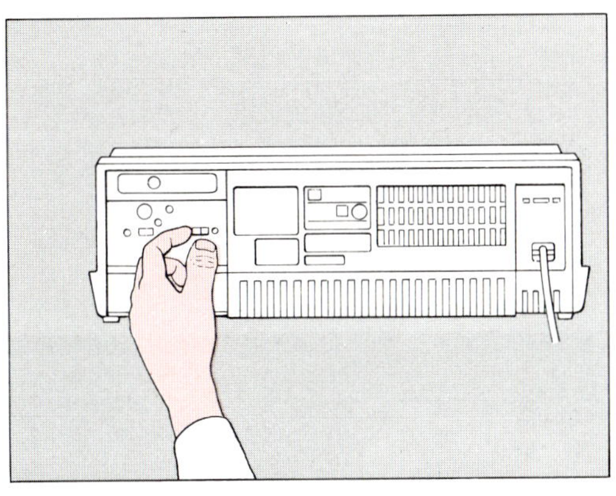

To produce a test signal to which to tune, either put the VCR's test signal switch to 'Test', or play a pre-recorded test tape, such as the one supplied with Grundig's VCRs, or any recorded tape.

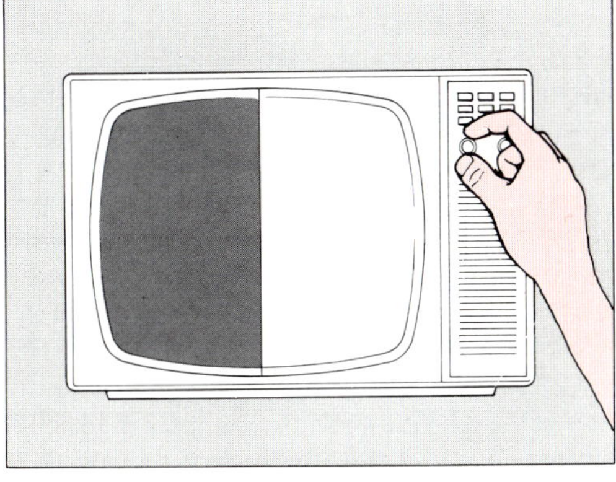

Adjust the tuning knob for your TV channel until a clear tuning signal, usually a pattern of black and white stripes, appears on the screen. Then switch the VCR back to its normal (colour) setting.

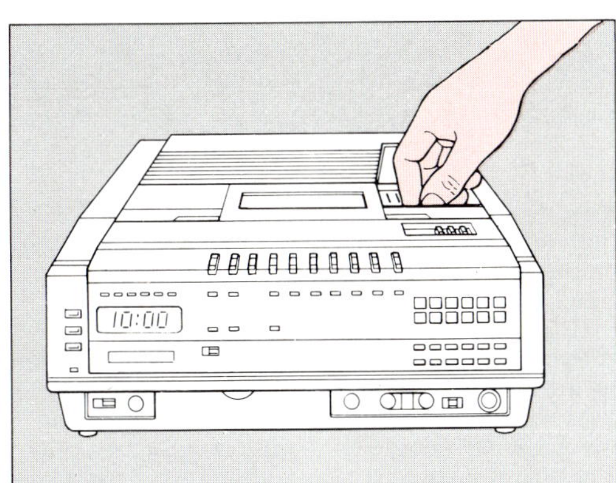

You can now tune each channel on your VCR to a TV station. Compare the picture on the VCR channel with your normal TV picture. If it is inferior, try retuning.

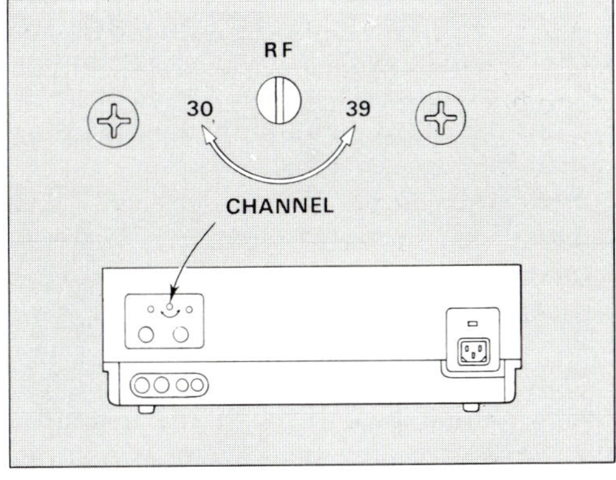

If you get herring-bone interference, or another station on the screen, you can often retune the VCR's RF reception by means of a screw on the back of the VCR.

The video cassette

Video cassettes work in basically the same way as audio cassettes, in that a tape made from plastic film and coated with a magnetically sensitive material is passed over a recording head. The signal fed into the head causes a changing magnetic field around the tape. Particles in the tape coating align themselves with the field and so reflect what the magnetic field at the recording head was like when that portion of the tape ran over it.

If the tape is passed at the same speed over a playback head, this is able to read the magnetic information on the tape. The information can be erased by passing the tape over an erase head, which uses a strong magnetic field to realign the particles of magnetic material in the same direction, wiping out the recorded pattern. The video tracks are about a thousandth of an inch (0.025 mm) apart.

Early professional video recorders recorded video tracks across a 2-inch (5-centimetre) wide tape. Since then, tapes have grown narrower and narrower, down to CVC format quarter-inch (6-millimetre) tape. This has been made possible by helical scan, the process of recording tracks at an angle across the tape, done by a pair of recording heads on a high-speed rotating drum. If the heads were fixed, as on an audio recorder, the tape would have to move impossibly fast for enough information to be crammed on it.

The tape follows a complicated path over a series of guides and rollers. Early video tape machines were reel-to-reel models, and threading the tape from the input reel through the machine, around the drum and on to the output reel, was a tricky business. It was only the invention of the pre-packaged cassettes which made loading and removing the tape from the recorder a simple process, but this depended on building the threading system into the recorder.

Sony introduced the cassette in their three-quarter-inch (19-millimetre) U-matic format, where the tape is taken from the cassette and fed into the machine in a path which roughly resembles the letter U. Their half-inch (5-centimetre) Betamax format follows a similar path through the machine. The VHS threading mechanism winds the tape around a more complex path, resembling a letter M; so the tape has to be wound clear of the record/playback path for fast wind or rewind.

The tape follows a complicated path through rollers and capstans for accurate positioning on the video heads. This is done by hand with reel-to-reel machines. Cassette recorders do it automatically.

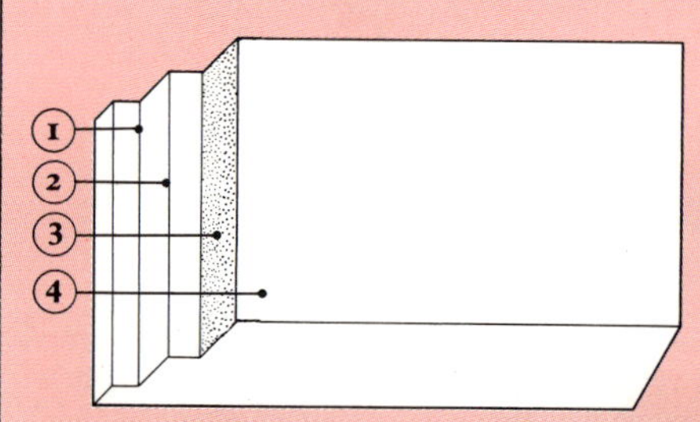

Video tape starts off as a clear polyester base, **2**, which must be thin, flexible, strong and non-elastic. An anti-static carbon backing **1**, prevents static eletricity from building up on the tape as it runs through the recorder. The metal oxide particles, **3**, which store the recorded magnetic fields, are bound to the other side of the base. Finally, to offer the smoothest possible surface to the video heads, the metal oxide is polished, **4**.

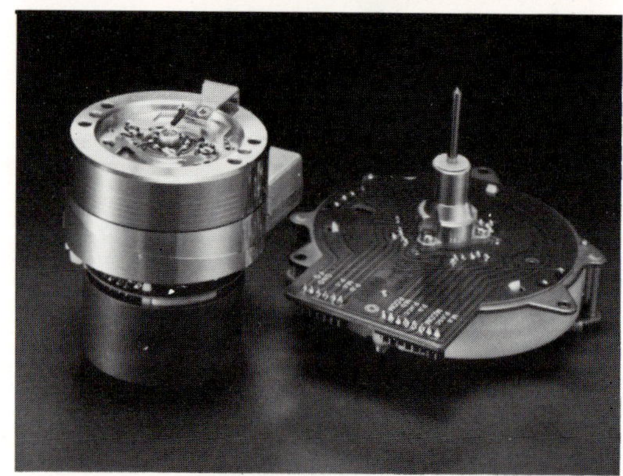

The drum in which the video heads are mounted, *above*, is tilted with respect to the tape, so that the video tracks are recorded and played back diagonally across the width of the tape. The drum normally holds two heads. On each rotation of the drum, one head records the even lines and the other the odd lines of the picture, so that a single rotation records one frame. A tracking control on VHS and Betamax VCR, allows the video drum angle to be changed to track the tape more accurately. Video 2000 drums move automatically for better tracking, especially in slow motion and fast picture search.

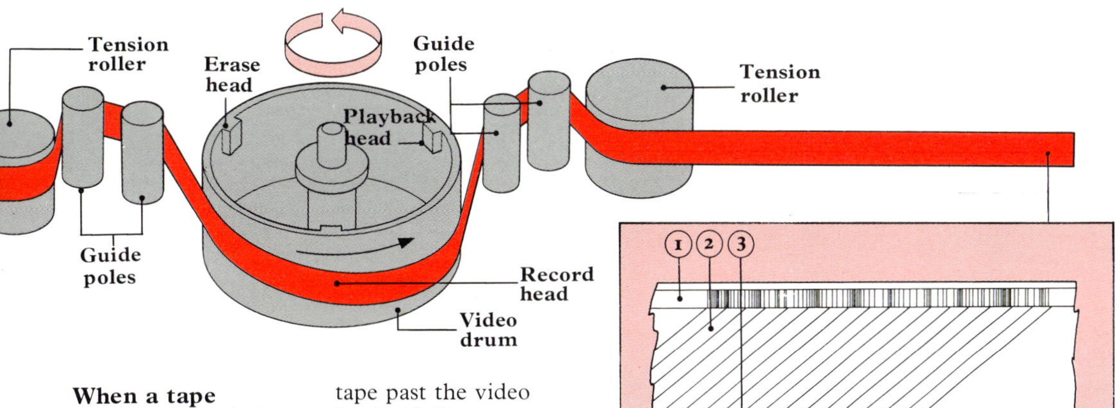

When a tape cassette is loaded into a VCR, guide poles pull a loop of tape out and wrap it round the video drum. This is called lacing. Tension rollers take up any slack.

At the end of the tape path, just before the tape re-enters the cassette, it is squeezed between a pinch roller and a capstan. The pinch roller is driven by a motor to pull the tape past the video heads. A Betamax cassette is laced up as soon as it is loaded, but a VHS cassette is only laced up when a playback control is operated. Different tape formats lace the tape in different ways.

During playback, the drum spins at 25 r.p.m. (UK) or 30 r.p.m. (USA), producing 25 (UK) or 30 (USA) complete TV pictures per second.

The sound-track, 1, is recorded along one edge of the tape. Video information is recorded diagonally across the tape, **2,** the full width in VHS and Betamax, or half the width in Video 2000. Pulses recorded on the control track, **3,** ensure that the tape is recorded and played back at the same speed. They are the magnetic tape equivalent of cine film sprocket holes. The quadruplex tape used on reel-to-reel video recorders has a cue track along one side which can be used to store pulses for electronic editing.

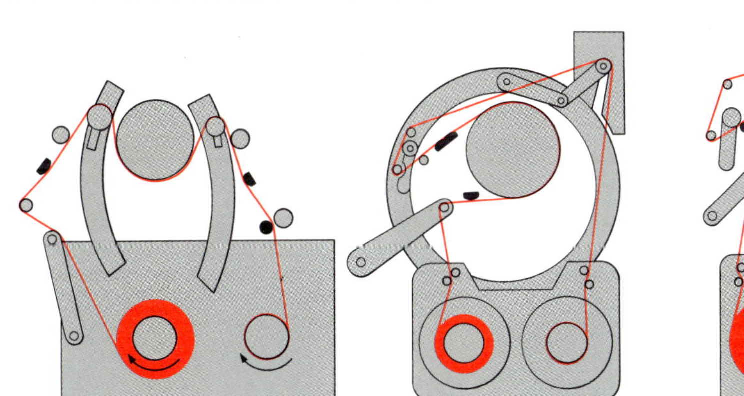

When a VHS tape is inserted in a recorder, the spool locks are released and two guide poles pull a length of tape out of the cassette. Because the tortuous tape path would place severe strain on the tape during fast wind, it is only laced up during normal-speed playback. M-load (or M-wrap) takes its name from the shape of the tape when laced up, *above*.

In the semi-professional U-matic format developed by Sony, the tape is laced up as soon as the cassette is inserted in the recorder. Capstans on a circular guide track catch the tape and pull it round the video drum. Tension arms take up any slack in the $\frac{3}{4}$-in (19-mm) tape. U-matic format has become the standard for the first cable TV stations in the UK.

Betamax, developed by Sony, is a smaller version of U-matic, using narrower $\frac{1}{2}$-in (12.5-mm) tape. Lacing operates in the same way as U-matic, although a smaller mechanism is used. The tape remains in contact with the video heads for all playback and winding functions until the 'Eject' button is pressed. Despite this, there is no excess head wear.

Formats

Video tape is no more than 20 microns thick (a micron is a thousandth of a millimetre). It has to stand up to winding on and off high-speed rotating drums and threading and unthreading hundreds of times during its lifetime, without breaking or distorting by so much as a thousandth of an inch.

The polyester plastic tape which carries the specialized coatings is prone to shrinkage if it gets too hot. Although the tape is stable over normal temperature variations, some shrinkage and stretching can happen with a brand-new tape, so try winding it fully forward and backward before using it.

Video tape records information magnetically on the ferrous oxide layer; so stray magnetic fields can erase recordings. In fact, a tape eraser (called a bulk eraser) works by using a strong magnetic field to realign the magnetic particles on the tape so that they all face the same way.

Keep cassettes clear of any other magnetic fields, such as those produced by magnets in audio speaker cones, in toys or even in magnetic document clips.

The worst effects are produced by collapsing magnetic fields. Switching off a vacuum cleaner close to a television can cause the colour guns to become misaligned, which may result in coloured patches being produced on the screen.

In time, the magnetic particles of the tape coating will be dislodged, causing white streaks on the picture. This is called dropout, and it is usually more common in the first few minutes of a tape. One way of postponing this problem is to begin new recordings at different times over the first five minutes of tape to reduce wear at any given point.

Tape development is now focusing on two areas: cramming the necessary information on to narrower tape, such as the quarter-inch (6 millimetre) Compact Video Cassette (CVC) tape, or the twin-track Video 2000 cassette (the only video cassette on which you can record both sides); and making the tape thinner so that longer tapes can be fitted into cassettes. The latest VHS cassette provides four hours of playing time, but the tape is so thin that some of the older VCRs with heavy threading mechanisms can stretch it. Sony's Betastack gives 13 hours of unattended recording by cycling automatically through four ordinary Betamax cassettes, stacked one on top of another.

Video cassettes have a safety tab to prevent accidental erasure of recordings. When a cassette is loaded into a VCR, a probe detects the presence of the tab. If it has been removed, the tape cannot be erased. If it is necessary to record on the tape at some future date, the tab hole can be covered by a piece of adhesive tape. VHS, Betamax and U-matic cassettes each have one tab. Video 2000 cassettes have two.

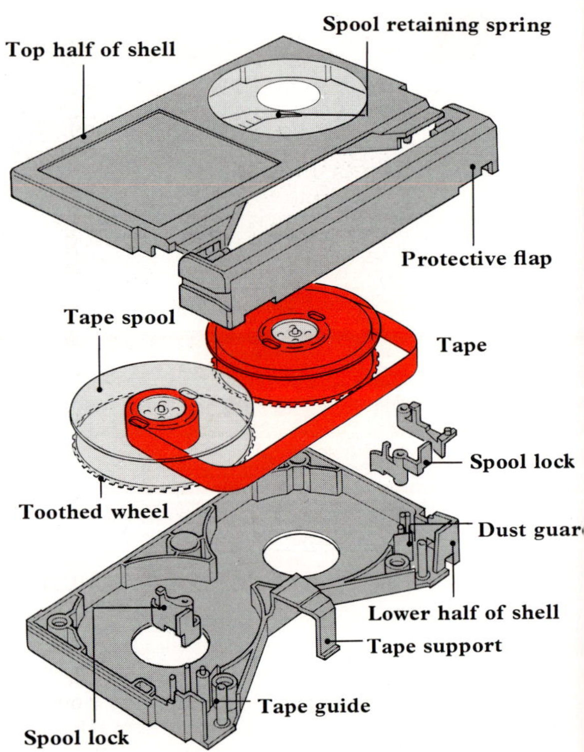

A video cassette might look like an audio cassette, but inside it is very different. The tape spools are mounted on toothed wheels, which are released only when the tape is inside the recorder, to prevent it from slackening. The two halves of the plastic shell are held together by screws (five for domestic VHS cassettes and six, in a stronger arrangement, for industrial machines).

Loading a cassette
Before a cassette can be inserted into a top-loading VCR, the cassette drawer must be opened by pressing the 'Eject' control. Hold the cassette with the protective flap farthest away from you and with the loose spools facing downward. Push the cassette into the open cassette drawer. The cassette drawer must then be closed by pushing it down until it locks into place. The recorder is now ready for playback or recording.

It is not necessary to prepare a front-loading model before inserting a cassette. The cassette is held as before and pushed into the loading slot. When it is about half way in, the motorized mechanism takes over, pulling the cassette away from you and down into the machine.

When loaded, a cassette can be seen through a window in the top loader's cassette drawer. This is not so with an all-enclosed front loader, and in some machines it is possible to insert a second cassette on top of the first. VHS, Betamax and U-matic cassettes can be inserted only if they are the right way up. Video 2000 cassettes can be inserted either way up, but the protective flap must always be farthest away from you.

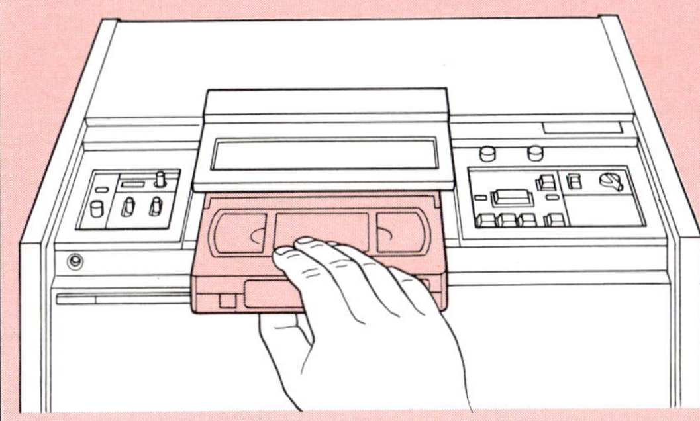

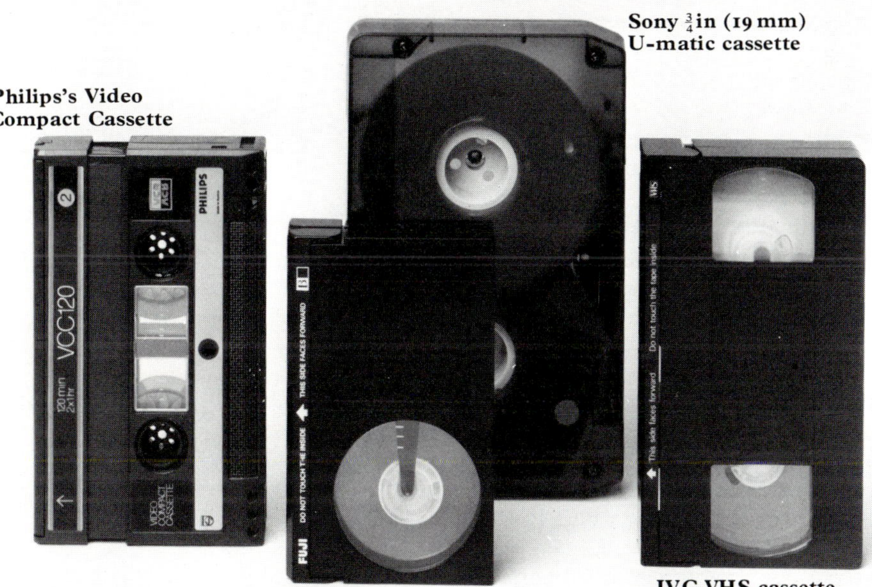

Sony $\frac{3}{4}$in (19 mm) U-matic cassette

Philips's Video Compact Cassette

Sony Betamax cassette

JVC VHS cassette

There are five widths of video tape in use today. In high-quality professional video, including broadcast, 2-in (50-mm) open-reel tape is the norm. In one 2-in system, Quadruplex (Quad) the video heads write almost at right-angles to the edge of the tape.

One-inch (25-mm) tape is now taking over. There are three formats: A-format was pioneered by Ampex in the sixties, B-format is used by Bosch, IVC, RCA and Philips, while the newer C-format is used by Ampex, Sony, RCA and Marconi.

Sony's U-matic, *top*, is the only $\frac{3}{4}$-in (19-mm) system, and was the first video cassette format. The High-band version gives broadcast-quality recordings. It uses the same cassettes, but will only play back on ordinary U-matics in black and white.

There are two $\frac{1}{2}$-in (12.5-mm) systems, Sony's Betamax, *above*, the smaller of the two, and JVC's VHS, *above right*.

The Video 2000 format uses the Video Compact Cassette, *above left*, by Philips, which has a tape width of $\frac{1}{2}$-in (12.5-mm) but is a $\frac{1}{4}$-in (6-mm) format as it only writes over half the tape width.

The Funai Microvideo system uses a $\frac{1}{4}$-in (6-mm) format tape the size of an audio cassette.

Recording off air

To save needless plugging and unplugging of antenna leads, manufacturers have ensured that as long as there is power running to the VCR, the video signal received by the antenna will pass through to your television set.

It is better to leave the VCR power on all the time as this saves the trouble of resetting the digital clock. There is no electrical hazard associated with such a practice. The machine is over-abundantly supplied with internal fuses and it uses about as much power as an electric clock.

On VHS and Betamax machines, you can use the tuner in the VCR for normal viewing. This means that if you leave the television set tuned to the VCR channel, you can change the channel by pressing the 'Channel Selector' buttons on the VCR. The advantage of doing this is that you can utilize a remote control (most VCR manufacturers offer an optional one if they do not supply one with the machine) to change TV channels – and the VCR usually has more channels available than a television set.

Prepare to make your recording well in advance of the time the broadcast begins. Make sure the VCR is switched to the tuner and that the television is set to the VCR channel. Manufacturers label their switches differently but the principles remain the same; consult your user's manual.

Insert a tape. Check that it is rewound and set the memory counter to zero. Make a short test recording and adjust if necessary.

You can usually hear the tape turning, and on most top-loading machines you can see it rotating. If it does not move, check that you have not accidentally pressed the 'Pause' button. If no picture appears the safety tab at the back may have been removed.

Rewind the tape, watching the memory counter, or using the VCR's memory facility, if the tape is already partly used. Be ready to record when the show begins. Remember to watch for the finishing time of the recording or the VCR will continue to record until the tape ends. Only by using the timer can you program the recording to stop on time.

While recording is in progress you can watch the show, or, because you are not recording from the television receiver but directly from the antenna (or aerial), you can switch the input select mode to 'TV'; the television receiver to another channel, and watch a different show.

Knobs and switches

You can tell a middle-aged VCR by its piano key controls. Pressing a key moves a series of metal levers and links to operate the appropriate solenoids (magnetized wire coils) and relays. Built-in safety checks to ensure that two keys cannot be depressed together exacerbate the slowness of the operation. Each change of function requires the stop mode to be selected in between.

The application of microchip technology has changed all that. The mechanical keys have been replaced by touch-sensitive controls. Only a featherlight touch is needed for these to send a minute electrical current to the microprocessor to instruct it as to which function is required. The microprocessor then operates the necessary switches and solenoids, ensuring that the correct sequence is followed, so you can now switch between functions without having to select the stop mode, knowing that the VCR's brain will sort things out. The removal of mechanical switching has also speeded up the operation of the various functions, drastically reducing the delay between, say, rewind and play.

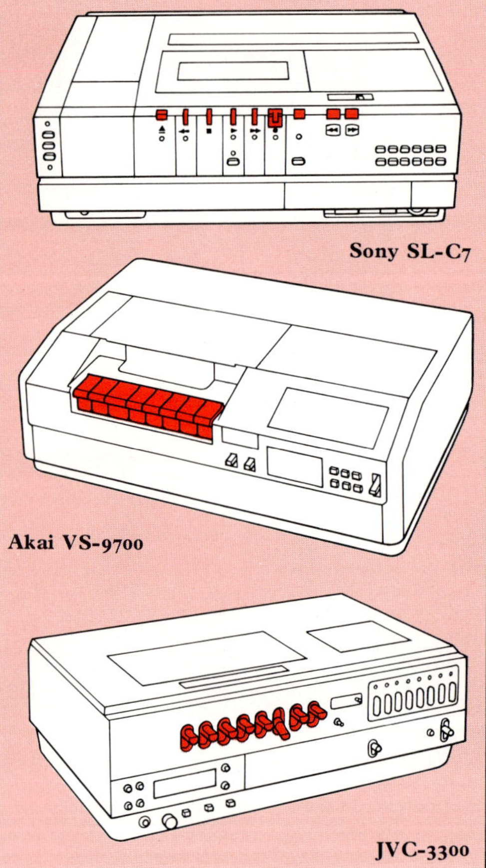

Sony SL-C7

Akai VS-9700

JVC-3300

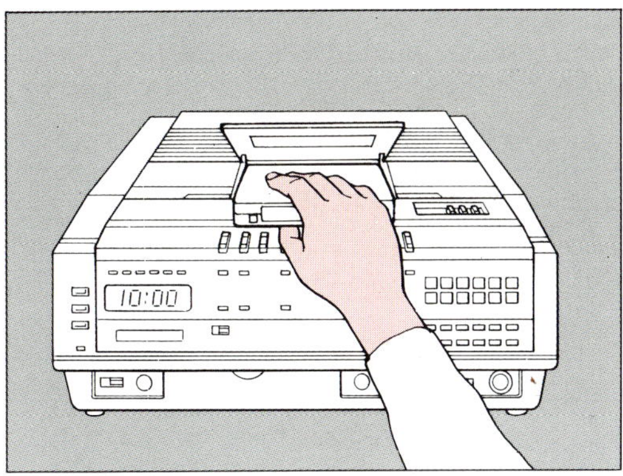

Make sure the VCR is switched on at the wall socket and the VCR 'Power' switch, and that it is not in the timer mode, or the cassette compartment may not open. Insert a tape and press 'Rewind'.

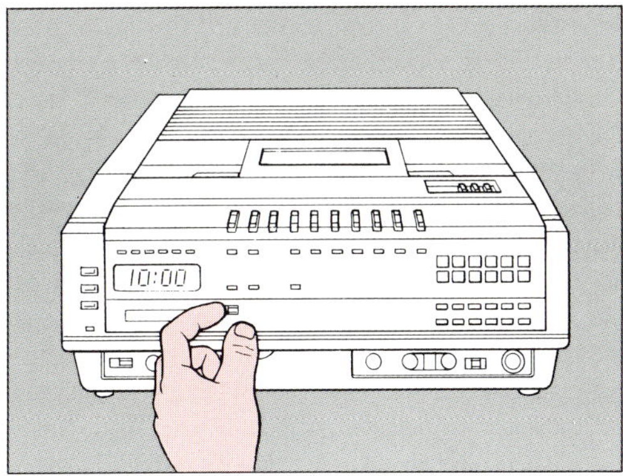

Set the 'Input' or 'Record Selector' or the 'TV/VTR' switch to TV: VCRs with a 'Tuner/Timer' switch should be set to 'Tuner'. On some machines the input selection is made automatically.

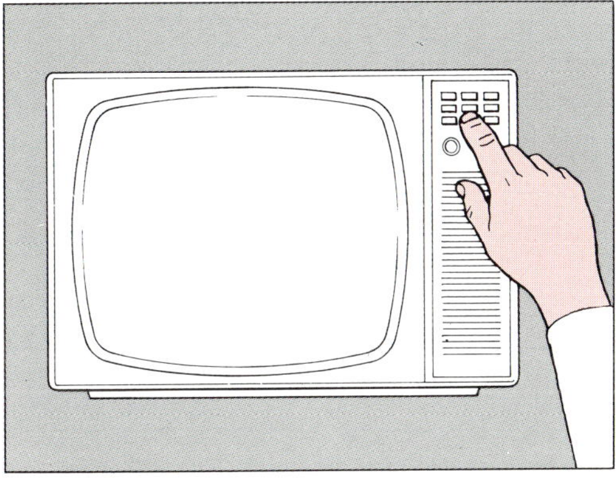

Select the channel you wish to record on the channel selector. The channel number will light up. Check that the correct picture is appearing on the screen, and make any adjustments.

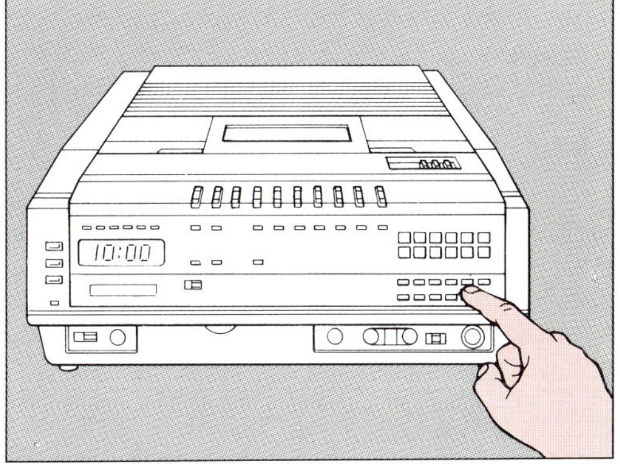

Turn on the TV set and select the 'Video/VCR/VTR or 'RF out' channel on the TV receiver.

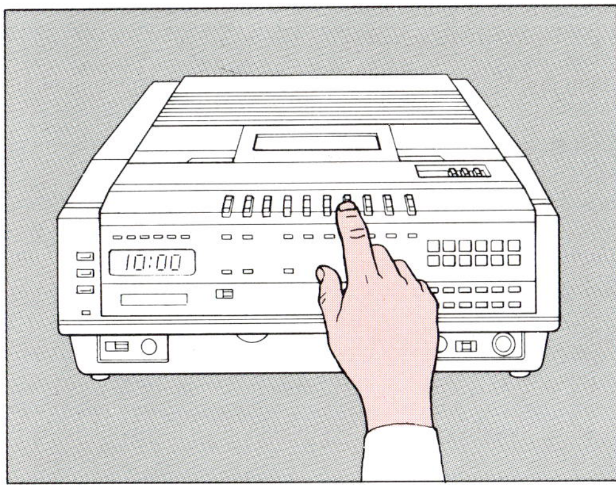

Press the 'Record' button or, on some VCRs, the 'Record' and 'Play' buttons.

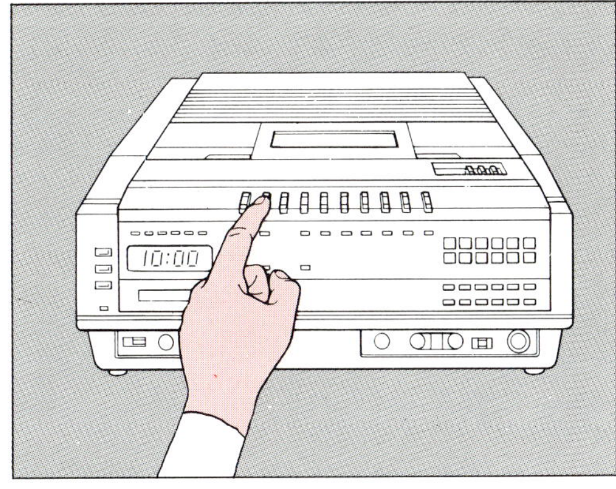

Press 'Stop' and/or 'Rewind' when the recording is complete. Some VCRs automatically rewind when the cassette comes to an end. Remove the cassette.

Time-shifting

In 1972 Philips launched the N1500, the first home video recorder to incorporate a tuner-timer, making the video recorder saleable on the domestic market. Until then, users had to rely on crude mechanical timers; finding the right starting point was something of a hit and miss affair.

These early timers were accurate only to the nearest minute. Modern VCRs have electronic timers accurate to the second.

The timer's job is to switch the VCR on to record at a predetermined time. The simplest timer will record only one such 'event', but most have a repeat facility to record at the same time daily or weekly. Sony's Betastack mechanism allows a number of tapes to be stacked up, giving 14 hours recording time.

Most budget models have a seven-day timer, with a single event capability plus a daily or weekly repeat. Luxury models usually offer eight events over 14 days, but any number of permutations – up to 99 days on the Grundig 2 × 4 Super – are available.

If you use your VCR mainly to record feature films a single event timer is all you will need. It is also easy to set up. The more event capabilities and the longer the timing period, the more complicated is the timer.

It is essential that the digital clock displays the correct time. It may use a 24- or a 12-hour display; so make sure it is not showing 6.00 am instead of 6 pm or 9 instead of 21.00 hours.

A single event timer is programmed with the start time, stop time, day of the week, and repeat facility. A timer that can be programmed over several days will show the days either by name or number (0 = today; 1 = tomorrow, and so on).

Remember to select the correct channel on the VCR before setting the timer, because the tuner will not operate, once the VCR is set to the timer mode.

Multi-event timers need considerably more information, so they are more difficult to program. Each model differs in the method of programming the timer, but the principles remain the same. To save confusion, make a list beforehand with all the information required for each event: 1 Event no.; 2 Channel no.; 3 day.; 4 start time; 5 duration or stop time. If you treat complex programming as a simple question and answer sequence following a logical progression, you will not miss out an essential step.

The timer setting procedure soon becomes second nature, but it is nevertheless easy to leave out an important step, so preventing the machine from making the recording. To stop this happening, before leaving the recorder to make an unattended recording, ask yourself these questions:

1. Is the AC supply switched on?
2. Is the recorder loaded with a cassette?
3. Is the cassette rewound to the correct starting point?
4. Is there a safety tab in place on the cassette? If not, has the hole been blocked by a piece of adhesive tape?
5. Is the digital clock showing the correct time and day?
6. Have the correct recording start time and length, or stop time, been programmed?
7. Is the VCR switched to timer mode?
8. Is there enough recording time left on the cassette for the total length of the show.
9. Should you have pressed the 'Record' button? Check your manual.

For accurate time setting, advance the clock to 1 minute after the correct time. Release the clock button on a radio time signal.

88:88

If the clock display flashes, there has been a power-cut and it must be reset.

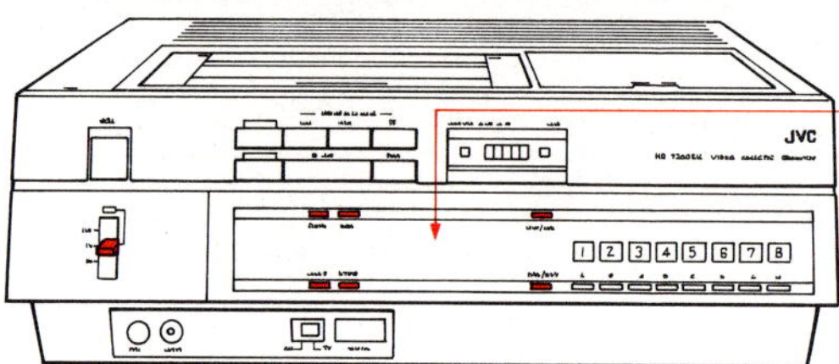

Japanese VCRs, such as the JVC HR7300EK *above*, feature a function switch in a prominent position on the front panel. This is normally set to the 'On' or 'Operate' position for recording and playback. In the 'Off' or 'Standby' position, power is cut off from the recorder section. For unattended operation, once the recording times have been keyed in using the timer setting-buttons, the function switch must be set to the 'Timer' position. The recorder will then appear to switch off. It remains apparently dormant until the start time, when recording begins automatically.

Recording continues until the stop time, when the power is once again cut off.

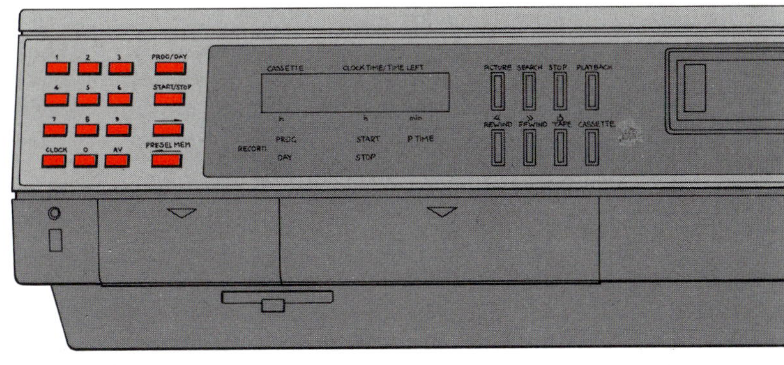

Grundig's 2 × 4 Super offers one of the most advanced timers available to domestic video users. Its tiny computer memory can hold all the information necessary to record up to five events on different channels, up to 99 days later. The method of entering information into this European machine is very different from that of Japanese VCRs.

The computer must first be told that you want to key in some details. Once the correct button has been pressed to start the sequence, the next nine steps necessary to enter all the information to make a future recording must be completed within 15 seconds. If you take longer, the memory is wiped clean. The whole procedure must then be started from the beginning again.

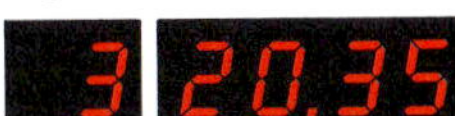

1 Press the top 'Presel. Mem' button to start the timer-setting sequence.

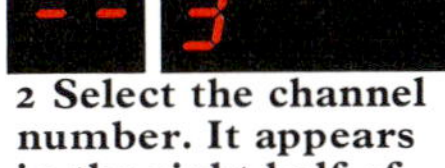

2 Select the channel number. It appears in the right half of the clock display.

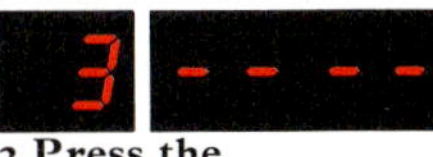

3 Press the 'Progr/Day' button to confirm the entry of the channel number.

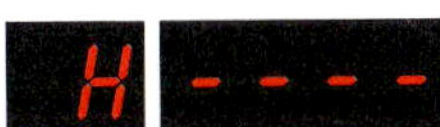

4 Key the start time using the numerical keypad. Remember, this timer uses a 24-hour clock.

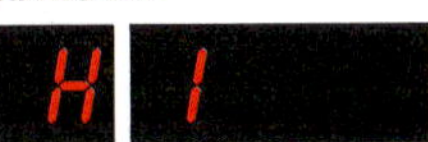

5 Pressing the 'Start/Stop' button confirms entry of the start time into the timer memory.

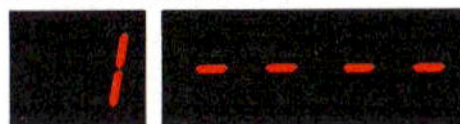

6 Press the top 'Presel.Mem' button again to switch the timer from time to day setting.

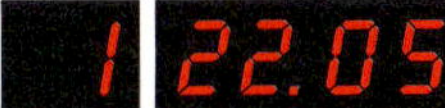

7 The 'H' on the display is German for today (*heute*). Press 1 for tomorrow, 2 for the day after and so on.

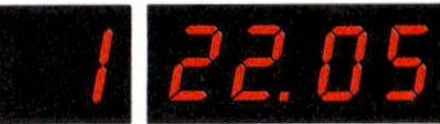

8 Press the 'Progr/Day' button to enter the day of the recording into the timer memory.

9 Finally, enter the stop time on the clock, and press the 'Start/Stop' button to enter it into the timer memory.

Auto timer recording

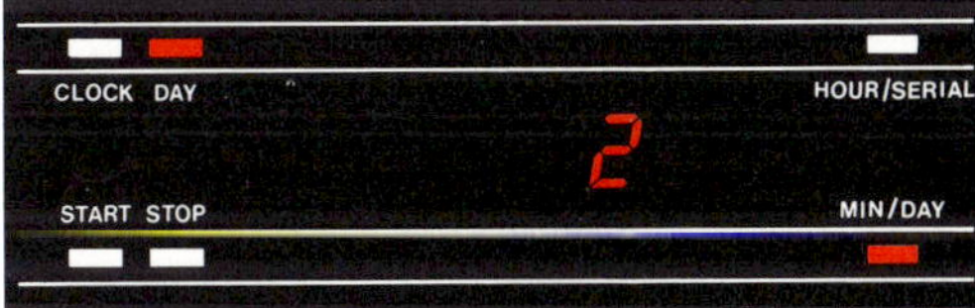

1 JVC's timer setting is typical of Japanese recorders. Hold the 'Day' button down and press the 'Min/Day' button until the recording day appears. Zero means today.

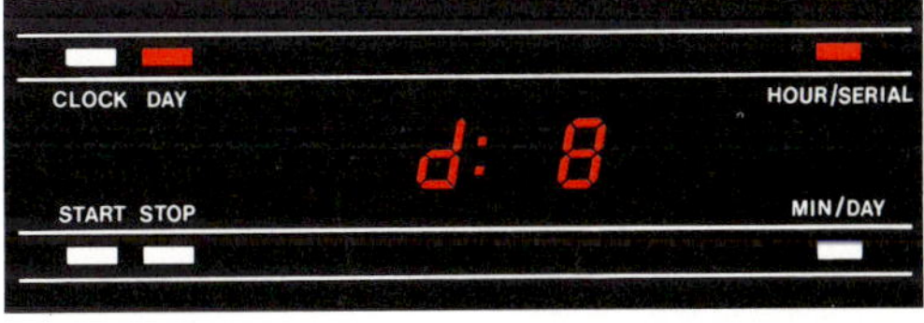

2 If the 'Day' and 'Hour/Serial' buttons are pressed at the same time, a 'd' appears on the clock display, and recordings are made daily at the same time. Most recorders have this function.

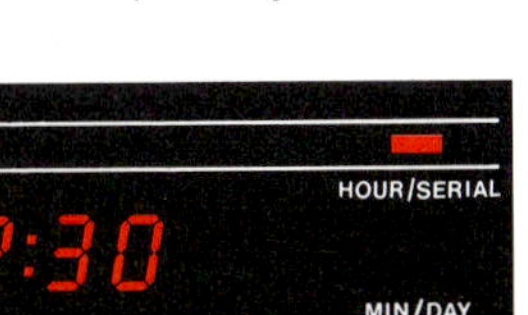

3 Next set the start time by holding down the 'Start' button, while pressing the 'Hour/Serial' and then the 'Min/Day' buttons. The hours and then the minutes will cycle round. Release the buttons when the correct time appears on the display.

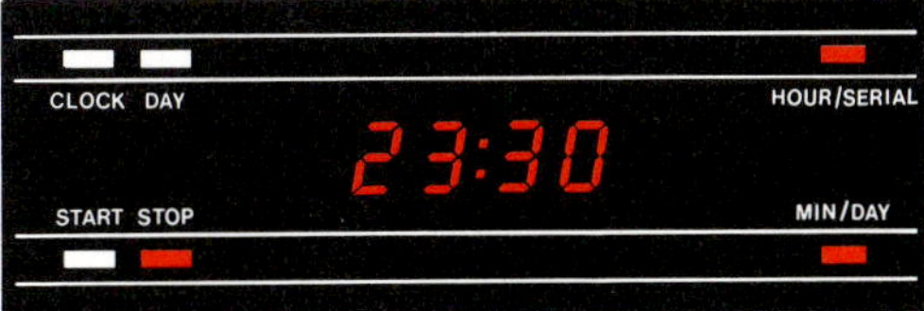

4 Set the stop time by holding down the 'Stop' button and pressing the 'Hour/Serial' and then the 'Min/Day' buttons until the desired stop time appears. Select the timer position on the function switch, and press the 'Record' button.

Use your memory

The simplest VCR memory facility takes the form of a switch labelled 'Counter Memory'. When this is switched on, the tape automatically stops in fast forward or rewind when the counter reading is zero. It is very useful for finding the beginning of a recording, provided you remember to set the counter to zero when you start recording.

More sophisticated recorders incorporate a system called 'Automatic Program Location' (APL). At the beginning of every recording an inaudible pulse is recorded on the tape. In fast forward and rewind the machine can be made to look for these pulses on the tape and to stop when it finds one.

A further development allows the memory to be used for playback. You can set a start and finish point, so that the machine will play a sequence, rewind and play it again as many times as required – to display advertisements in a store, for example.

The 'Go To' system on Video 2000 recorders makes it possible to program-in a tape location to which the machine will automatically go. To locate a recording which is 1 hour 10 minutes into a tape, punch in the coordinates on the numerical keypad and the machine will do the rest.

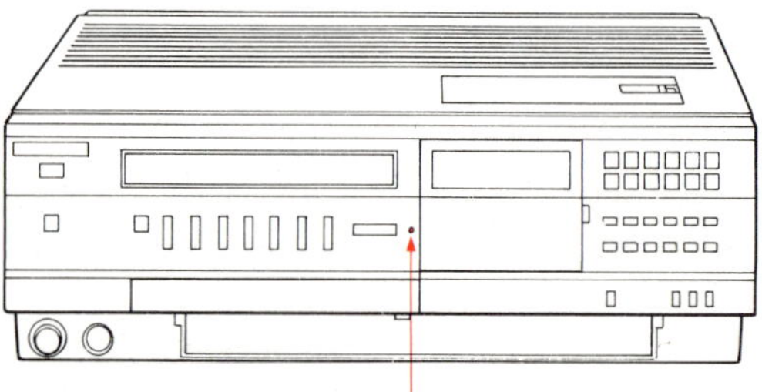

Most Sharp VCRs, *above*, feature an APL device. It scans the tape in forward or reverse playback, looking for inaudible marker tones, which are recorded on the tape every time the 'Record' or 'Pause' button is pressed, or when recording with a camera begins.

All recorders have a counter, *above*, with three or four digits. The 'Reset' switch sets the counter reading to zero. When the 'Memory' switch is on, the tape automatically stops when the counter reading reaches zero in both fast forward and rewind.

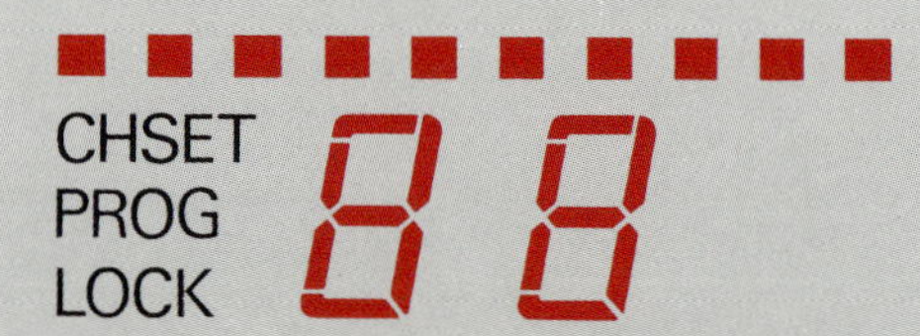

The 'Tape Remaining Indicator' on a machine such as the JVC HR-7700EK gives a rough guide to the length of tape left on a cassette. It takes the form of a series of illuminated bars. The bars are turned off one by one, from left to right, every 10 minutes until, during the last 5 minutes, the last bar flashes. It is certainly a useful extra: a glance at the display shows whether or not there is enough time left on a cassette to make the desired recording, but it is by no means essential. Several top-of-the-range models feature a tape remaining indicator, that on the Grundig 2 × 4 Super is the most sophisticated, with a full clock display.

When a cassette is loaded into the Grundig 2 × 4 Super, *right*, the display shows the length of the tape in hours, and the playing time already elapsed. In this case an 8-hour cassette (4 hours each side) is in the machine and 58 minutes have been used; 182 minutes remain free. Unfortunately, this very useful feature is unique to this VCR.

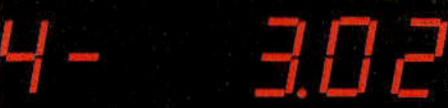

Pressing the 'Time left' button displays the running time remaining on that side of the tape.

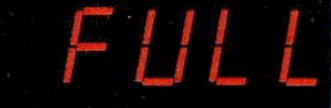

The display shows 'Full' when there is not enough tape left for the programmed instructions.

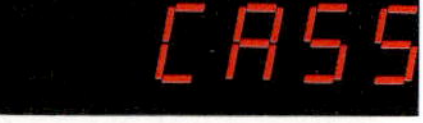

If the display reads 'Cass', it is necessary to insert, change over, or release a cassette from the holder.

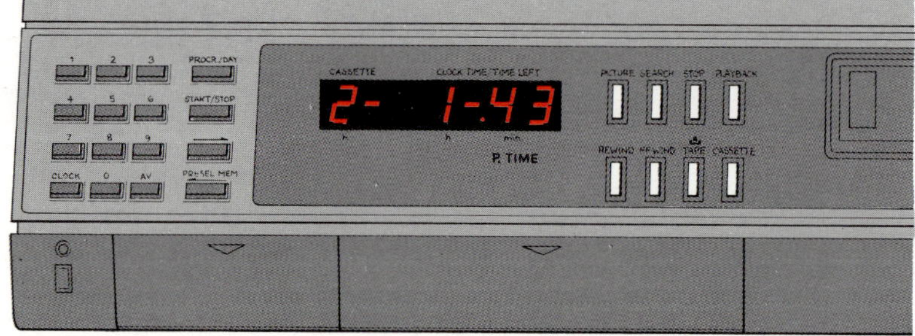

Picture search

Probably the most useful of the so-called trick frame facilities, fast picture search, allows the viewer to speed through the tape forward and (usually) backward at between five and ten times normal playback speed, while retaining a picture on the screen. The image can sometimes be almost unintelligible, particularly on the more advanced machines which give more than 20 times normal speed search. Used with discretion, this is a very valuable asset to any VCR, and it is pleasing to see the facility appearing on new models, even at the budget end of manufacturers' ranges.

Video 2000 picture search gives much better picture quality than either VHS or Betamax, because of different head-mounting methods. Indeed, Video 2000 picture search quality on both Philips and Grundig recorders is as good as normal speed playback quality. VHS and Betamax picture search is characterized by bars of white dots (noise bars) stretching across the screen.

Different manufacturers have different names for picture search: visual search, shuttle search or cue and review. Some machines display only a black and white picture in picture search.

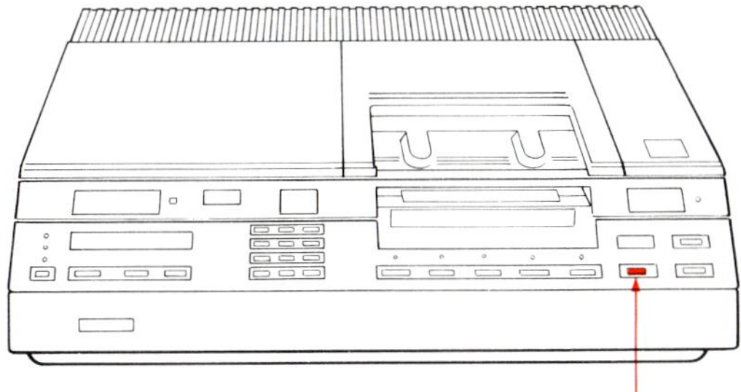

Tape movements are controlled by a series of buttons on the front panel. These feed control signals to a microprocessor. However quickly or randomly the buttons are pressed, the microprocessor is programmed to make sure neither heads nor tape are harmed.

The useful 'Go To' function has so far been limited to Video 2000 recorders, *above*. On Philips machines, you press the 'Go To' button and enter a counter number on the keypad. The tape is then automatically wound on or back to the required counter reading and stopped.

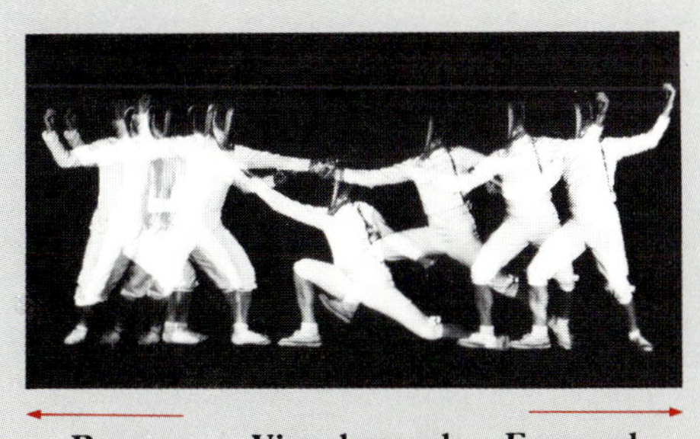

Reverse **Visual search** **Forward**

Visual search, also called picture search, cue and review and shuttle search, is the most useful of all the trick frame facilities. It provides a quick and easy way of skipping through unwanted sections of recordings, commercials perhaps, to find the beginning of a film. The tape is played faster than normal speed and a picture is shown on the television screen, but there is no sound.

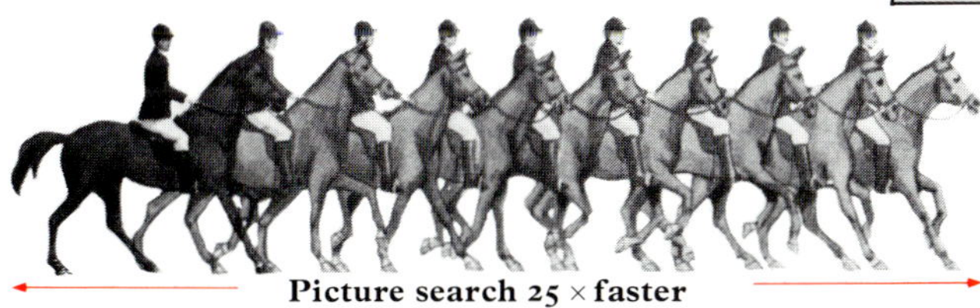

Picture search 25 × faster

The Toshiba 8600 (Europe) 8500 (USA) has a two-speed picture search. Touching the 'Cue or Review' button plays the tape (without sound) at 7 times normal speed. Pressing it down selects 25 times normal speed but it is hard to follow the action at this speed.

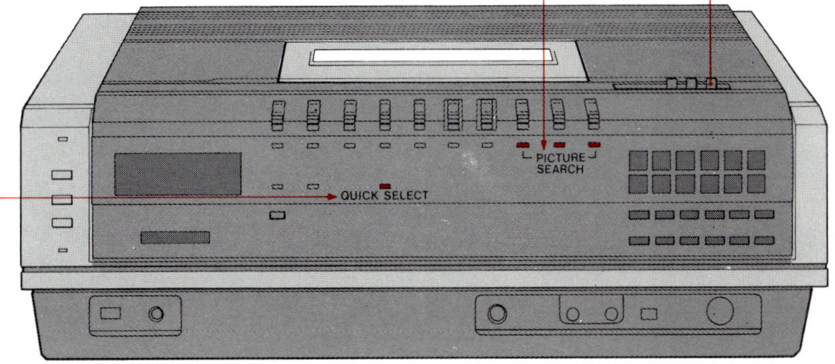

Toshiba's APL system is called 'Quick Select'. When activated, it automatically stops the tape at breaks between TV shows in fast forward or rewind. This is very useful for finding the beginning of a show.

Fast and slow

All but the most basic domestic VCRs now offer a range of playback speeds. These vary from slow motion, which may be a simple half-speed, on-off switch or a continuously variable effect controlled from the remote handset, to high-speed playback, with sound that is just intelligible.

Most common, and perhaps most useful, is picture search. As the name suggests, this offers a picture at, usually, five to ten times normal speed as a means of searching through tape to find a particular sequence. Sound is muted. Some VCRs will search only in the forward direction. Others will search only in black and white, but as search is not used for general viewing, this is not a serious drawback. The fastest search speeds are very difficult to watch.

The pause control usually doubles as a still picture control: the tape is stopped, but remains threaded round the video drum which continues to spin, and the video circuitry remains live. The information on one small section of tape is therefore continually scanned. As this would soon lead to excessive wear of that one piece of tape, a safety mechanism is usually tripped after several minutes to limit the length of pause possible.

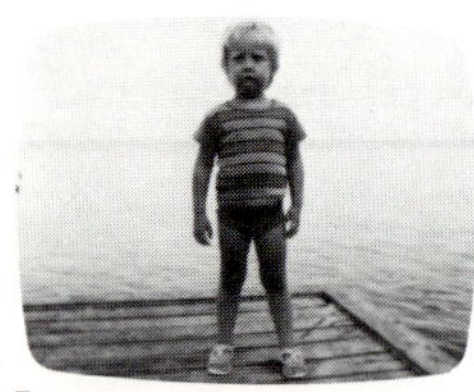

1

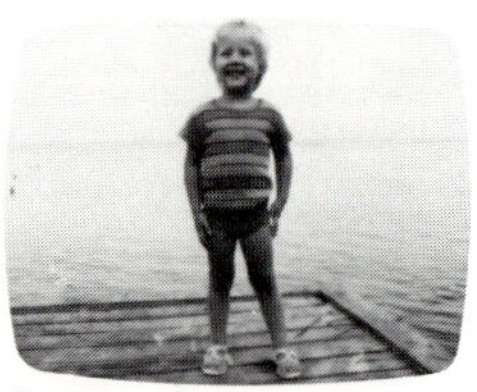

2

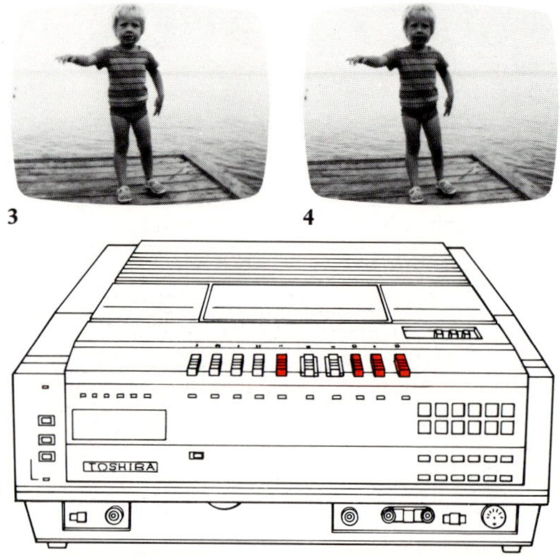

3

4

Picture search, **2**, *above*, can be used to find a particular place on a tape, either skipping forward, **1**, through unwanted material or backward, **3**, using fast wind, if you overshoot the desired point. It can be used to wind past commercials.

Pressing the 'Pause' control of a VCR stops the tape and produces a still picture on the screen, **4**, *above right*. Still picture quality varies enormously. It is generally accepted that Video 2000 is the best.

Running the tape at a slower than normal speed causes irritating noise bars across the picture. Some models give a better performance than others, and some very sophisticated models incorporate extra video heads to reduce this problem to a minimum.

The JVC HR-7700 has an unusual slow motion speed control. When slow motion is selected, the playback speed can be varied continually using two press buttons, one to raise the speed, the other to reduce it from normal through slow to freeze frame.

The Sony C7 features triple-speed playback in both forward and reverse directions. It is operated by first selecting playback at normal speed and then pressing the 'X3' button. In this mode of playback, there is no sound. Any noise bars (white dots and streaks) which appear on the screen can be removed by adjusting the tracking control, which should be returned to its usual central position for normal speed playback. The search speed of three times normal is slow, but perfectly adequate.

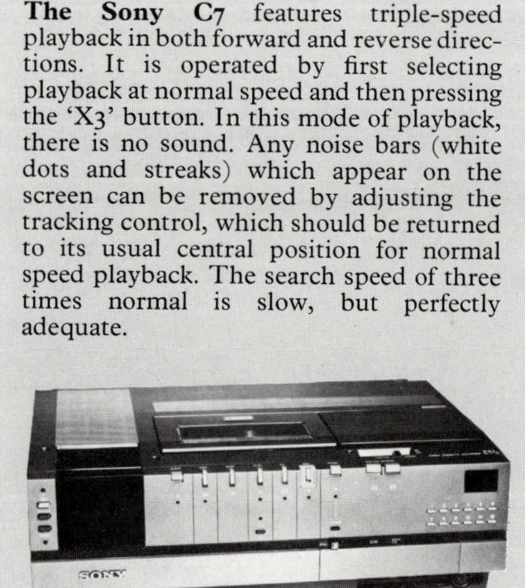

Sound

Sound has never been a strong point of domestic television sets and the reduction of home video recorders in size and quality from professional U-matic down to the domestic half-inch (12.5 millimetre) format has compounded the problem. There is not enough room on half-inch (12.5 millimetre) tape to store all the information needed for high-fidelity sound.

As home video expands, more and more users are becoming dissatisfied with the quality of the sound on their television and, in response, manufacturers are searching for ways to improve VCR sound quality by adding noise-reduction facilities. The most famous and well-proven noise-reduction system is Dolby, designed to solve problems in audio recording.

Other systems, such as Dynamic Noise Suppression, or Beta Noise Reduction, are attempts to circumvent the Dolby patents.

Many home video users plug their video recorders into their existing audio systems to try and improve the sound. Some hi-fi systems incorporate filters to cut out hiss at high frequencies, or booming at the bottom end, but most merely amplify any weaknesses already present on the recording.

An audio dub facility, *below*, is almost universal on home decks. Simply pressing the 'Audio dub' button at the same time as the 'Play' button erases the existing sound-track. A new sound-track can then be recorded using a mike or other external sound source. Several different shots can be held together by over-dubbing with a new sound-track.

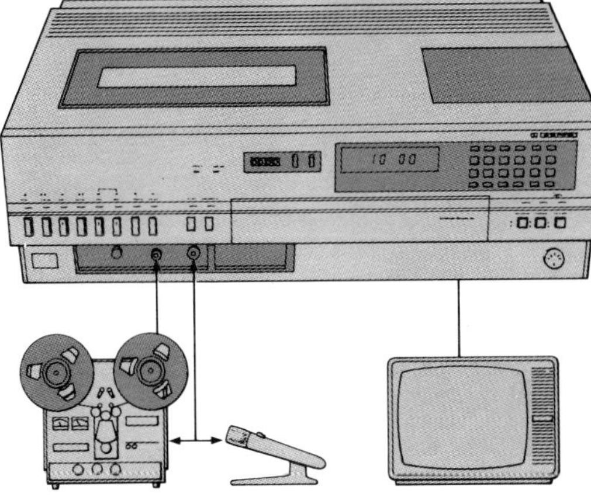

The sound quality of a recording is improved on VCRs incorporating the Dolby system. This effectively reduces tape's hiss.

Betamax VCRs use a system similar to Dolby to reduce the hiss on their sound-tracks. It is known as Beta Noise Reduction, or BNR.

A headphone input socket is useful, not only for monitoring the sound quality of a recording but also to maintain control over simple mixing effects.

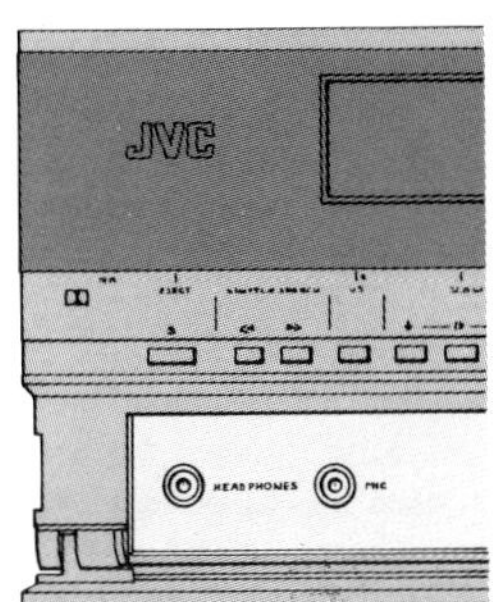

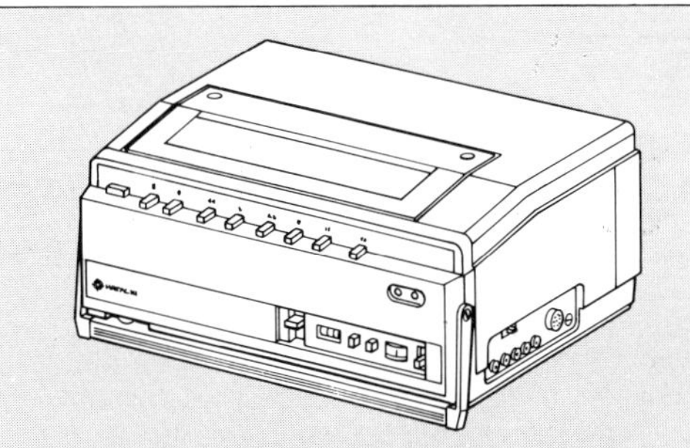

The Hitachi 6500 portable VCR, *above*, features a facility called sound on sound, to record a sound-track without erasing the existing sound-track. This basic mixing is useful for adding music to an otherwise dull tape, or for recording a commentary.

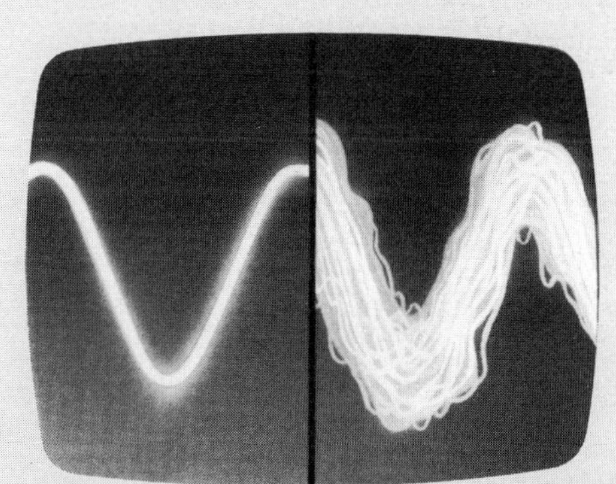

Improved sound quality on Video 2000 VCRs such as Philips VR2022 is achieved by the Dynamic Noise Suppression facility. For any noise-reduction system to work at its best, it must be used in both record and playback modes. It works by boosting high-frequency signals during recording and cutting them (and the hiss) during playback.

Editing

When a VCR plays back a recording, it locks on to a control track which was laid down on the tape as the recording was being made. Any interruptions in this control track will cause picture break-up until the VCR locks on to the new control track.

On the newer models of VCR, particularly on portable versions, a simple form of editing called back-space editing overlaps the new with the old recording, allowing sequences to be assembled one after the other, with clean cuts or edits between them.

More sophisticated models can memorize the control track and align the new recording with the old even if the VCR has been switched off for a while between shots. To insert a picture into a recorded sequence calls for a machine that will cut it in cleanly, and cut it out again to restore the original picture, equally cleanly.

By using a tape with a control track, it is possible to assemble a sequence and cut in further shots without disturbing the original control track. The development of this facility in home portapacks makes them more versatile: you can now shoot a sequence and then cut in further shots without having to use a professional edit suite.

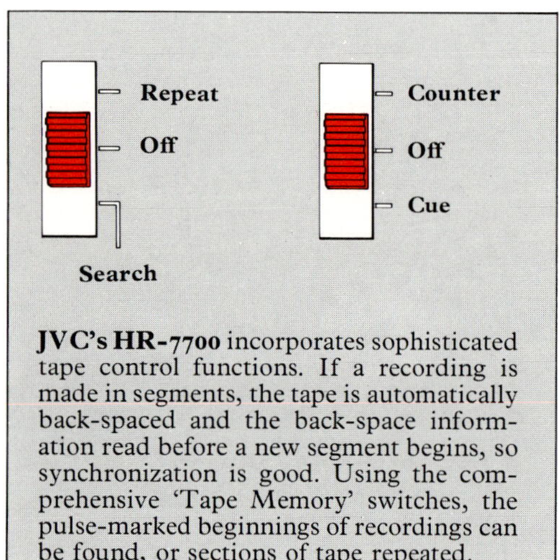
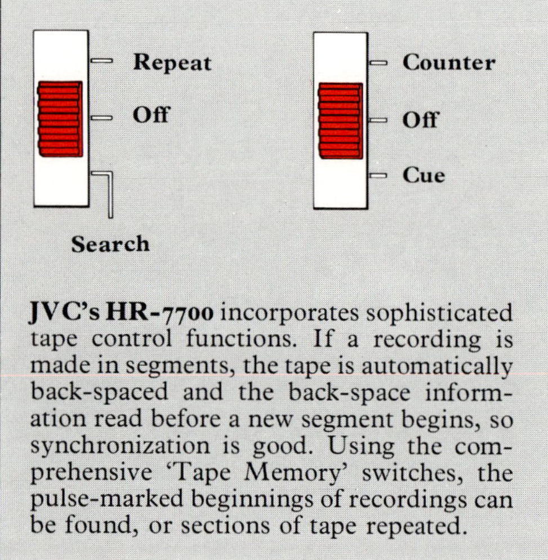

JVC's HR-7700 incorporates sophisticated tape control functions. If a recording is made in segments, the tape is automatically back-spaced and the back-space information read before a new segment begins, so synchronization is good. Using the comprehensive 'Tape Memory' switches, the pulse-marked beginnings of recordings can be found, or sections of tape repeated.

In auto-rewind, when the tape reaches the end it will stop and rewind for the recording to be viewed, *below*. If the tape memory is used with auto-rewind, and the counter set to zero before recording begins, the tape can be made to rewind to the point where the counter reads zero. This is useful for finding the beginning of a recording which is not at the beginning of the tape.

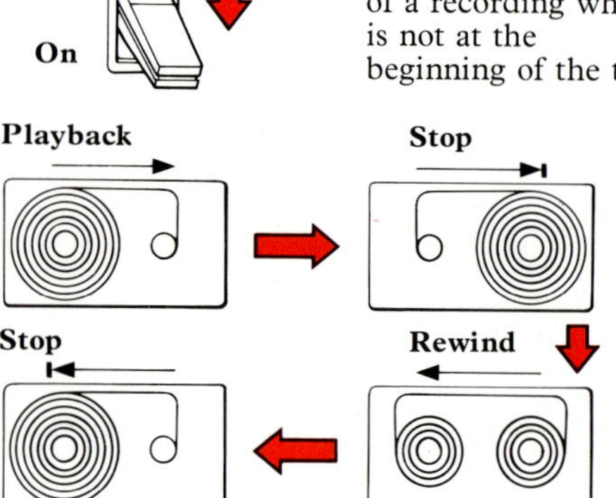

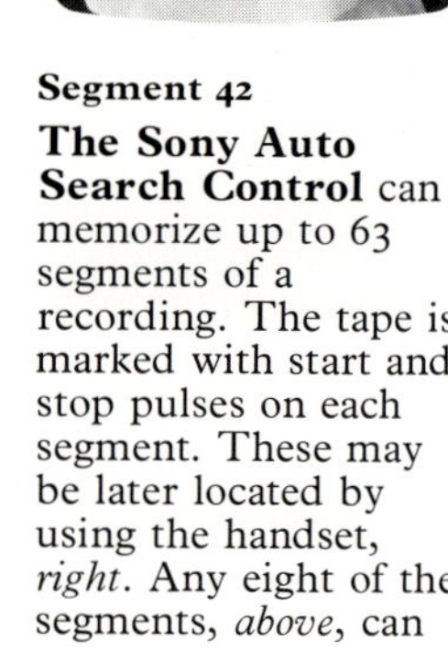

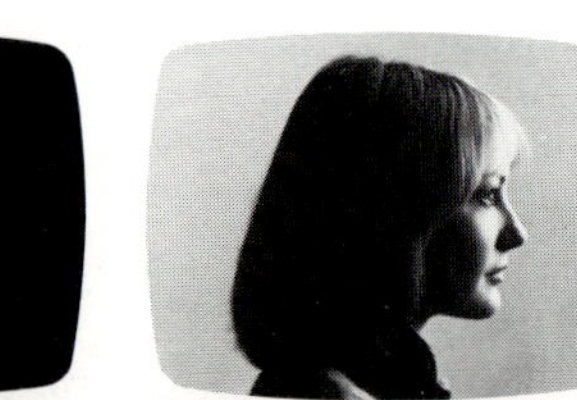

Segment 1

Segment 5

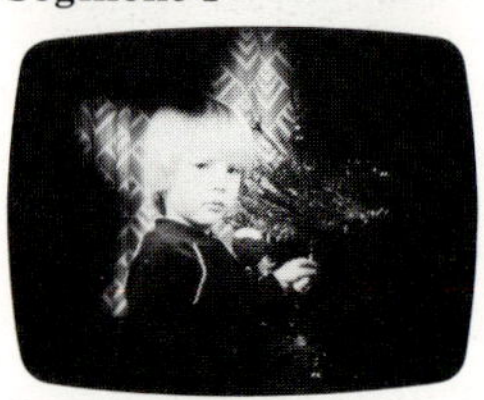

Segment 18

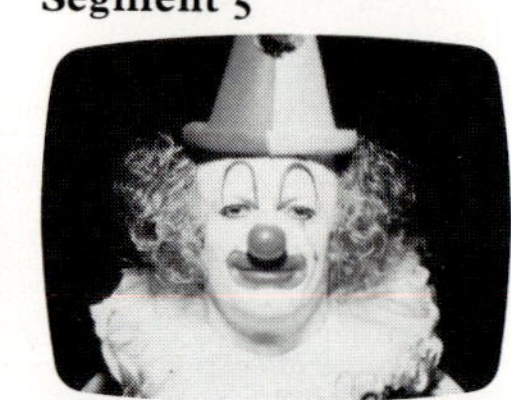

Segment 19

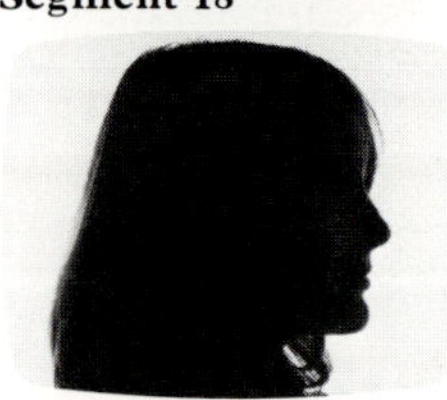

Segment 28

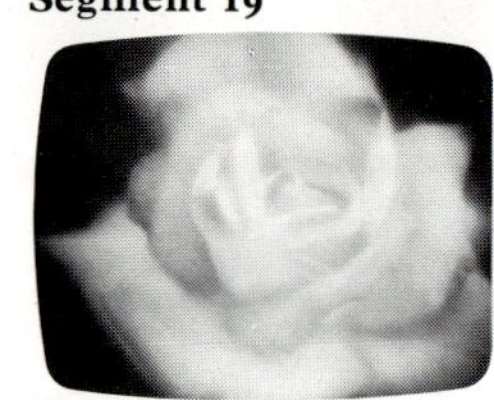

Segment 35

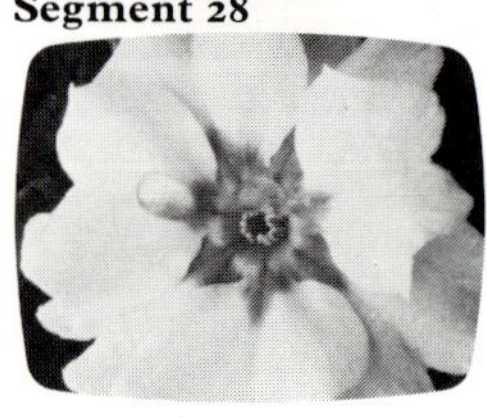

Segment 42

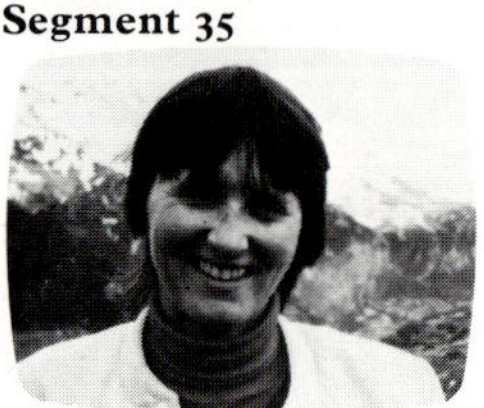

Segment 60

The Sony Auto Search Control can memorize up to 63 segments of a recording. The tape is marked with start and stop pulses on each segment. These may be later located by using the handset, *right*. Any eight of the segments, *above*, can be programmed to play in any desired order. The cycle can also be interrupted at any point to play a different segment, then return to the cycle. The unit can be used to add a professional touch to presentations or exhibition displays.

Remote control

Most VCRs are supplied with some sort of remote control. The simplest is a pause switch connected to the VCR by a length of cable. Although it is undoubtedly useful for stopping the tape temporarily, it has two major disadvantages. First, the viewer must sit within the cable's reach of the VCR. Second, a trailing cable is a hazard. Some basic or mid-range domestic VCRs use a more sophisticated cable handset with control switches for most of the VCR functions.

Infra-red remote control overcomes many of the problems of cable control. There is no connection between the VCR and handset; commands are transmitted using an infra-red data link, which is an invisible beam. Unlike cable remotes, infra-red units need a power source: a set of batteries. Because the standby current of these units is so small, typically only several millionths of an amp, they can be left on all the time. In fact they cannot be turned off. They should be removed if the remote is going to be inoperative for a long time. When a button on an infra-red remote is pressed, an indicator lights to show that a signal is being transmitted. If the indicator does not light, replace the batteries.

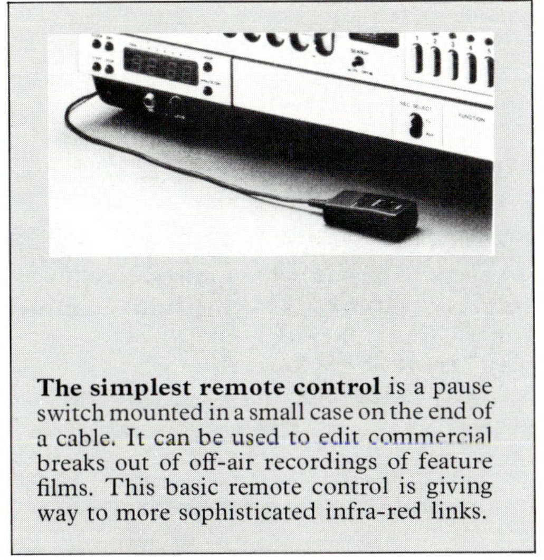

The simplest remote control is a pause switch mounted in a small case on the end of a cable. It can be used to edit commercial breaks out of off-air recordings of feature films. This basic remote control is giving way to more sophisticated infra-red links.

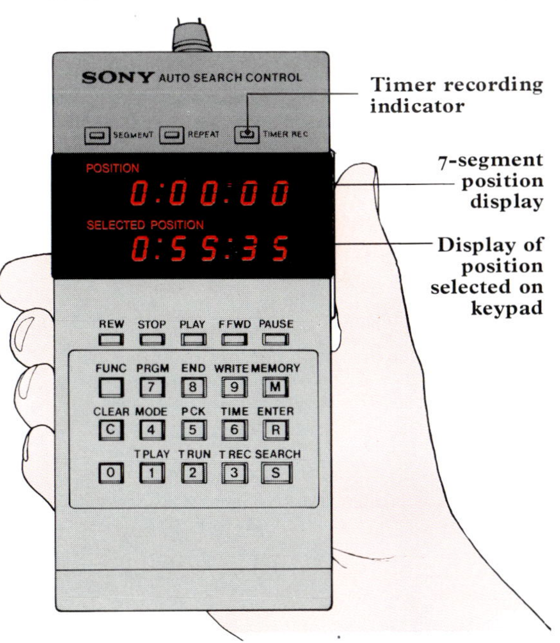

The Sony C7 is controlled by a remote control commander: a handset with press switches for all the main tape transport functions. Information is transmitted to the VCR using an invisible and harmless infra-red beam.

Although the beam is directional, the handset can be used up to 30° either side of dead ahead, and up to 30 ft (10 m) away from the VCR. In practice, the beam will reflect off smooth objects such as mirrors and window panes.

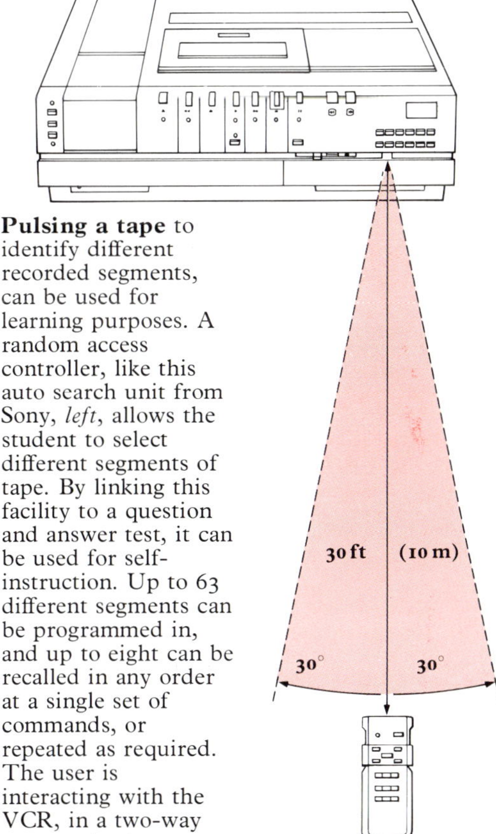

Pulsing a tape to identify different recorded segments, can be used for learning purposes. A random access controller, like this auto search unit from Sony, *left*, allows the student to select different segments of tape. By linking this facility to a question and answer test, it can be used for self-instruction. Up to 63 different segments can be programmed in, and up to eight can be recalled in any order at a single set of commands, or repeated as required. The user is interacting with the VCR, in a two-way communication.

Faults and how to fix them

The VCR you have bought or rented is likely to be the most complex and delicate piece of engineering in your possession. Indeed, because VCRs are so specialized, any servicing must be left to professionals.

You do not have to be an expert to get the best from it, for the design of a modern VCR is increasingly foolproof, but it is also increasingly complex, and there is a real chance that the time will come when you press a switch and nothing, or the wrong thing, will happen.

Before you call your dealer or reach for a screwdriver, remember that in a survey carried out over thousands of reported video faults, it was established that fully ninety per cent were not faults at all. The errors were in the way the owners had set the controls.

The first rule for trouble-free video is: read your manual. Run through all the processes you are going to use with any regularity and familiarize yourself with them. That way, you will not be baffled if something unexpected happens and, above all, you will not be billed for expensive, unnecessary service calls.

Check list

Some of the most annoying and frustrating faults are the simplest. If your VCR will not work, go through this deceptively simple check list and see if there is anything that you have overlooked.

If the VCR shows no sign of life:
Is it plugged in?
Is the plug properly connected?
Is the wall socket switched on?
Is the VCR switched on?
Has the plug/socket/VCR fuse blown?

If the VCR will not record:
Is the record tab still on the cassette?
Have you released the pause control?
Has the tape come to its end?
Is the dew warning light still on?
Are all the connections plugged in?
If your VCR has a TV/VCR select switch, is it correctly switched to VCR?

If the VCR will not play back:
Is the television switched on?
Has the tape come to its end?
Is the television tuned to the VCR's output channel?
Is the cassette properly loaded?
Is the TV/VCR select switch set to 'TV'?

If the tape will not fast wind:
Has the tape reached its end?
Is the VCR properly connected and switched on?
Has the tape memory stopped the tape?
Is the tape loaded properly?

If the tape will not eject:
Is the VCR in its timer/record mode, waiting to do a pre-set recording?

Connecting up and tuning in

The first and most complicated process you perform on your video is connecting it up, tuning it to your antenna and tuning the television to the VCR. Although manuals are now easy to follow, connecting and tuning need care and concentration. If you miss a step, you may find you have to retune either straight away or, if the tuning drifts out of alignment, in the future. Here are some common tuning problems with their symptoms and solutions.

There are broadcast pictures on the television but no replay picture from the VCR.

The television needs to be tuned to the VCR. If tuning to a pre-recorded tape, check that you are receiving the tape, by operating the pause briefly.

Video tapes play back perfectly but the VCR will not record off air.

The VCR's receiver needs to be tuned to the antenna. Check the VCR tuning sequence (different from the television tuning sequence) in the manual. Although some VCRs have auto-tuning, the picture can sometimes be improved by hand-tuning.

The VCR is connected but the television gives no picture.

The connection between the television, VCR and antenna is wrong. The antenna must be plugged into the VCR, and the lead provided used to connect the VCR and the television. Double-check the correct inputs in the manual.

The television gave a good picture when first tuned in to the VCR, but the picture has since deteriorated.

You have either forgotten to switch off the Automatic Fine Tuning (AFT) when you tuned in the television, or you have forgotten to re-engage it when you finished the tuning.

Picture faults

Once a picture has been established, you may find that it is not perfect. Here are some common picture faults, their causes and solutions.

Short, fine, white horizontal lines keep darting across the screen.

This is tape dropout, which means that flecks of oxide have fallen off the tape's recording layer. This can happen if the tape is worn, or of poor quality, or if it has been left in 'Pause' too often. This is particularly likely to happen to the first few minutes of a video tape, since these are played most often. Worn tape cannot be repaired, but the ends can be spliced. If dropouts appear on good-quality tapes frequently, and for no apparent reason, call an engineer to look at the VCR.

The picture has become progressively more grainy and the colour and audio poorer. This can happen quickly or gradually.

The video heads need cleaning. For the best results, call an engineer. New heads sometimes get dirty unusually fast. This is because they need to be worn in, and they will settle after a few plays. You can run a blank tape a few times to save wear on recorded tapes.

The recording or timer controls were set carefully, but the VCR did not record.

If nothing was recorded, the VCR was probably not switched to its timer mode. The switch may have been accidentally knocked to 'Off.' If audio and video noise was recorded, the channel selector switch may be switched to a channel not in use. Check these controls when setting up a recording.

Thick white lines appear on the screen on playback, and the VCR makes a noise.

The tape has been creased or crumpled. Do not continue to use the tape, for it may damage the heads. If this happens suddenly and without reason to a known tape, consult an engineer. The VCR may be faulty.

The playback picture is poor, perhaps with marked noise bars. It may break up altogether.

The video heads are mistracking to some degree. If the tracking control does not help and the fault persists, call an engineer. The video heads may be misaligned as the result of a knock to the VCR.

If the tape is borrowed, the VCR it was recorded on was mistracking.

On playback, the top of the picture bends to the right.

This is 'hooking'. It occurs when the television cannot synchronize immediately to the VCR's output. Most televisions can be suitably adjusted by an engineer.

There is herring-bone interference on the screen.

The VCR may be too close to the television, or the VCR's RF output too close to a broadcast channel. Some RF outputs can be retuned by the owner, if the manual gives instructions. Otherwise, call an engineer.

The VCR gives only a monochrome picture.

This may be a sign that the VCR or television needs retuning. Otherwise, the VCR may have a 'Monochrome/Colour' switch which should be set to 'Colour' (or 'Auto') for colour and 'Mono' or 'B/W' for black and white.

White sparks appear on the screen during playback.

There is static electricity on the head drum. An engineer can demagnetize it.

Small, almost static black lines appear on the screen when a pre-recorded tape is played.

This is a recording fault on the tape. Ask your dealer to exchange it.

Regular white streaks deface the video replay picture every few seconds.

This may be radar interference. The VCR's receiver unfortunately amplifies radar signals. You can get a radar filter fitted to the antenna, but it is cheaper and usually more effective to retune the VCR to a new reception channel away from the radar band.

Faults and how to fix them

You are, by now, an expert in the art of using and enjoying your VCR. Your antenna, television set and recorder are all adjusted for the best picture reception. You know your way round the controls and are familiar with the manual. If your VCR fails to record you know how to check thoroughly that you have operated the controls correctly; that your television and VCR are tuned to best advantage; and that all the connections are in the right place before you make an expensive appointment with your service engineer.

This expertise has not been too hard to acquire and you feel at home with your VCR. Now is the time to acquaint yourself with some of the less routine surprises which it might spring on you. The potential problem areas can be divided into cables, tape and breakdowns of the VCR; if you use a portapack, batteries can cause some problems.

Listed below are some symptoms you may encounter, and what to do about them.

Cables

Cables are most likely to become worn or damaged at the point where the cable joins the connector; sometimes the internal cable cores can be broken if they are folded sharply or trodden on. This can be hard to detect from outside. About eight per cent of all video problems are caused by faulty cables.

It is worth learning to do a simple continuity check, using a multimeter or continuity tester. If you know how to splice and solder, repair the cable yourself; if not, it is easier to buy a replacement.

If you are using a screw-together power plug, it is common for one of the wires inside the plug to become disconnected. It is a simple job to open the plug and reconnect it. When doing so, ensure that the metal strands at the end of the wire are twisted neatly together and screwed down tightly, and that the cable is firmly clamped in the plug's cable collar.

Make sure the live wire is as short as possible, so that it will be the first to pull loose if the cable is strained, removing the threat of electric shock should the earth or neutral wire be disconnected.

If you have a sealed AC plug, you may have to replace the cable and plug. However, sealed plugs do not often fail.

Cable problems

The VCR works well near to the television but poorly at a distance.
The extension cable may not be the full-specification UHF coaxial cable which is absolutely necessary for transmitting signals to and from a television receiver. If in doubt, ask your dealer's advice.

You want to use a hi-fi lead for audio dubbing from one VCR to another. The connectors fit perfectly, but you can get no sound from the monitor.
Some audio and video leads look alike, but the connectors are wired differently inside. Ask your dealer for the correct audio lead, or have one made up.

Tapes

Video cassettes are fairly robust if kept inside their cases when not in use. Video tape is very delicate, and small amounts of dust, grease, damp or heat can damage it.

The tape appears to play for a shorter time than is stated on the label.
This is a very common complaint. It is more than likely that the tape memory on the VCR is switched on, and that the tape is not winding back to its beginning.

The picture from one tape is getting progressively worse.
The tape may be wearing out. If a tape is well treated there will be wear eventually, owing to the abrasive effect of the video heads. Tapes used to show signs of wear after about fifty plays, but a good-quality tape with a properly adjusted VCR should last for hundreds of plays in good condition.

If your video heads get dirty quickly, this could be a sign that you are using a cheap brand of tape which sheds its oxide easily, clogging the tape heads.

Once worn, a tape cannot be revitalized. It can, however, be copied: the result will be poor but it will last longer.

A tape has snarled up inside the VCR.
The tape may be caught round the head drum. Call an engineer.

The tape is broken or creased.
You can salvage a broken or partly creased tape by cutting out the creased section and, using a video tape splicing kit, joining the end to one end of the leader tape. Never splice any part of the tape that will contact the video heads.

The cassette is jammed in the compartment.
Cassette housings can become invisibly warped if left near a heat source. Call an engineer.

A cassette housing is broken.
Since cassette housings are screwed together, they can be dismantled and, therefore, it is possible to swap housings with those on a less valuable tape. This requires great manual dexterity: if you handle the tape, you will damage it beyond repair. If the tape is valuable, or you are in doubt about the procedure or its results at any stage, consult your dealer.

If possible, both cassettes should be fully rewound before opening. To open them use a Phillips pattern (cross-headed) screwdriver to undo the screws which hold the housings together. Lay each cassette down with the screw-holes underneath and press the tiny release catch by the hinge of the protector panel which covers the front of the cassette. Hold this panel open and carefully lift off the top half of the cassette.

Gently unthread the tape and lift the loaded spool into its new housing without touching the recording tape. Clean rubber or surgical gloves can be worn as a precaution. If the tape is not rewound, both spools will have to be transferred.

Rethread the tape, replace the top of the new housing, turn it over and replace the screws.

Batteries
Battery problems usually concern recharging. Rechargeable batteries are expensive, so it pays to care for them properly.

The battery pack has suddenly stopped working. They are lead acid batteries.
If lead acid batteries are allowed to go completely flat, they will cease to work.

Tired batteries can cause the tape threading mechanism on the portapack to malfunction, trapping the tape.

The battery pack will only work for twenty minutes. They are nickel cadmium batteries.
It is harmful to nickel cadmium (NiCad) batteries to recharge them when they are only partly used. If done repeatedly, this will inhabit the batteries' ability to accept a charge. Use them until the battery warning light comes on and then recharge them. If stored, discharge and recharge them every six months.

VCR malfunctions
Apart from minor tracking faults and dirty video heads, real problems with the VCR are rare. On older VCRs, damage could result from hasty and incorrect use of the switches. Modern logic-controlled switches are less susceptible, but it is wise to treat electronics with some gentleness. Operate all switches with a brief pause between each operation, and do not be tempted to test the system to its limits.

There are traces of either the audio or video from a former recording on a new one.
The VCR is not erasing properly and requires professional attention.

The cassette keeps ejecting from the VCR.
You may be pressing buttons too fast or in the wrong order, causing the VCR's logic to malfunction.

If you have an infra-red remote control, flashes of sunlight or heat sources can cause it to function unexpectedly. Ultrasonic remote controls can sometimes be set off by whistling or sibilant sounds.

There is a persistent buzz from the VCR.
This can be the first sign that it is drifting out of tune, and needs retuning.

Some people are troubled by traces of audio noise from the VCR. If you are sensitive about sound quality, you need a VCR with a noise-reduction system.

Rewind varies in speed and sometimes stops.
The VCR's drive mechanism may be malfunctioning, or the cassette may be faulty. Consult an engineer.

Understanding the recorder

From the moment you load a blank cassette into the machine the VCR is ready to record signals off air, or from a camera, and to carry out a whole series of other actions in response to the touch of a button.

The VCR's logic circuits are behind the translation of commands into actions. Supposing you select 'Record', the circuitry triggered by the switch sets in motion a mechanical threading system which opens up the guard flap on the cassette, and draws a loop of tape out to a set of moving rollers. These draw it into the machine along a highly complicated path which takes the tape past the fixed audio heads (which record and play back the sound-track); the erase head (which wipes out any existing recording on the tape passing the heads when 'Record' is selected) and the control track head, which writes a regular series of pulses on the tape to calibrate it for accurate timing control. It also winds the tape round the heart of the machine, the helical scan head drum, which records and plays back the video information.

Because of the amount of video information which has to be recorded on the tape, these heads are not fixed, but the drum on which they are mounted spins very fast past the moving tape. The tape transport mechanism swings the tape around the drum at an angle, so that the spinning heads write a series of very narrow tracks at a shallow angle to the length of the tape. This is the helical scan system which all video cassette recorders use. The audio head records the sound-track along one edge of the tape, and the control track is recorded on the opposite edge.

When the recording is finished you touch the 'Stop' button and the machine responds, not only by stopping the tape, but by reversing the laborious tape-threading mechanism, so that the loop inside the machine at the time you stopped the recording is threaded back into the cassette, and the guard flap is ready to unload the cassette.

If all you want to do is wind the tape back to the beginning to replay the recording, this is a more simple process. All the machine has to do is to wind the spools so that the tape is fed back in the reverse direction. This it does by connecting the drive motor to the tape spools rather than to the pinch wheel which pulls the tape loop past the heads for recording and playback. This means that the tape has to be held clear of the heads to allow a fast winding speed: on Betamax machines the tape is retained within the VCR for fast wind and rewind; on VHS machines it is wound back into the cassette first, and on Video 2000 machines the tape can be wound quickly inside and outside the tape path within the VCR.

On selecting 'Play', the tape is threaded back around the rotating head drum and past the other fixed heads. The other functions affect mainly the speed with which the tape follows this path.

Most other features on your VCR will be refinements of these basic systems, or a sophisticated timer-control mechanism, which increase their versatility.

Another option is the automatic program search facility. This is a circuit which records an invisible and inaudible 'blip' on the tape at the beginning or the end of each recording. A control then lets you tell the recorder to fast-wind through a tape, and stop only when it recognizes one of these signals. This will put you at the end of a particular recording, and the beginning of the next, quite automatically.

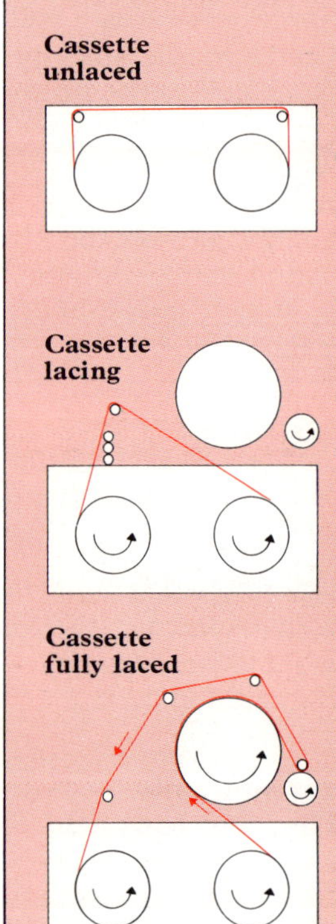

When the 'Play' button is operated, a series of pins rises up and draws the tape against the head drum, *left*.

The tape is moved along by the pinch wheel, a rubber wheel which pushes it against a turning capstan.

In 'Fast search' the capstan moves the tape over the video heads faster than in normal playback, and 'Slow motion' slows the tape down. Extra electronics are needed to get a stable picture.

In 'Pause' the pinch wheel releases its contact with the capstan, so that, although the capstan turns, the tape does not move along.

On most VCRs with twin recording heads, 'noise' bars cross the top of the screen when pause or freeze-frame is selected. The Toshiba V-8600 (Europe), V-8500 (USA) eliminates these by doubling the number of recording and playback heads. Instead of tracing two recorded tracks, they trace only one track accurately. As a result, Toshiba offers 'Super Still' and 'Super Slow Motion' facilities on this machine, and also provides fast picture search at up to 40 times normal speeds.

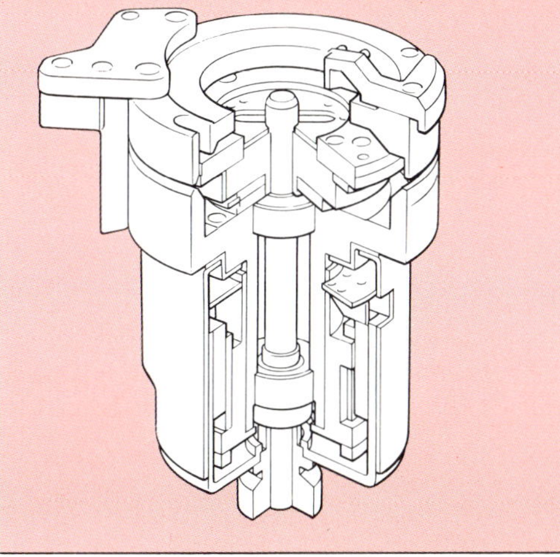

The neat, unruffled exterior of a modern VCR covers a mass of electronic and mechanical machinery which carries out a wide range of delicate tasks with precision and reliability.

The main mechanical part is the head drum which can be seen, ringed by the tape-lacing arm, in the centre of a Sony SL-C7, *below*.

The majority of components, however, are electronic and are mounted on printed circuit boards (PCBs). One PCB may carry several hundred components.

A large VCR factory can turn out over 20,000 machines a day, each one with over 3,000 separate parts. All are tested for a period. VCRs are very reliable; however, the delicacy of some adjustments make them vulnerable to rough treatment. Some parts are set to tolerances 50 times finer than a hair.

The minute slits in the video heads which record the video signal, are so small that particles of dust can clog them, so VCR factories are kept as free of dust as possible. In large factories, much of the work, including PCB assembly, is done by machines.

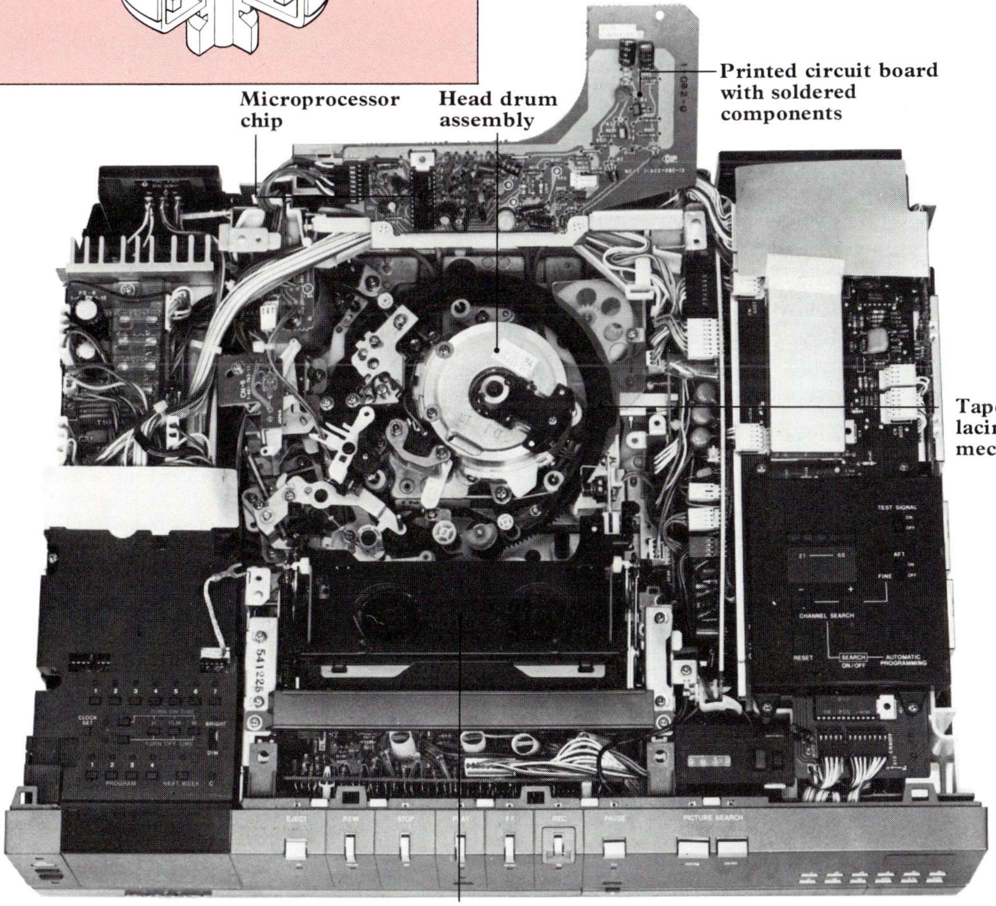

Printed circuit board with soldered components

Microprocessor chip

Head drum assembly

Tape-lacing mechanism

Cassette compartment

Caring for the recorder

What keeps a VCR working well? It needs the right environment to work in: not too cold, nor too hot, humid, dusty or prone to vibrations or magnetic fields. It needs what many engineers call mechanical sympathy on the part of its owner: do not hit it, drop it, shake it, stand heavy objects on it, or spill things into it. Use the controls gently and do not confuse it by trying to give it too many contradictory commands at once because you are unsure about what to do next.

The only repair attention it will need is general cleaning and demagnetizing every so often. It is difficult to be specific about how often chores like these need to be carried out. Most experts say that preventive maintenance (doing the job before the need becomes obvious) is to be avoided.

It is better to watch out for picture distortion or signs of partial erasure on some of your tapes, which are indications that the recording heads are retaining some of their magnetism, before calling a service engineer to demagnetize (or de-gauss) them.

Much more common is the need to clean the heads of particles of magnetic oxide which flake off the tapes in regular use. It is difficult to give an accurate figure of how often this is needed. Manufacturers' and testers' estimates vary from 'up to every 1,000 hours of use' to 'as few as every 400 hours or so' depending on the cleanliness of the atmosphere in which the machine is kept, and the quality of the tapes being used. Poor-quality tapes may shed more oxide particles than the latest, best-quality tapes do.

Once again the answer is to watch for the symptoms. Progressively noisier pictures (that is, grainy and speckled with snow) on all your tapes during playback will indicate that the heads are becoming clogged and that it is time to clean them.

Opinions differ sharply too on whether you should do the job yourself, or call in a service engineer. The problem is that the heads are extremely delicate, and the slightest slip in cleaning them could easily cause damage which would cost far more to repair than the price of having the heads cleaned professionally in the first place. Add to that the fact that most manufacturers' guarantees are rendered invalid by opening up the machine to perform an operation like this, and it hardly seems worth while trying to undertake the work oneself.

Care and maintenance check list

1. Make sure your VCR is horizontal before you operate it.
2. Never stand a VCR on the floor, or on a carpet.
3. Keep your VCR well ventilated at all times. Never cover the ventilation grilles.
4. Do not expose your VCR to direct sunlight or heat; keep it in a cool, dry place.
5. If you have disconnected the machine for some time, do not use it immediately after reconnecting it. Give it time to warm up before playing a tape.
6. Do not expose your VCR to moisture from spilled liquids, sudden changes of temperature, or the damp air near a window.
7. After playing a cassette, remove it from a VHS or Betamax machine, and never transport the machine with a cassette in place.
8. If a fault occurs, do not try to repair it, and do not continue to use the machine. Call a service engineer.
9. Keep the VCR's original packing in which to transport it.
10. Do not use solvent cleaners on a plastic VCR case. Use a little isopropyl alcohol (white spirit) to remove stubborn or oily marks.
11. Wipe the VCR over regularly with a damp cloth, using clean water, or an anti-static cleaning fluid obtained from a video dealer. Dust or vacuum carefully around and behind the VCR to prevent dust and fluff accumulating. Keep the cassette compartment closed when cleaning.
12. Do not use a tape-head cleaner unless you suspect that your video heads are clogged, and then use it sparingly. If in doubt, call a service engineer.
13. Run your VCR in record mode for an hour every couple of weeks, even if you do not wish to record. If you use it only for playback, the heads will gradually become magnetized, causing a loss of picture quality which is a symptom of permanent damage to the recordings. The strong RF signal used for recording reverses this process. If you suspect your tape heads need demagnetizing, call your service engineer.
14. Take care not to drop or knock your VCR. This could cause the video heads to fall out of alignment, a condition which can only be repaired in a video laboratory.
15. Run trailing leads round the edge of the room, or cover them with a mat or rubber electrical cord protector, so that people cannot trip over them or damage them by treading on them.

Although you should never attempt to open your VCR, you should keep a basic tool kit for servicing other parts of your system. As well as standard and cross-headed screwdrivers, you will find that you need Allen keys or hexagon keys or ball-tip drivers, *below*, for hexagon socket screws on tripods, light-stands and video furniture.

For the very small screws on some photographic equipment, jeweller's screwdrivers, **2**, are the right size. When assembling cables or repairing connections, clean wire ends with emery cloth, **1**, and use a soldering iron, **4**, which gives a controlled heat and can take alternative bits, **5**. For stiff tripod joints, use household oil, **3**.

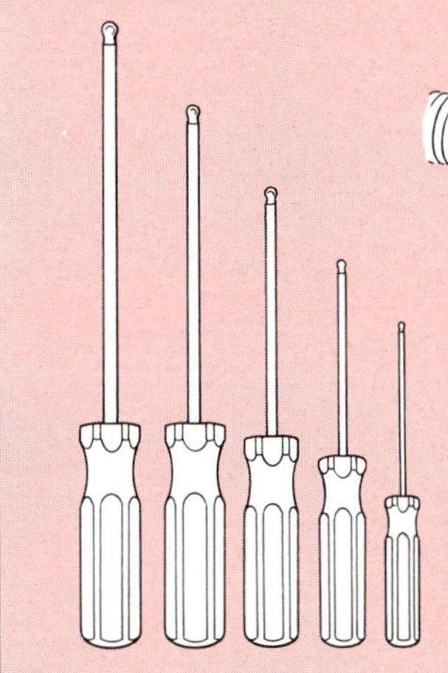

Hexagon socket screws, *above*, can be operated by a hexagon ball-tip driver at angles of up to 25°.

Hexagon ball-tip drivers, *left*, are an alternative to hexagon keys for screws in awkward locations.

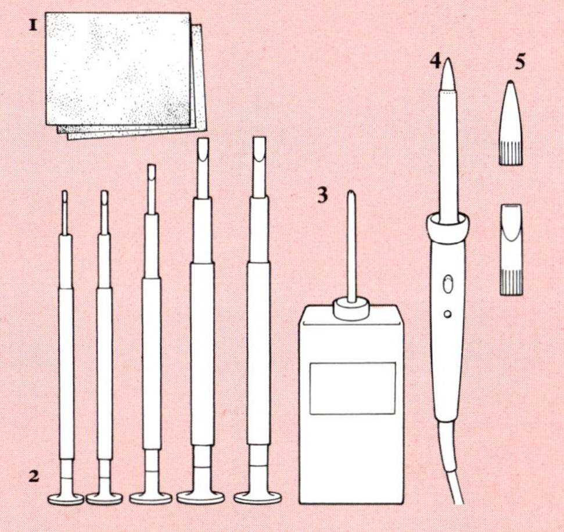

Dust is a bitter enemy of the VCR, so much so that it has to be assembled in a constantly filtered, dust-free factory environment, *bottom*.

Dust clogs the video heads, so when your VCR is switched off, put a dust cover over it, *below*. Be careful not to switch it on with the dust cover in place. This can cause overheating.

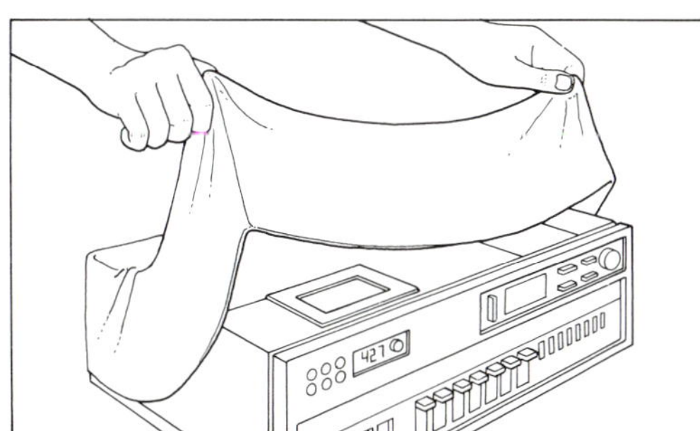

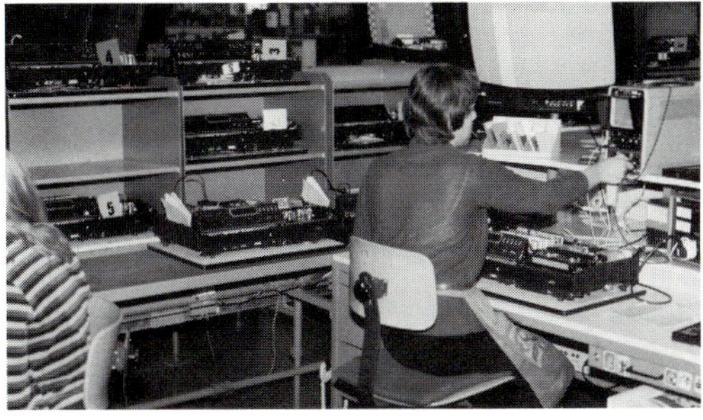

A head-cleaning cassette is a welcome alternative to the use of swabs and sprays to clean clogged tape heads. It obviates any need to open up the machine, and is unlikely to cause damage. All you do is load the cassette in the normal way, and play it for a short time: between 10 and 30 seconds, depending on the manufacturer's instructions. The tape cleans the heads and tape guides by direct abrasive action, so take care not to overdo the treatment.

One type of cassette eliminates the extra wear caused by using abrasives. It has one spool of felt, which has to be dampened with cleaning solvent before use, and one spool of a drying material. The problem is applying the right amount of solvent to the cleaning strips. Dry, non-abrasive, cleaning cassettes use direct contact with a soft material.

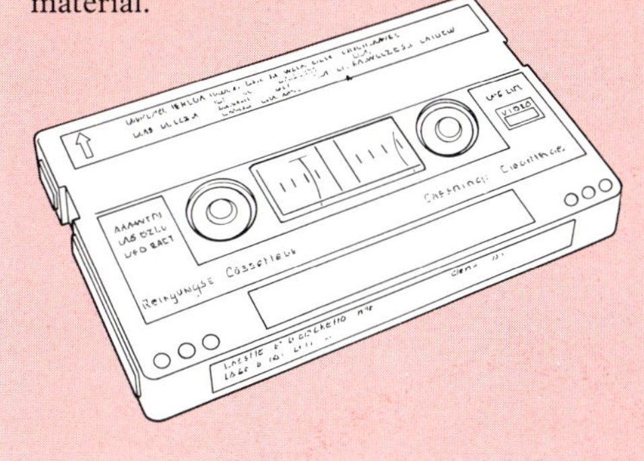

Building a video library

Many of the recordings you make will have a short life. These will be the shows you record because you will be out when they are screened, or because two movies you want to see coincide on different channels.

Yet there will always be some tapes you want to preserve, to play again and again, or to consult at intervals. Some of these you will probably buy. The choice of videograms (pre-recorded video cassettes) has for some time been broadening from the original base of full-length movies. They have penetrated into fields as varied as cookery, cricket and rock music.

However, the price of pre-recorded cassettes makes buying one an investment, worth while only when this repeatability factor comes into play. Classic movies which are only occasionally given an airing on the networks, such as *Casablanca* or some early Hitchcock masterpiece, might be well worth the cost, if you can take them out for a private showing whenever the spirit moves you to renew the acquaintance of such old and valued friends.

Tapes may also merit a place in your library. There are whole series of specialized tapes on diving, dinghy-sailing, rally-driving, fishing, tennis, rugby, football, badminton, soccer, squash and skiing, and the list is growing year by year. Here, the extra facilities on your VCR, such as freeze frame or slow motion, can be of the greatest help, enabling you to study the movement involved in a tennis service or a golf drive really closely, in a way that would not otherwise be possible.

The desire to see over again some of the greatest and most dramatic moments in a truly classic contest may motivate you to expand your library of sports tapes. Here, too, there is a steadily growing choice: provided agreements can be hammered out with the professional bodies involved, sports fans can look forward to being able to watch whole cricket series from the past, closely fought football finals, or Davis Cup tennis.

If you try listing your ten favourite movies, you will probably discover that you remember most clearly their great moments. Few people think of Steve McQueen in *Bullitt* without instantly picturing the amazing car-chase sequence, and perhaps the chess game. Similarly, the final sequence on the vertical cliff-faces of Mount Rushmore is what most people remember about *North by North-West*. Moments like these can form the basis for a specialized collection which will withstand repeated viewing.

How do you build up such a collection? One way is to set aside one cassette to begin a special collection. Every time a movie or a show with a contribution to make appears in the schedules, you watch it through until the moment you want comes up on the screen, and record just that fragment. Unfortunately this has a two-fold problem: first, you cannot pre-set the machine unless you know the exact timing of the segment you want to record; and second, the excerpts will be recorded in the order that the fragments are broadcast on television.

The easiest way round this difficulty is to tape the movie, and the show containing the fragment, in their entirety. Then, wind through the tape to find the parts you want to preserve, and copy them on another tape which you keep purely for these edited highlights. To do this, you will have to borrow another VCR for a few hours from a like-minded friend, for whom you may later be able to return the favour.

The list of edited library tapes you can build is endless: great moments from football games; appearances of your most popular rock band or whole series of comedy shows which may never be repeated.

One word of warning: it is all too easy to lose track of an important recording; just when you remember it, you find that something much less important has been recorded over it. For this reason alone, a proper labelling and indexing system is essential. If you label and index each cassette as soon as you have recorded a show, you will never have to face a dozen unlabelled tapes, having no idea which is the only surviving blank one on which to record a show which is starting in ten seconds.

Finally, as an alternative or in addition to building up your own library, consider joining one. Video libraries are springing up all over the country, with different terms for membership and borrowing cassettes. Joining a good library with a wide selection of tapes of movies and a few special-interest subjects, is a splendid way of widening your choice of tapes in the areas into which your own collection cannot venture: tapes that will be of interest for a single showing at a time.

Copying a video tape

It is quite simple to copy a video tape, but unfortunately a copy is always slightly less perfect than the original. There are ways of minimizing this drop in quality.

In order to copy a video tape, you have to link two VCRs. Do not, however, do this by plugging a VCR/TV coaxial cable into the RF input socket in the second VCR. If you do that you will be playing back the copying signal, converting it to an RF signal in the first VCR, passing this across to the second VCR and there converting it back into video and audio signals.

Instead, buy a pair of video and audio leads, with the correct connectors, to link up the 'Video out' and 'Audio in' of the first VCR to the 'Video in' and 'Audio in' of the second. You will then be able to play back the recorded tape on the first VCR and simply switch the second machine to the record mode for the duration of the piece you want to copy.

By this method you can even copy a cassette on another of a different format, but make sure the copy is made on a cassette which will fit your machine.

Tape care check list

1. Buy a cassette only if it is in a sealed library storage case.
2. Do not touch the tape; it is protected by a flap on the front of the cassette.
3. Never leave cassettes lying on the floor; they may collect dust or be damaged.
4. Do not subject cassettes to strong physical shocks.
5. Always wind cassettes on to the end or back to the beginning before storing them.
6. Store cassettes vertically.
7. Keep cassettes dry: do not have your library near a window. If you take cassettes out of the house, carry them in a waterproof bag or case; do not put them down on damp grass.
8. Keep cassettes cool. Never leave them where sunlight streaming through a window could damage them, such as on the seat of a car, and do not have your library near a fire or radiator.
9. Keep magnetic fields away from tapes.
10. Remove the recording tabs at the back of the cassette from any tape containing recorded material that you wish to keep.

Step-by-step guide to cataloguing

Step 1: Give each tape a number when you buy it, not when you want to use it. All tapes come with stick-on labels and numbers, so stick them on before you need a blank tape in a hurry. Label the first 01. It looks more professional and is easy to identify at a glance. Remember to number both tape and box, and never allow the two to become separated. Make sure you write all labels legibly.

Step 2: Buy yourself a notebook or a set of cards on which you can write the details. Allocate a page for each numbered tape. Only write the details on the labels provided with the cassette when you know it is something you will want to keep. If you have continually to relabel a tape you use over and over again, it will soon become unreadable. If this happens, relabel with self-adhesive labels.

Step 3: On each card, write all the details you need to identify the tape. You will need to note down what excerpts show, and the footage numbers at which they start and finish. Not only does this help you to find sequences quickly, but if you also enter each time a tape, or part of a tape, has been shown, you can work out how much life a recording has had when it begins to lose playback quality.

Step 4: If you remember to reset the counter to zero every time you load a new tape, and to rewind each tape to the beginning after use, you can record footages accurately.

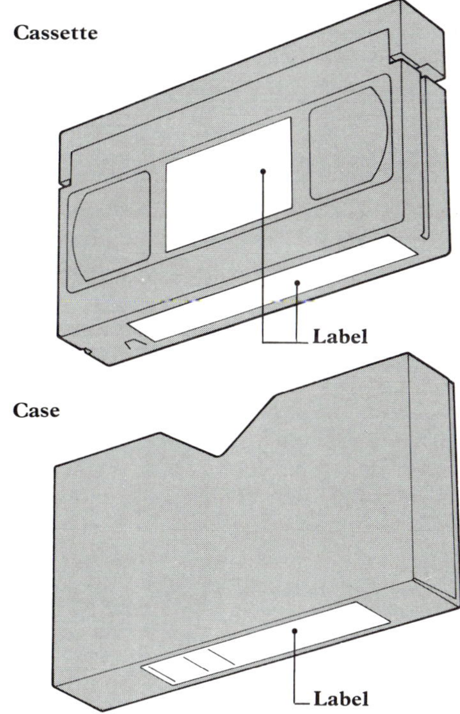

Video and the law

Most people who own or rent a VCR are aware that they are, in some way, affected by the law of copyright. There is, however, enormous confusion in the public mind about the extent and implications of that law and the legality of video recording.

The whole principle underlying the complex law of copyright is that creative people should be allowed to control the exploitation of their work, to enable them to secure a proper reward for their creative efforts.

There are two popular misconceptions about the effect of copyright law on video: first that a private individual can record any television programme without obtaining anyone's consent, provided that the recording is intended for home use only; and second, that permission is required for the recording of all television broadcasts. On a strict interpretation of the law, neither is entirely correct, but the second of these two propositions is, in fact, the one nearer to the truth.

A television broadcast usually consists of a number of elements, each of which gives rise to rights belonging to various people as creators of those elements. If you record a television programme, you may breach the copyright of one or more of those elements and, for that reason, break the law. While the copyright of a programme (owned by the broadcasting authority) is not infringed by a video recording made for private, domestic use, the rights of the author of the book on which the programme was based, for example, the composer of the featured music, or the record company who manufactured featured records, would be infringed.

As well as copyright, the 'performing rights' of people such as actors, singers, musicians and dancers are involved. Under the Performers Protection Acts 1958 to 1972 it is a criminal offence knowingly to make a record or film, other than for private and domestic use, from the performance of a musical work, without the prior written consent of the performers. This does not cover live sports and public ceremonies, but there is some doubt as to whether it covers activities such as figure skating and ballroom dancing and the performances of variety artists such as jugglers, magicians and acrobats.

From a strict legal point of view, the main reason why it is illegal to record the vast majority of television programmes off air is that they constitute what the Copyright Act 1956 calls 'cinematograph films', simply because they are pre-recorded on video tape before transmission. The copyright owner of a cinematograph film is the film maker, and so is usually (unless they are broadcasting a commercial feature film) the broadcasting authority. The copyright owner is entitled to control the making of copies of the cinematograph film, and there is no provision in law for copies to be made for private use.

Television programmes, when broadcast live, do not constitute cinematograph film, but often they contain some copyright element, such as the music featured in ice skating events, which would require consent for recording. Even live broadcasts of sporting events contain copyright elements such as theme music or the performance of a brass band at half time. Or they may contain cinematograph film in the form of action replays or other pre-recorded items. The copying of any such film (a cup final is a good example) would, without consent, amount to a breach of copyright.

Even if approached, the BBC and Independent Television companies are usually unable to give consent for copying in respect of all the elements that comprise the majority of their general programmes. The reason is that the BBC, for example, usually obtain, from the owners of the rights in each element, only sufficient rights to enable them to produce the programme and broadcast it. They do not generally acquire the right to authorize others to make recordings of the programmes that incorporate other copyright elements.

There are, however, some exceptions to the rigorous copyright laws which enable educational establishments to make use of certain copyright works. These do not allow schools to record television and radio broadcasts, nor to copy commercial films, but arrangements have been made so that educational establishments can legally record BBC educational programmes. (The exception is Open University programmes which can be recorded only by registered Open University students; otherwise a special licence must be obtained from the Open University.) Even so, a number of conditions must be complied with, and further details can be obtained from the BBC Copyright

Department. ITV will issue a licence to schools to record educational programmes for use in schools. But the limited exceptions for recording educational programmes do not extend to documentaries such as 'Horizon', however educational their content, and do not apply to the VCR owner.

Despite this strict legal interpretation, in practice neither the BBC nor the Independent Television companies complain of recordings made off air for private and domestic purposes. Largely, this is a recognition of their inability to prevent people making recordings in their own homes. Also, the broadcasting organizations seem to recognize that 'time shift' recording harms their commercial interests little, if at all, as long as such recordings are 'wiped' and the tapes reused for further recordings. This said, they cannot specifically authorize any such private recording.

While turning a blind eye to purely domestic off-air recording, as long as the recordings are not copied, or used for commercial gain, the broadcasting organizations and film distributors recognize the absurdity of the present law, which renders illegal the vast majority of home taping off air, and finance a pressure group to campaign for a change in the law. They would like home taping to be legalized in return for the imposition of a levy on the sale price of blank tapes and VCRs. The money collected would be distributed among the owners of the rights comprised in the programmes. This levy is not intended to legalize the subsequent sale of tapes, even for domestic purposes, nor the copying of prerecorded commercial tapes from one VCR to another.

Although a commision was set up to examine copyright law, and recommended in 1977 that a levy should be made on recording equipment only, as it is in the German Federal Republic, the British government has decided not to take any immediate steps and has invited further representations from the public and interested parties. In the meantime, the VCR owner will, it seems clear, be taking no risk in taping programmes off air for his own private and domestic use.

Any VCR owners, who produce multiple copies of their recordings and attempt to sell them, could end up at the wrong end of a legal action brought by the film distributors. Large-scale copying of film amounts to fraud, and the court may permit a person's premises to be searched if it is believed that he or she is copying pirate tapes, or storing them. Any counterfeit material is liable to be siezed and used as evidence.

Having made an off-air video tape, is it permissible to show that tape to an audience at home or elsewhere? The answer is that it all depends whether the showing amounts to a 'performance in public'; which depends, in turn, on the nature rather than the size of the audience. If the showing is essentially domestic it is not 'in public'. Otherwise it is.

The fact that some video films are legally made and distributed does not mean that their content is censored, and such films are available to children of all ages. Thus it is important for parents to be careful about vetting films intended for adult audiences only before permitting their children to watch them.

Many video owners make use of goods and services offered in classified advertisements. Most advertisers require full payment in advance, and the would-be purchaser could be left without an effective remedy if an advertiser subsequently went bankrupt without having supplied the ordered goods. Some magazines, however, and certainly those published by IPC, operate a Mail Order Protection Scheme under which the magazine undertakes to consider the customer for full or part reimbursement in the event that the advertiser went bankrupt without supplying the goods, subject to certain conditions being complied with. Otherwise, there is little the customer can do except make a formal claim against the liquidator with little prospect of getting the money back.

If you own a video camera you are, in general, free to film whatever you like, particularly if the film is intended for domestic viewing, and also to sell your movies. You should, however, remember the Performers Protection Acts if you intend filming concerts or the like, other than for home use. Although there is no right of privacy in the UK, some people may object to being filmed, so it is prudent to stop filming if you come up against this problem.

For further information on the legalities of video recording, contact the Video Copyright Protection Society Limited, London NW10, or the BBC Copyright Department.

The video buyer's guide

Acquire a video recorder now, the time is right. Years of competition have brought the prices down, and development has reached the stage beyond which only details have still to be perfected. Eventual obsolescence is unlikely: the VCR's growing role as the heart of the home computer guarantees its survival.

The projected arrival of the video disk, and its likely rivalry with video tape, has caused confusion in buyers' minds. However, you cannot record on a video disk; its potential lies in markets other than those where tape presides. Eventually, each will take its place as an interactive piece in the home computer jigsaw.

Choosing a VCR from among the dozens on the market involves, first, selecting a format. There are three widely available home video formats. VHS (Video Home System), a Japanese system developed by JVC, has so far proved the best seller in the European market. Betamax, another Japanese system developed by Sony, has been more popular in the USA. Video 2000, the late contender from Europe, has yet to make a strong impact in the USA. Complicated though it may seem, multi-format video is likely to persist unless buyers and manufacturers decide to opt for one system. Current trends suggest the opposite.

The choice of format will inevitably be dictated by personal requirements, such as where you live, for example. Each VCR format will play only cassettes designed especially for it: you cannot play a VHS cassette on a Betamax or Video 2000 VCR. If you want to exchange tapes with friends, choose a VCR in the same format as theirs.

If your VCR is to be used mainly for playing pre-recorded tapes, in the UK you will find most material available on VHS; in the USA, on Betamax. However, manufacturers of all three formats are improving market penetration into areas where they are weak, so the balance may be redressed.

In the following pages the differences between the leading formats are discussed in detail. Broadly speaking, however, there is currently the widest choice of machines in the VHS format, while Video 2000 offers the longest-playing cassettes and the most sophisticated timing facilities. Betamax cassettes are the most compact; Sony's portable VCR is smaller and lighter than its VHS competitors. All three have proven quality and reliability, and there is little difference in price between them. Each format offers low-priced basic machines, and more expensive models featuring advanced facilities.

Weighing up the relative value of sophisticated features takes time. You need to visit stores and showrooms, to see demonstrations and compare models, to read brochures and magazines and talk to friends with VCRs.

Your choice should be based on how you want to use your VCR. If you are often away, you will need a model with a sophisticated timer which takes long-playing cassettes. A machine with remote control is a must if a member of your family is bedridden – or if you just hate getting up to change the television station. Should you want to collect excerpts, or to store off-air broadcasts for educational purposes, you will need a machine with an auto tape-search facility.

Slow motion, combined with frame-by-frame advance and freeze frame, is essential if you want to analyze a recorded sequence – a golf or tennis stroke, for example. If you are training seriously for a sport, consider the possibility of hiring a video camera to analyze your own stroke and compare it with that of a professional. To do that most successfully, you will need a VCR with good roll-back editing facilities.

Consider buying a portable system, if you want to make home movies. With the addition of a tuner/timer it will double as a home deck. The best have all the advanced facilities.

If you want to exchange tapes with friends overseas, you will need a triple-standard VCR. Different countries use different television systems, and so whatever the format, VCRs are designed to record from and play back through the television system of the country in which they are sold. If you want to play a tape of a baseball game recorded in the USA on a French television set, you will need a triple-standard VCR and the television must also be a triple-standard model, which is expensive.

Finally, the enthusiast with professional ambitions might want to invest in an industrial-standard VCR (U-matic, Betamax or VHS). These are more expensive than the domestic formats, but give a higher quality of sound and picture and can be used with sophisticated editing controllers to produce professional-standard video movies.

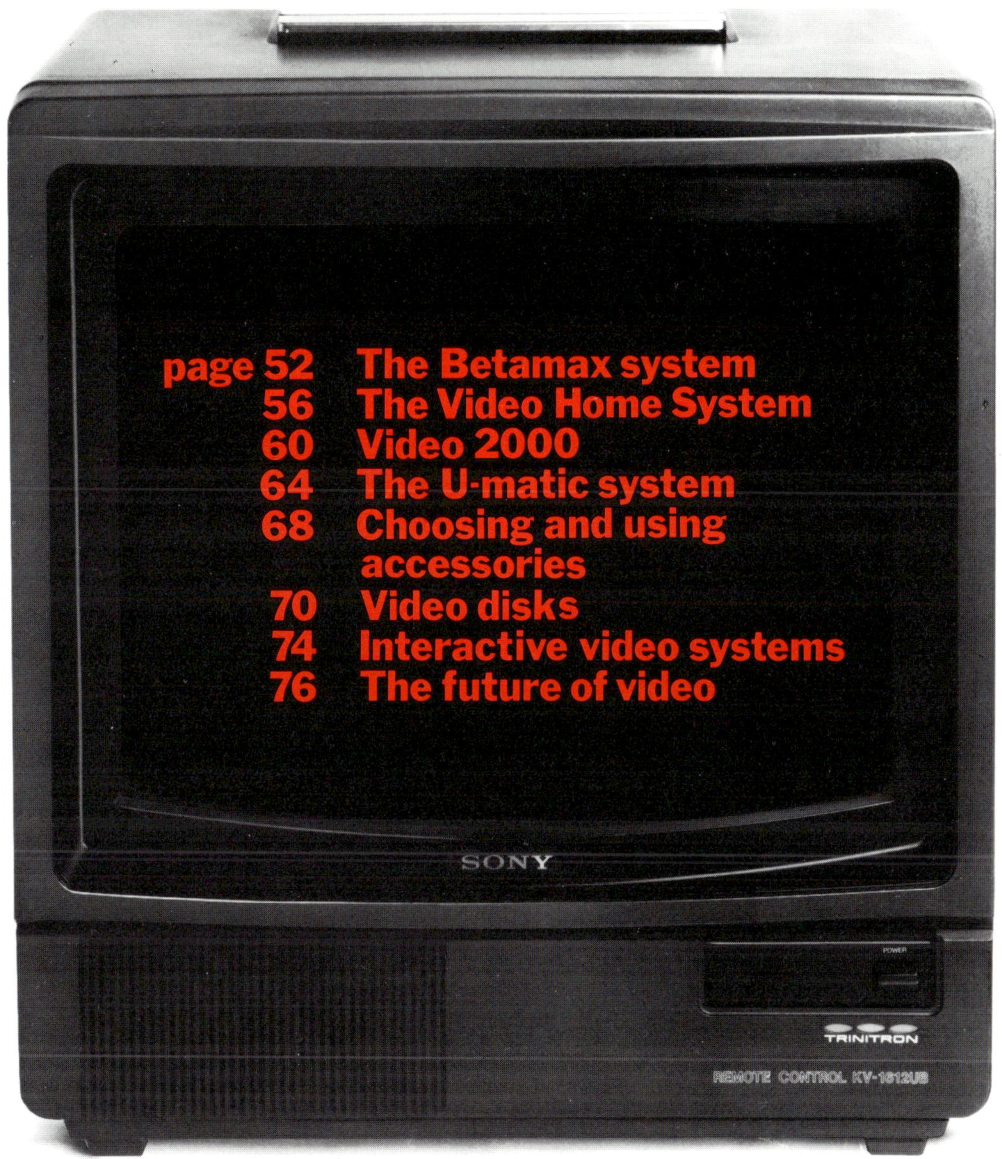

Before you buy, shop around. Prices fluctuate widely. If you are undecided rent a machine for a period. This way, you pay for the VCR in about three years and lose any resale value; but in return for a commitment of about one year the rental company bears all servicing and repair costs. Portapacks and cameras can be hired for short periods.

You will often find second-hand VCRs for sale. Some will be obsolete formats for which you cannot buy pre-recorded tapes, but blank tapes may still be on sale. Check that servicing is still possible. Even if it is only one year old, discarded in favour of an updated model, run a tape through it to check that it works properly. Later, when you come to update your own machine, think about keeping the old one. It could be useful for editing tapes.

The following pages guide you through the range of video recorders and their accessories currently on the market, and the video disk players you may soon be able to buy. Then, since innovations reported in today's video magazines will be on sale sooner than you think, the last pages take a look at video R & D, to give you a glimpse of what the future may bring.

The Betamax system

The Betamax system was developed by the Japanese electronics giant, Sony, in response to the now-obsolete 'VCR' format developed by Philips. Philips was the first company to enter the domestic video market, but the VCR cassettes were too expensive to attract enough buyers to become well established.

Until that time, Sony had concentrated on its industrial U-matic format. Open-reel VTRs were still the norm for professional video, and cassettes were an innovation. U-matic, with its automatic tape-threading system, called U-loading, became the basic blueprint for Betamax. However, where U-matic uses three-quarter inch (19 millimetre) tape, Betamax uses the half-inch (12.5 millimetre) tape used now by all home decks. This gives a compact cassette only 6 × 4 × 0.9 inches (155 × 95 × 25 millimetres).

Starting from an already-viable format, Sony were able to release Betamax in 1975, a year before its main rival, VHS. Consequently, in the areas where video was marketed vigorously and was quickly adopted – the USA and Japan – it long remained the favourite. In many other parts of the world the VHS system has made more impression. The main result of this is that prerecorded tapes will be easier to obtain in the locally-favoured format, although, because they are now simultaneously released in both formats, your supplier should be able to order tapes to the format you require.

Betamax has a reputation for technical excellence. The Betamax head drum, with a diameter of 2.9 inches (74.5 millimetres) spins at a speed of 19.2 feet (5.83 metres) a second, enabling the heads to write over a larger area of tape than in VHS. More information can be stored on the tape, and under optimum conditions this should give a better video picture and better audio.

Betamax cassettes are slightly smaller than VHS and Video 2000. This means that portapacks can be smaller and lighter, but also that the longest Betamax tapes give a shorter playing time than the longest VHS, since they are physically smaller. They tend to be cheaper. On the other hand, Betamax is still the only format with an optional tape-stacking mechanism which allows 14 hours of continuous playing.

The U-shaped tape path is thought to place minimum strain on the tape, particularly during on-screen picture search.

Unique to Betamax is Sony's Betastack tape-loading attachment. This is the video equivalent of a disk auto changer; up to four cassettes can be loaded into the cartridge, giving a playing or recording time of 14 hours at normal tape speed. As each tape finishes, it is ejected and a new tape inserted by the autochanger.

An attachment like this is at its most useful with a VCR like the C7, *below*, which has a multi-event timer, so that a long series of features can be recorded. This may account for the fact that video auto changers have not yet become common.

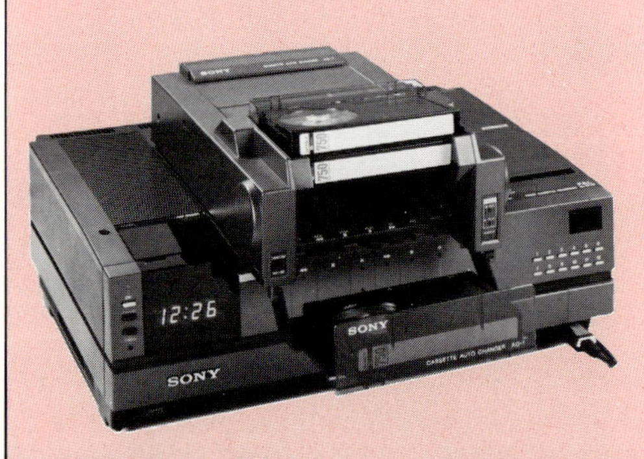

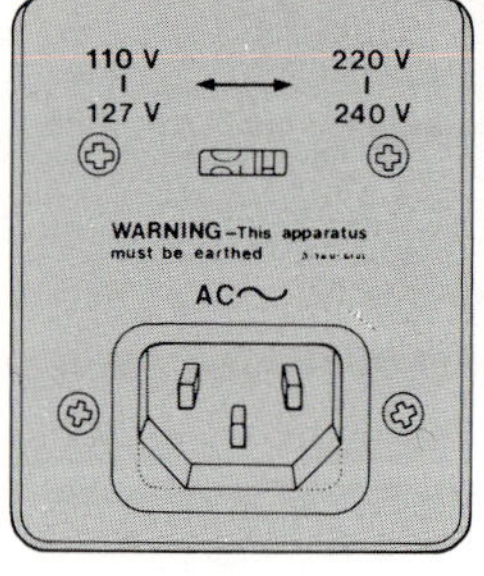

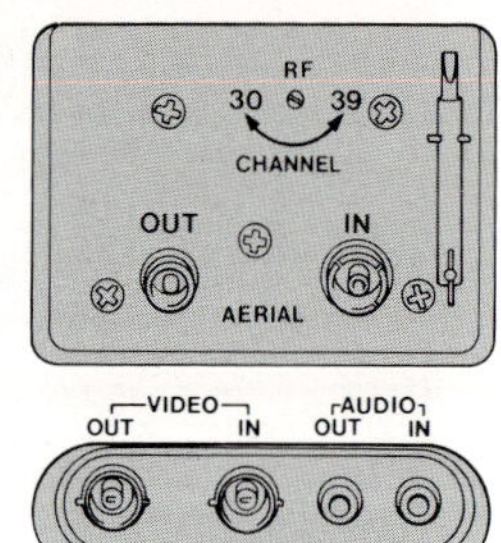

The voltage selector on the back panel of the Sony C7, *above*, gives a choice between 220 and 240V and 110 and 127V. Some voltage selectors give a choice between three or four single-voltage settings.

Methods of tuning a VCR's RF-decoder to the TV vary. In this case, a test pattern is generated by the VCR, and a screw is turned until the pattern shows clearly on another channel.

New from Sony in the USA, the SL-2500, *right*, is the first VCR to be designed to stack with hi-fi components. Modular stacking with other equipment is a natural progression for video, and this is its first realization.

The 2500 has the advanced features of larger VCRs: high-speed tape search (called Betascan), with random access to nine points on the tape, and a direct camera input with roll-back editing.

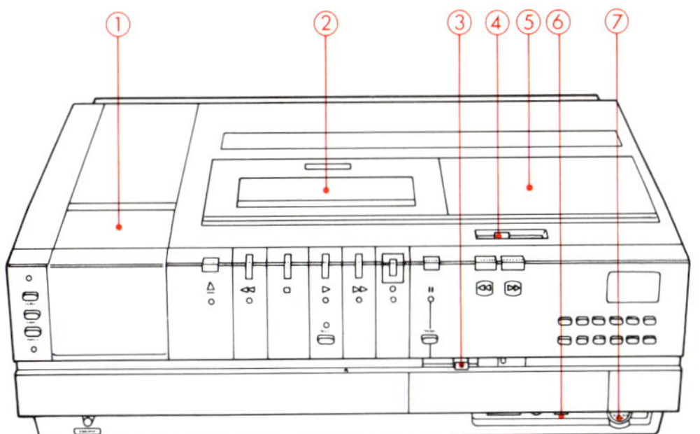

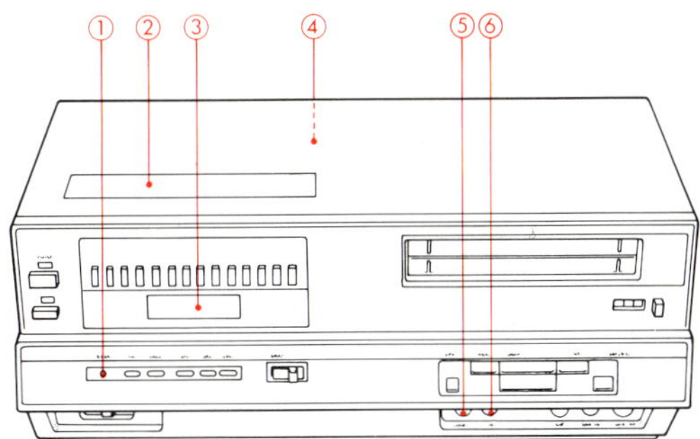

Sony SL-C7, *above*, has been Europe's top-line VCR since 1980. The top-loading cassette compartment, **2**, tuner panel, **5**, and timer panel, **1**, distinguish it from newer front-loaders.

Some older VCRs only have a 3-digit tape counter because they pre-date longer tapes. The C7's tape counter, **4**, has four digits, typical of modern VCRs.

A special tracking control, **3**, can be adjusted to give a good still frame and slow-motion picture.

The C7 was one of the first VCRs to provide a direct camera input, **7**. An input select, **6**, switches between the camera, the TV receiver, and a line 'Video in' input on the rear panel.

The American SL-5000, *above*, is the most advanced of the series which includes the 5400 and 5800.

The 14-switch channel-select panel, **3**, is typical of American VCRs, and will become more widespread as TV stations proliferate. The tuner controls, **2**, include a selector switch for VHF and UHF.

The tape speed select switch, found on many American VCRs, is at the rear of the SL-5000, **4**.

A cable remote control input, **5**, and separate camera pause, **6**, audio and video inputs make this VCR a mixture of advanced and basic features. The clock-set switch, **1**, is operated by a fine-pointed tool.

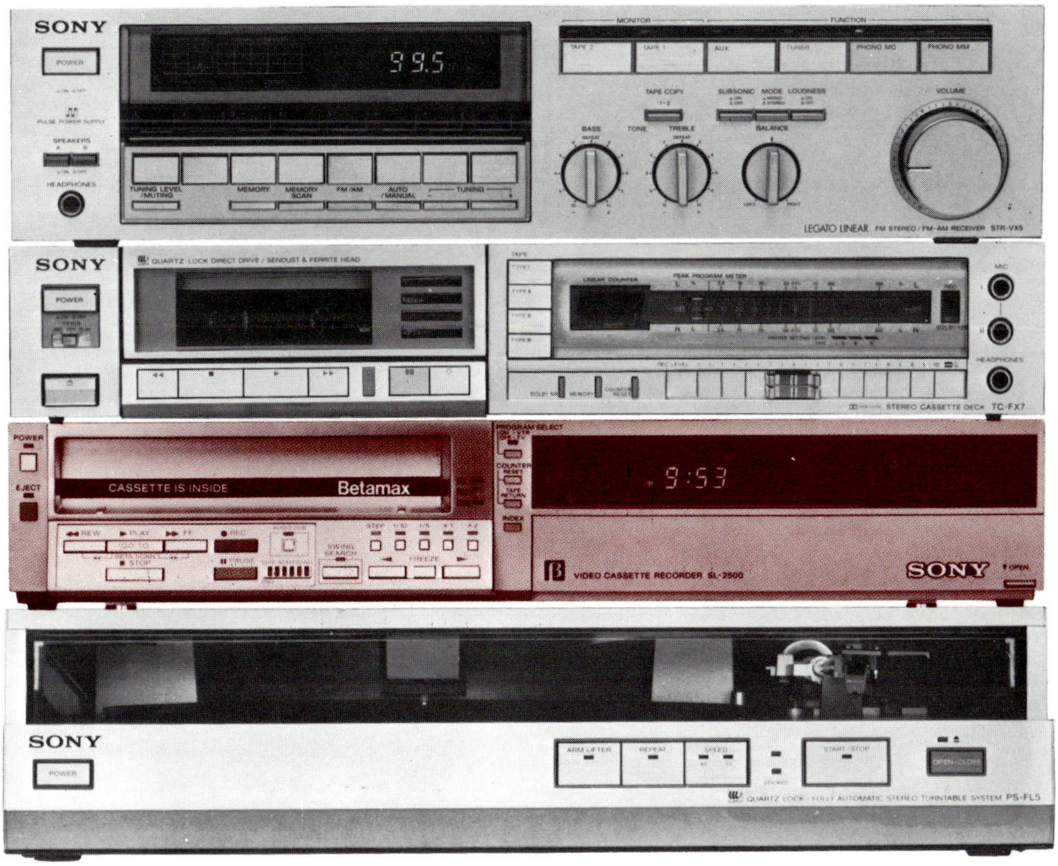

On the market/Betamax recorders

The prospective buyer will notice immediately that there are far fewer Betamax models than VHS to choose from. This is not a disadvantage; VCRs today vary so little from one brand to another that some people may even consider it an advantage.

VCRs fall into three categories: 'budget', 'de luxe' and portable. The lower-priced, or budget, models have few facilities, to minimize costs. The most obvious limiting factor will be the timer, which will allow you only one or two pre-programmed recording events, or one event at the same time on several consecutive days. Many people find that a one-event timer is quite sufficient.

In the second category are the models with more advanced features and a higher price tag. Some manufacturers have several machines on their current catalogue, each slightly more complex and expensive than the one before, so study the most advanced model and then work down through the price range until you find one that suits you.

A portable is built for lightness. Betamax portables are potentially slightly smaller and lighter than VHS, since the cassettes are smaller. However, the choice of Betamax portables is much smaller than for VHS.

Prices tend to reflect facilities rather than picture quality: the best way to study this is to try a model in your own home. This may be hard to arrange if you are not renting; so check whether the store you are dealing with will give you another model if you are not happy. Check that the focus is crisp and the colour true, and that the sound quality is satisfactory for your needs.

All Betamax machines are manufactured under license to Sony. There are fewer manufacturers offering Betamax than VHS: the Sony, Sanyo and Toshiba companies are the primary manufacturers, while others market Betamax machines under their own labels. The choice of facilities and the standards of maintenance and service are much the same for Betamax and VHS.

One point that may decide you between VHS and Betamax is Betamax's faster write speed, which gives it better sound reproduction. However, if audio is an interest, be sure to look for a machine with a noise-reduction system. Sony has its own system, Beta Noise Reduction, but this is apparently less effective than the Dolby system, which VHS has been the first to adopt.

The superb design and engineering for which Betamax is known make it an attractive format. As an example, Toshiba's new model has four video heads instead of two, a facility uncommon at present, but likely to become widespread in the future.

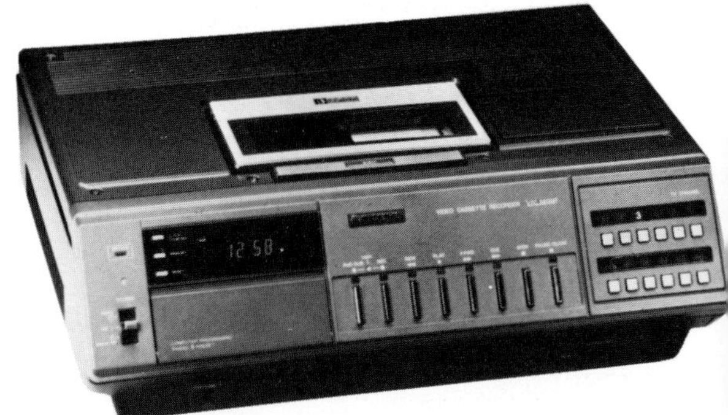

Sanyo VTC-5600 (Europe)

One of the new generation of Betamax, the Sanyo VTC-5600, *above*, has some advanced features at a medium price.

The soft-touch logic controls are a popular modern feature, and are complemented by a record-lock switch to prevent a chance touch from starting the record controls and erasing a tape. On-screen picture search is in forward mode only.

Sony's SL-2000, *below*, is the smallest and lightest $\frac{1}{2}$-in (12.5-mm) portable in the world. Combined with the TT-2000 tuner/timer, it gives many facilities associated with home decks.

The timer can memorize four events over two weeks, and there is a full set of fast and slow frame features.

Important and unusual is the SL-2000's ability to review just-recorded material, in forward or reverse, without disrupting the camera's recording mode. Normally this would happen if the portapack were taken off the 'Pause' or 'Record' setting, causing a burst of video noise on the picture.

Sony SL-2000

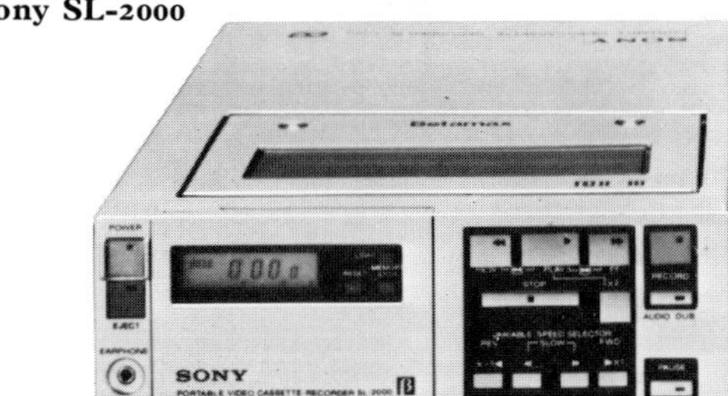

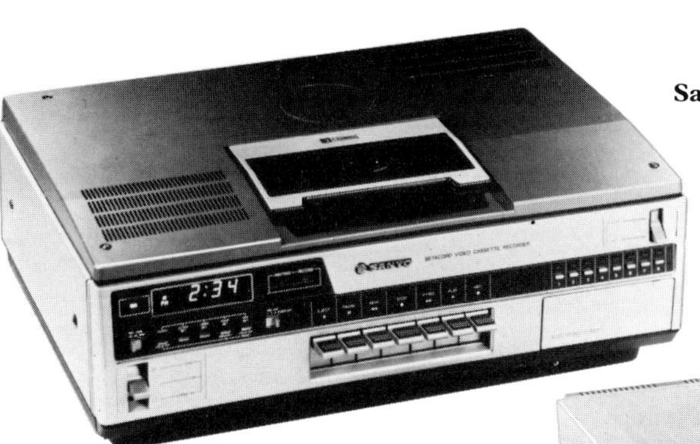

Sanyo VTC-9300 (Europe)

Sanyo VTC-9300, *above*, is old-fashioned, but an inexpensive and popular machine. It is identified as first-generation by its piano-key style controls. The switches are mechanical, and 'Stop' must be selected between other modes, or the controls may suffer damage or strain.

Typically on old VCRs, there is no on-screen picture search.

The VTC-9300 has held its ground as a budget VCR because it offers similar quality to more expensive machines.

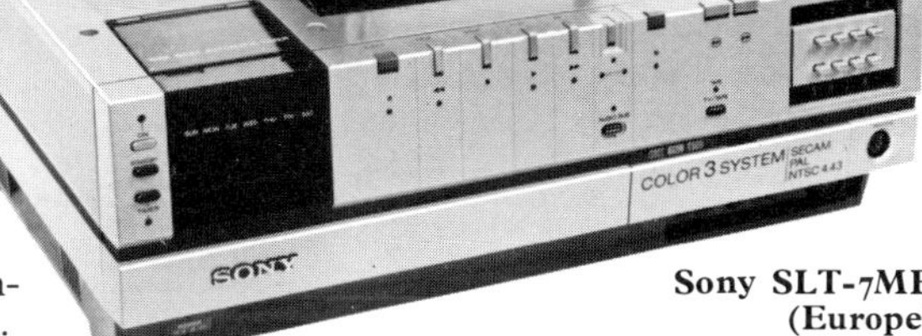

Sony SLT-7ME (Europe)

Sony SLT-7ME, a European triple-standard version of the C7, is aimed specially at customers in the Middle East, where PAL and SECAM are transmitted, and NTSC recordings are on sale. Like most PAL-based multi-standard VCRs, it will play, but not record, in NTSC. Oddly, it will not record off air in the UK.

It is simpler than the C7, having monochrome freeze frame instead of colour, and a corded remote, instead of infra-red.

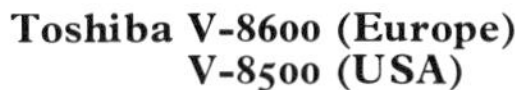

Toshiba V-8600 (Europe)
V-8500 (USA)

Toshiba's new model is one of the few VCRs to have four video heads. The extra pair are specially included to overcome tape distortion during freeze frame and to give a noise-free picture.

Tracking controls also allow the fast picture search functions to be noise free.

This VCR also uses the 'Beta Noise Reduction' system but this apparently is not as effective as the Dolby system, and the audio is not exceptional.

A cable remote control is standard, with infra-red optional.

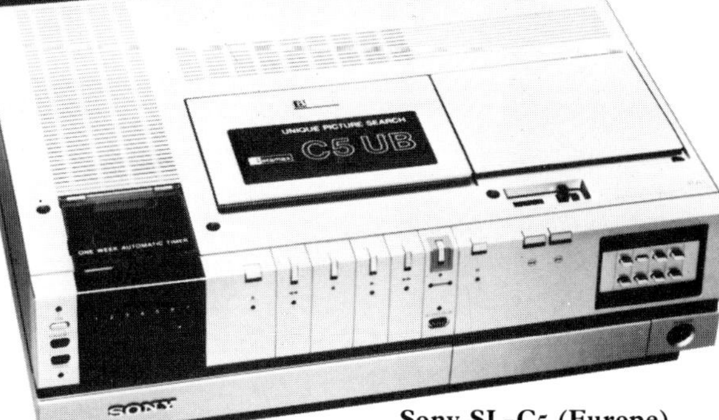

Sony SL-C5 (Europe)

A budget version of the Sony C7, the European C5, *above*, gives the same basic quality at a budget price. The timer allows only one event to be pre-set over seven days, and the freeze frame and picture search in forward and reverse are in monochrome.

Unusually for a budget model, the C5 has a direct camera socket, and an optional cable remote control.

The Video Home System

VHS (Video Home System) was announced by the Japan Victor Company (JVC) in 1976, a few months after the launch of Betamax. By 1978, home and portable machines were widely available, and VHS has become the market leader over most of the world.

For VHS, JVC developed a new tape-loading system known as M-loading. This uses a complex tape path, and the system has been criticized since it was believed to cause tapes to stretch during fast-winding. Therefore, early VHS machines fast-wound with the tape unlaced from the tape heads, so that no picture could be seen. Further developments have alleviated the likelihood of tape damage and given good on-screen picture search.

The VHS head drum is 2.6 inches (62 millimetres) in diameter and revolves at 7.3 inches (2.34 millimetres) a second, giving a slower write speed than Betamax and, therefore, in theory, a slightly inferior picture. In practice, however, if anything has seemed to suffer, it has been the audio quality. Sophisticated electronics have improved this to the point where only serious music lovers will be disappointed, but video sound is generally poor because so little space is allocated to it on the tape.

The VHS cassette is bigger than the Betamax at 7.3 × 4.1 × 9.0 inches (187 × 104 × 25 millimetres). It is much the same size and shape as a Video 2000 cassette. VHS tape has a longer playing time than Betamax, four hours for the longest tapes, and six hours if a long-play machine is used, but since long tapes suffer more from breakage and poor reproduction quality, this is not necessarily an advantage. Video 2000 tapes give a longer playing time as they can be recorded on both sides.

Factors affecting picture quality are so many, and so often external to the VCR (the television receiver and antenna used, for instance) that a sensible comparison can no longer be made between VHS, Betamax and Video 2000 on picture quality alone. Furthermore, fierce competition has ensured that the range of facilities any one format now has to offer does not differ much from that of another. The choice must ultimately depend upon the qualities of one particular model, and local factors (such as availability of video tapes and service), rather than a preference for one format.

JVC's flagship, the HR-7700, is one of the newest, most advanced and most expensive VCRs in any format. It is the ultimate in VHS and, in Europe, has become the successor to the Betamax SL-C7 as the state-of-the-art popular-format VCR. It is expensive, but a luxury machine.

The HR-7700 follows the trend of other, newer VCRs in having a separate RF (radio frequency) output for standard TVs and a video input and output for non-RF monitors and cameras. This, and the good back-space editing facility, make it an ideal VCR to use for simple editing.

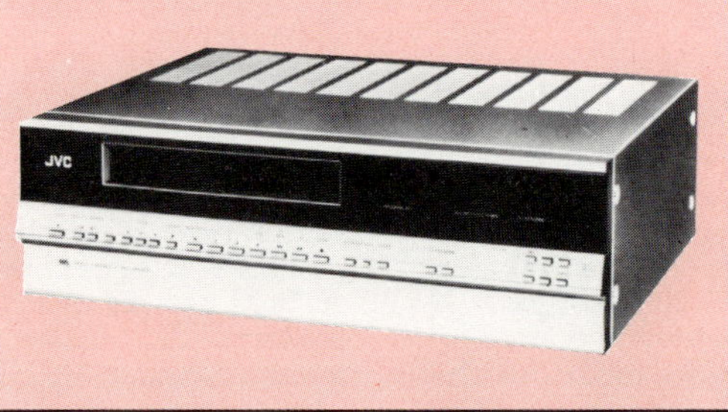

The HR-7700, *below,* has a catalogue of advanced features. For audio fans especially, the Dolby noise-reduction system will be attractive. Look for the Dolby logo, **1**, on the front panel.

The 'x2 with sound' facility, **2**, is not just a fast picture search: busy users should be able to watch recordings in half the usual time and still follow the sound-track.

Like nearly all new-generation VCRs, the HR-7700 has a front-loading cassette compartment, **3**.

Slow motion, **4**, frame-by-frame advance, **5**, and still frame, **6**, once considered gimmicks, are now found on nearly all VCRs. The HR-7700 gives colour freeze frame and slow motion pictures which, when the tracking is well adjusted, are virtually noise free.

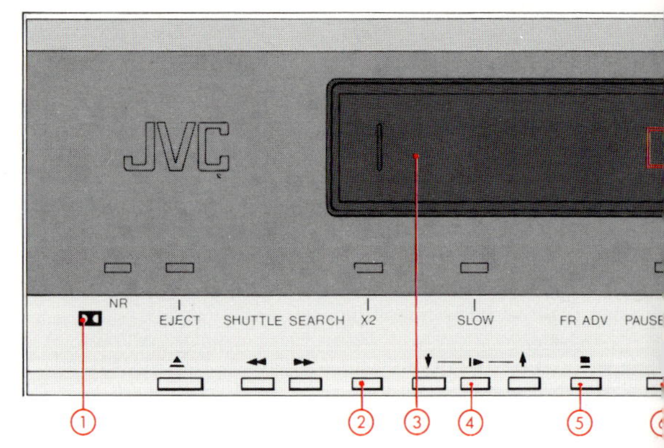

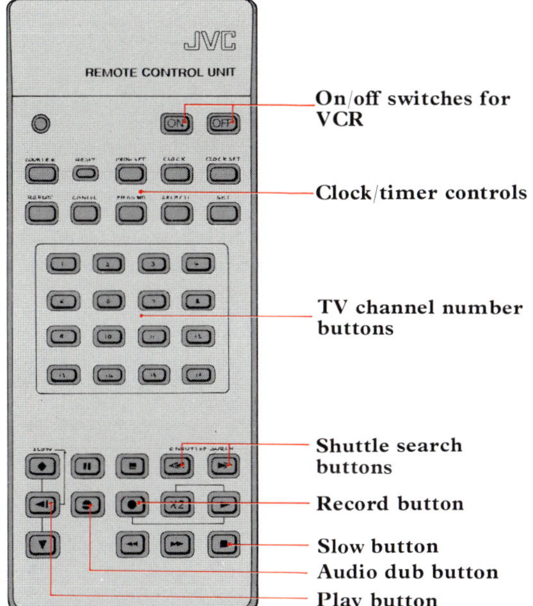

A full-function remote-control unit is standard with the HR-7700. This controller, *left*, is unusually comprehensive. It uses an infra-red signal, which is more flexible than a cable control, and less liable to interference than an ultrasonic signal.

Remote controllers vary in the distance and angle at which they can be used. Test the unit in the shop and try to find its limitations. You should expect to have infra-red on advanced models in any domestic format.

This unit has a complete set of clock/timer controls, channel selection buttons, variable-speed functions and audio dub. A light emitting diode (LED) lights up to show when the remote control is operating. One feature of an infra-red control is that is has no 'On/off' switch of its own, needing power only when in use.

The choice of a remote control is a matter of convenience. While a corded remote can be used anywhere within the cord's reach, an ultrasonic or infra-red remote must be aimed at the window on the VCR.

'Audio dub', **7**, is another facility appearing on more and more VCRs.

Concealed neatly by an aluminium hatch below the front panel, are headset and mike sockets; a tape memory switch; tracking controls for slow/still frame and x2/play modes; clock/timer controls; auto-sweep tuning; auto tape search (with which the VCR will home in on a key signal recorded along with a broadcast), and a direct camera input.

The soft-touch switches, **9**, are logic-controlled, so that any sequence of commands is stored in the VCR's microprocessor before it responds to them. This reduces the risk of damaging the VCR by pressing switches too fast, or in the wrong order.

The display-mode indicator, **10**, reads 'Clock', 'Prog' or 'Count', telling the user whether it is in a clock mode, timer-setting mode or tape-counter mode.

The clock display, **11**, has fluorescent numerals which give a 24-hour display. The remote control indicator light, **12**, comes on when the remote control infra-red is in use.

A line of ten fluorescent segments, **13**, can either light up to monitor the auto-tuner's progress as it sweeps the TV broadcast channels, or progressively light up from left to right to show what portion of the tape remains to play.

The TV channel indicator displays the number of the channel which is being pre-set or which is being held in the channel lock, **14**.

The sub-power buttons, **15**, send power to different parts of the VCR: the power supply is on if the recorder is being used and off otherwise, and the timer power stays on while the timer is set to record. Some power is used at all times to keep the clock running.

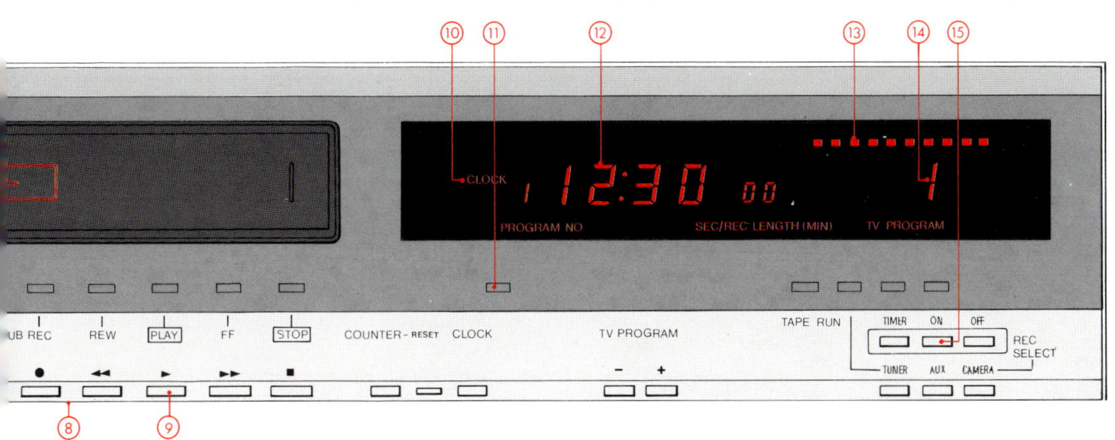

On the market/VHS recorders

A majority of manufacturers has opted to manufacture VHS machines under licence to JVC. Primary manufacturers are JVC, Panasonic, Hitachi, Mitsubishi, Akai, Magnavox and Sharp; certain other manufacturers package VCRs manufactured by JVC and Panasonic. As well as the VCRs on the current catalogues, there are often slightly older machines in stock, for which parts and service should present no problem but which may have been reduced in price to clear them. It is perfectly normal for companies to delete a particular model from their catalogue and to stop producing it, but to go on providing parts and supporting services for years afterwards.

VHS is the best-selling format over much of the world, including the UK, the USA and the Middle East. Pre-recorded tapes and to a lesser extent blank cassettes, may be marginally easier to obtain for VHS in these areas, but you can order tapes of other formats.

A good way of inspecting machines is to buy a cassette of the right format and take it to your video dealer. Ask the assistant to record something off air for you, and compare the results with the pre-recorded versions. This may also be a good chance to see in detail how the tuner/timer controls work. You could rent a recorded tape and use it for comparing VCRs in the same way. It is a good idea to try the same model in different shops, if possible.

Another advantage of VHS is that there is a considerably wider range of cameras and portapacks in that format. Cameras are not restricted to one format or another, but it tends to be true that a camera of a particular brand works best with a VCR of the same kind. This is not invariable, though, so test any equipment you buy thoroughly.

Although prices are now fairly consistent from one brand and one format to another, it is worth looking at both VCRs and cassettes in your area to see if one format gives a distinct price advantage. If you are going to build a complete tape library, you will want to choose the format with the cheapest tapes. VHS tends to be a little more expensive per hour (the only meaningful comparison) than Betamax, and Video 2000 is cheaper.

A hard-hitting marketing campaign by JVC back in the seventies accounts for VHS's current domination of the European market. Sony was much too slow.

VHS offers by far the widest choice of individual VCRs of any format. Although all VHS machines are made under license to JVC, other manufacturers have been innovative in their modifications to the system. For instance, Hitachi's VT-6500 portapack is one of few VCRs to offer 'Sound on sound', whereby an extra sound-track can be recorded over the original one without erasing it – music over commentary, for instance.

Although all VCRs are alike in many ways, it is worth while looking closely at several to see what features they offer.

Akai VS-5 (Europe)

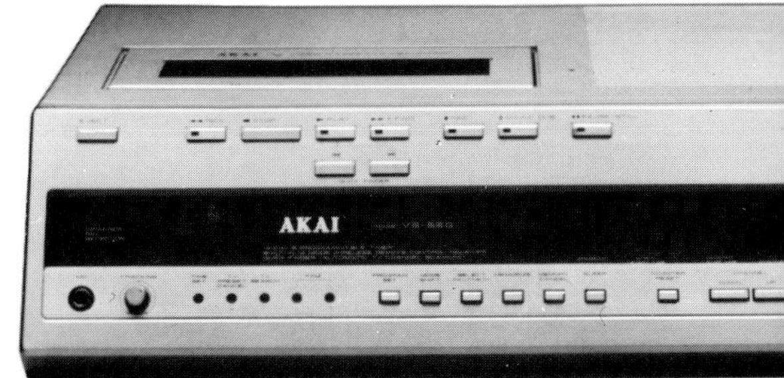

Akai VS-5, is a European top-loader. It has a steeply angled front panel making the controls easy to see from all angles: a neglected point with top-loading VCRs.

The VS-5's timer can handle nine events over 14 days. A full-function infra-red remote control is standard, and it has a safety lock to disarm the device while recording is in progress, or a tape is being played, to prevent accidental disruption.

If a malfunction should occur during recording, an 'error' buzzer sounds.

This VCR has a channel-tuner with one button to locate stations, and one for tuning.

JVC HR-7200

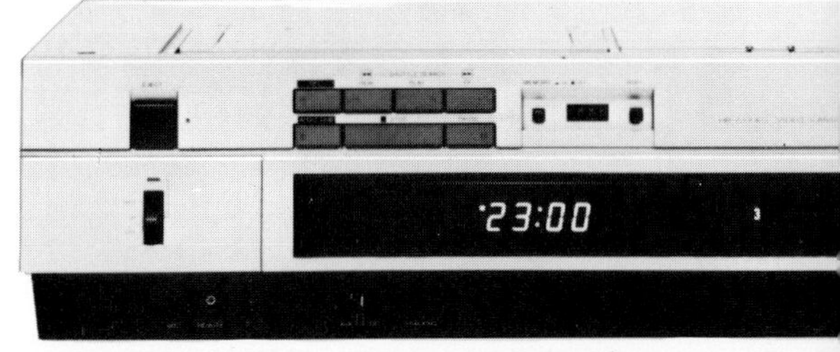

JVC HR-7200, is a good-value middle-market VCR.

The timer will memorize only one event, up to seven days ahead. However, there is a full-function remote control, and nine times normal speed in forward and reverse, automatic rewind, and tape memory, which return the tape to predetermined points.

Akai VPS-7350 (USA)

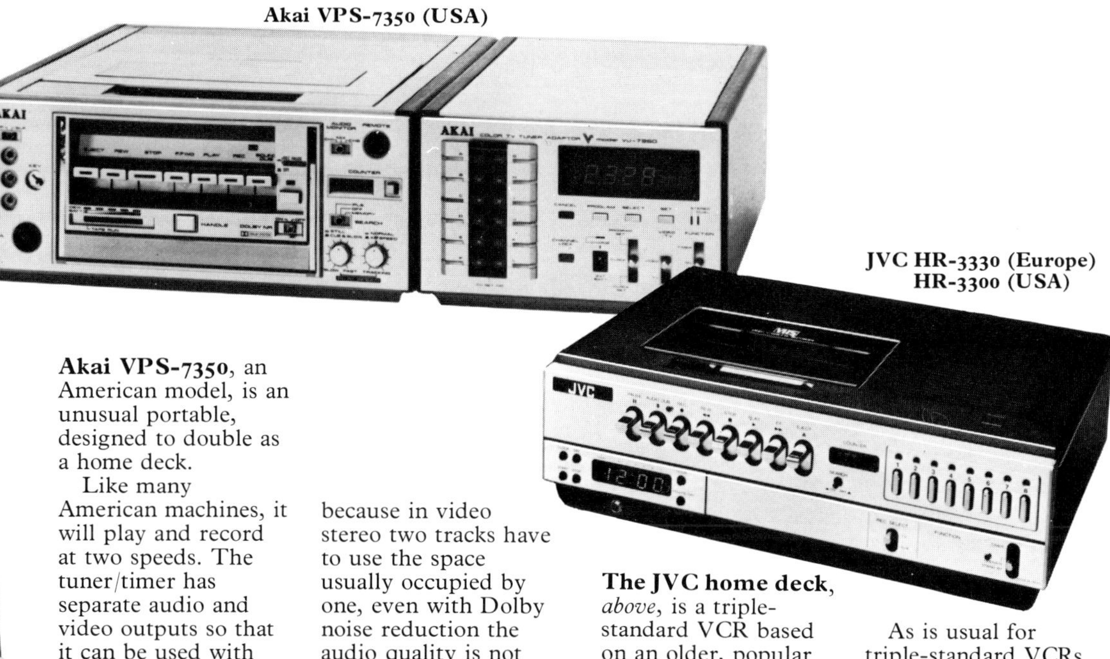

JVC HR-3330 (Europe)
HR-3300 (USA)

Akai VPS-7350, an American model, is an unusual portable, designed to double as a home deck.

Like many American machines, it will play and record at two speeds. The tuner/timer has separate audio and video outputs so that it can be used with older and industrial VCRs, which have no tuners.

There are two audio tracks, allowing stereo recording, but because in video stereo two tracks have to use the space usually occupied by one, even with Dolby noise reduction the audio quality is not particularly good.

The VPS-7350 can be locked, perhaps to prevent children from using it when parents are out.

The JVC home deck, *above*, is a triple-standard VCR based on an older, popular JVC machine.

Its facilities are basic, with no picture search, and a cable remote control with a pause function only.

As is usual for triple-standard VCRs and for some single-standard models, it is possible to select between 110, 127, 220 or 240V and 50 or 60 Hz AC supply.

Panasonic NV-7200 (Europe)

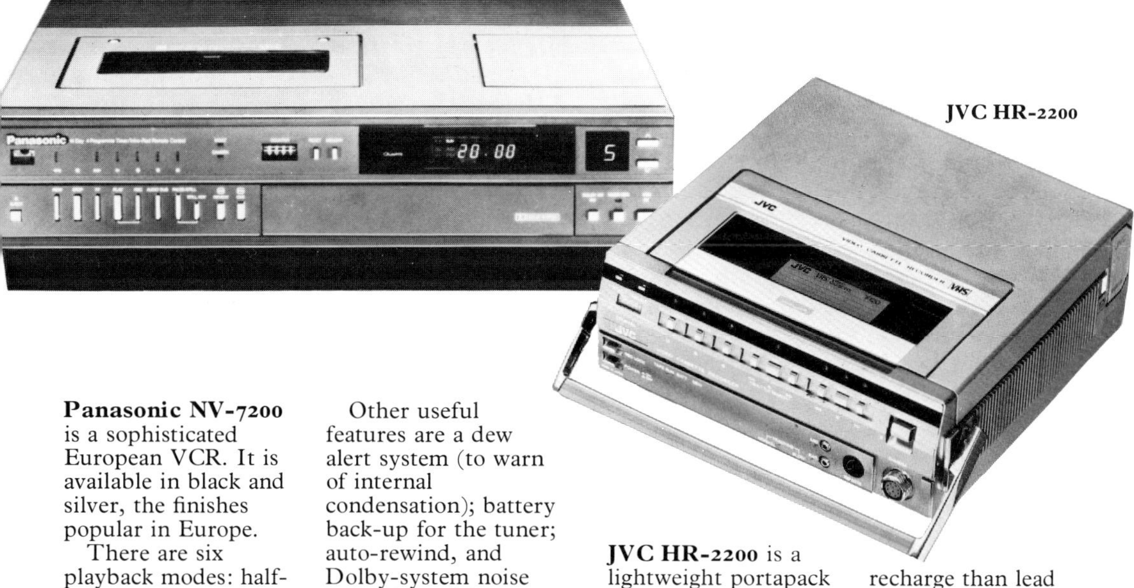

JVC HR-2200

Panasonic NV-7200 is a sophisticated European VCR. It is available in black and silver, the finishes popular in Europe.

There are six playback modes: half- and double-speed, normal, freeze frame, frame-by-frame advance, and nine times normal speed. All are operable from the infra-red remote controller.

Other useful features are a dew alert system (to warn of internal condensation); battery back-up for the tuner; auto-rewind, and Dolby-system noise reduction, giving better than usual audio quality.

Panasonic prides itself on building its VCRs on die-cast, lightweight aluminium chasses.

JVC HR-2200 is a lightweight portapack weighing just over 11 lb (5 kg). Older machines weigh around 19 lb (8.6 kg). It uses rechargeable nickel cadmium ('NiCad') batteries, which are quicker to recharge than lead acid.

The HR-2200 can be powered from its battery pack, an AC supply or a 12V car battery. A built-in RF unit allows connection with a TV.

Video 2000

Video 2000 is Europe's answer to the Japanese takeover of the home video market. Ironically, Philips, co-inventor with Grundig of Video 2000, were the first to market home video with their aptly-named VCR (Video Cassette Recorder) format, but they later lost ground to VHS and Betamax.

In the competitive world of video, as soon as a new model appears, it is safe to assume that a rival has a more advanced model in the pipeline. Outstripped by Betamax and VHS, Philips and Grundig had to produce something to challenge the market.

Video 2000, like VHS and Betamax, uses half-inch (12.5-millimetre) tape. The cassette, at 7.5 × 3.9 × 0.9 inches (183 × 109 × 25 millimetres) is very like a VHS in shape, but it is unique in having two tracks, like an audio cassette, so that the tape can be turned over to give twice as much playing time. (The compact audio cassette, developed by Philips and now the world standard, was the model for Video 2000 cassettes.)

Another major advance is Dynamic Track Following. This process uses a flexible video head which is able to detect the exact path of each video track and follow it precisely, a facility until now only available on professional recorders.

DTF means that the video tracks can be narrower, allowing a more economical use of tape and permitting the two-track format. It also gives very stable freeze frame and slow motion pictures. To achieve this picture stability, the other two formats have produced models with two extra video heads, but they do not give such a good result as the mobile heads. For this reason, there is no tracking control on a Video 2000 machine.

A problem facing Video 2000 is one of market resistance. The video-buying public has become familar with VHS and Betamax. Moreover, the ambiguous name chosen for the Video 2000-format cassettes, Video Compact Cassette (VCC), was intended to draw attention to the link between the Philips system and the successful compact audio cassette. However, the lack of direct reference to Video 2000 on the cassette has given rise to the rumour that Video 2000 cassettes are nowhere to be bought, whereas VCC is widely stocked.

It seems likely that Video 2000 will survive and prosper, and enthusiasts, looking for above-average quality, should consider it.

Video 2000's most advanced model, Grundig's 2 × 4 Super, arrived in time to save the format's reputation after poor performances by some of the earlier Video 2000 models. The unique Dynamic Track Following (DTF) system gives outstanding picture quality with still frame, slow motion, frame-by-frame advance, and on-screen forward and reverse and picture search functions, where the picture should be completely noise free: a great asset for anyone using slow motion for analyzing their tennis style, for instance.

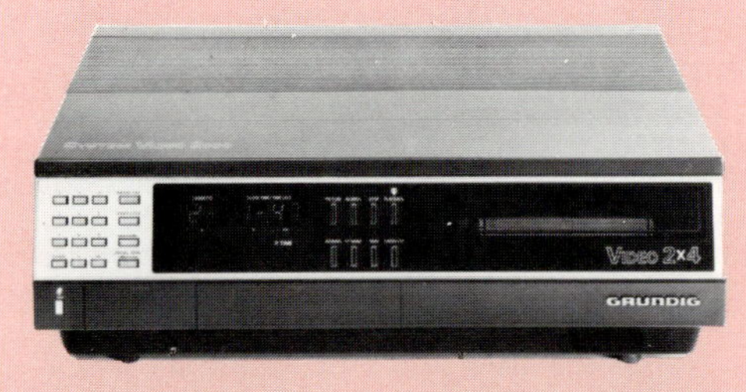

The Video 2000 system is different from VHS and Betamax in the style of some of its controls, as well as some of its central technology.

The timer keyboard, **1**, of Grundig's 2 × 4 Super will be familiar to anyone who uses a pocket calculator. It is simple to key in the start and stop times for pre-set recording without running through a sequence of hours and days. However, the VCR's own microprocessor will clear all commands that are not completed or continued in 15 seconds.

The Preselect Memory switches, **2**, are used to alert the microprocessor that you wish to key in a command, or to make the VCR display any instruction in its memory on the clock panel, **6**.

Tuning controls are concealed behind a panel under the keyboard, **3**. There are three channel-search buttons; in the UK only the one marked 'IV/V' is normally needed.

The two fine-tuning switches, **4**, are the equivalent of the AFT switches on some VCRs and radios.

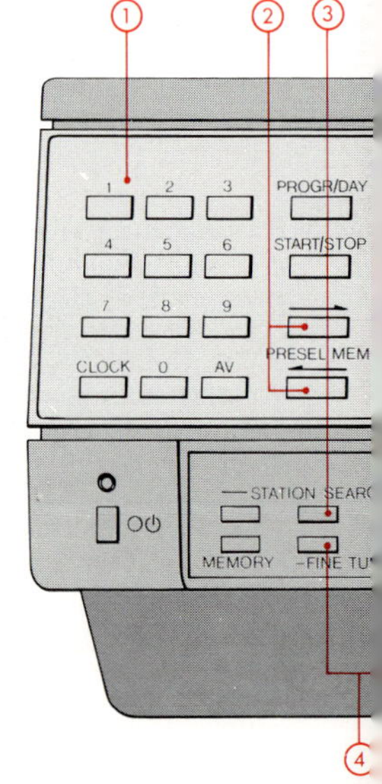

Dynamic Track Following (DTF) is unique in home video to Video 2000. DTF's secret is that the video head can alter its position to follow the video track it is reading. The tiny video head, *below*, is mounted on two bonded strips of piezo-electric ceramic. When an electric current is passed between the strips they bend slightly, moving the head up or down.

The diagram, *below right*, shows how the Video 2000 tape is recorded in two segments like an audio tape. The slanted video track covers less than half the width of the tape, the remaining space being occupied by two audio tracks for each video track, and two cue tracks in the centre of the tape, which could be used later in the format's development for carrying extra control signals. Each half of the tape is a mirror image of the other.

Because the heads follow each track so precisely, there is no need for the guard tracks used on VHS and Betamax to prevent the video heads picking up interference from adjacent tracks.

The video head stays in line by picking up minute traces of signal from the wrong tracks, analyzing and avoiding their source.

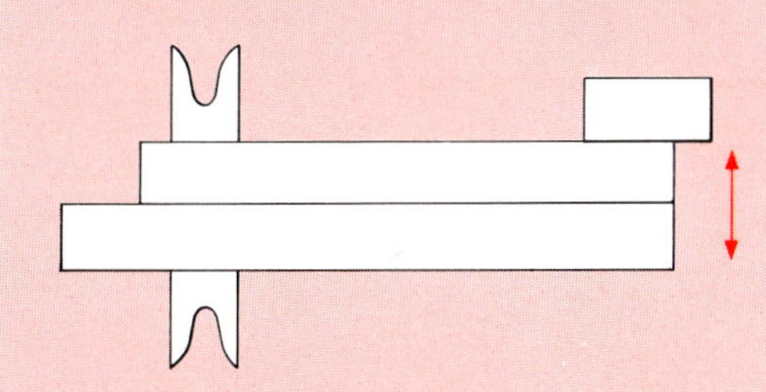

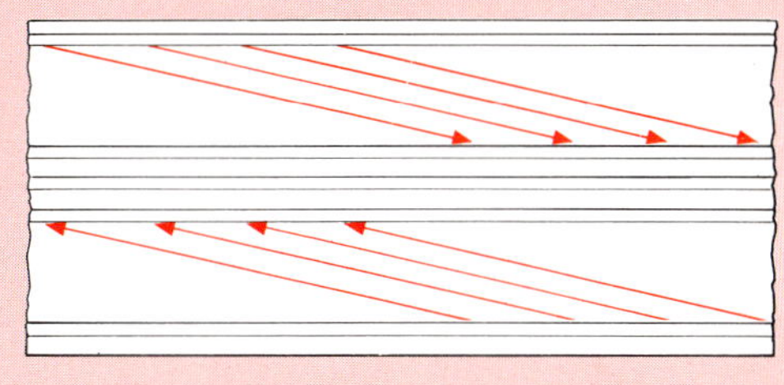

Although there is no tracking control, a crispener lever, **5**, gives the user a choice of sharp- or soft-focus picture. The microprocessor clears its memory after a 15-second delay if left standing, but there is also a manual clear control, **7**, in case the user wants to clear an error as soon as it happens.

The 'Go To' function, 8, is unique to Video 2000. If the user keys in a number of minutes from the start of a tape, the VCR will fast wind to that point. The 2 × 4 Super also has automatic picture search (APS), **10**, which finds the start and end of recordings by homing in on an audio tone.

If the 'Time Left' button, 9, is used, the clock display will show how much time is left for use on the cassette in the machine. It will automatically subtract any recording time stored in the VCR's tuner memory. With this VCR it is practical to base a tape-indexing system entirely on timing.

One confusing point about the 2 × 4 Super is the English wording on some of the switches. 'Stop', **11**, means 'pause', 'Tape', **12**, means 'stop' and 'Cassette', **13**, means 'eject'.

The 2 × 4 Super will also display 'Full' if you try to pre-set more recording time than is left on the tape being used.

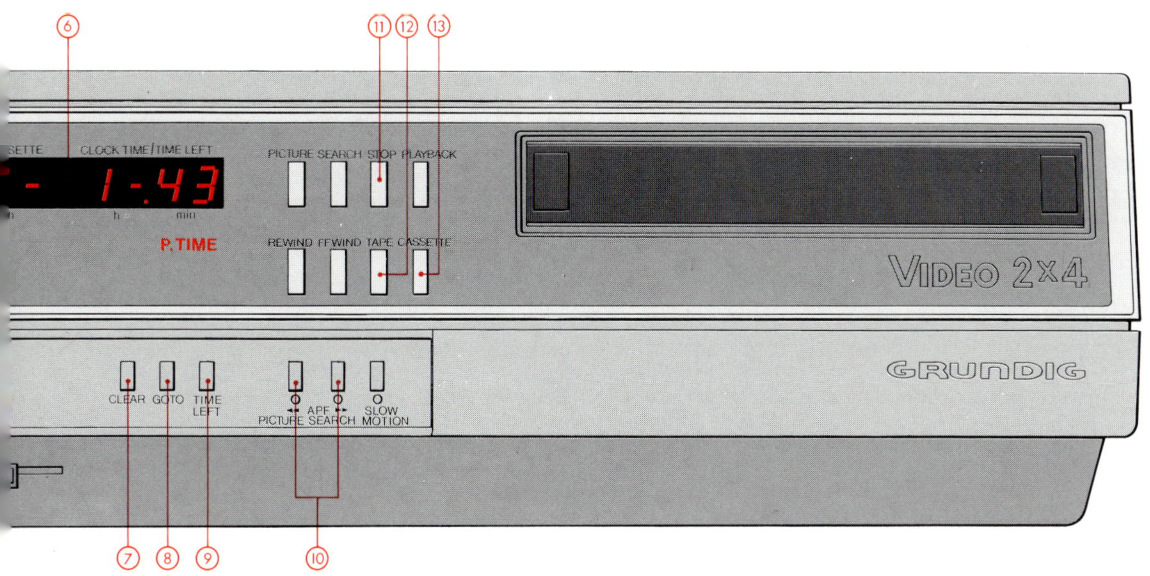

On the market/Video 2000 recorders

The late arrival of Video 2000 on the home video front explains why few models are yet available. The system was developed by Philips and Grundig, in cooperation; Philips machines are packaged by ITT, Pye, and Bang & Olufsen (the last being the only manufacturers to repackage the VCRs in their own style); Pye and ITT have simply relabelled the Philips machines.

Grundig, on the other hand, has designed its own version, aimed at the deluxe end of the market, and this is generally held to be the most advanced domestic VCR on the market today.

The late arrival of Video 2000 has given it advantages and disadvantages, but whereas the disadvantages are commercial (the reluctance of video retailers to commit themselves to stocking a third set of pre-recorded tapes) the advantages are technical.

Philips has been able to learn from the shortcomings of its two rivals. The design of the VCC (Video Compact Cassette) allows it to wind quickly in the housing, as do VHS tapes or, alternatively, in contact with the heads, as with Betamax. The capacity for a second audio channel was built into the VCC from the start, giving stereo or second track audio dub, whereas VHS and Betamax, with their fixed video heads, are limited in the extent to which they can compress the video track and, thus, in the extra space they can allow for audio recording. They can provide the same facilities, but are unlikely to match the quality.

Some models marred Video 2000's reputation because the incredibly sensitive DTF (Dynamic Track Following) system, which keeps the video heads in perfect contact with the tape, did not work properly. Since then, the shadow of these early troubles has been lifted by Grundig to reveal a picture of unsurpassed quality. Freeze frame, slow motion and picture search are especially noise free compared with other formats.

Will Video 2000 be the system of the future? Until it gets a larger slice of the world market it will be hard to tell. However, its popularity in Europe indicates that the enthusiast need not fear its sudden disappearance, and pre-recorded tapes are now readily available. Many enthusiasts today say that Philips's earlier formats, the VCR and VCR-LP are still giving top-quality pictures after many years.

Philips VR-2022 (Europe)

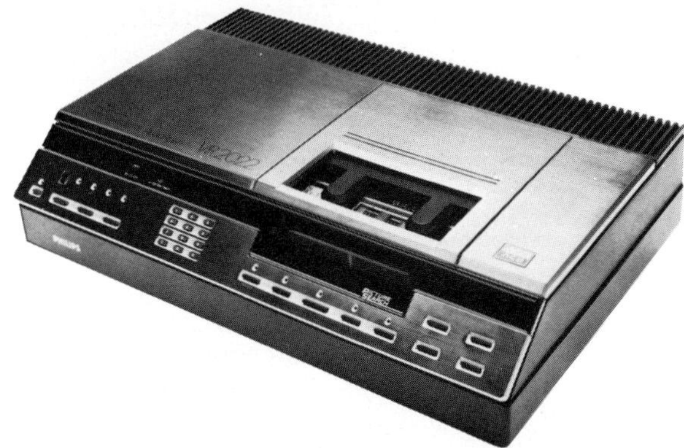

For Philips, the inventors of Video 2000, the VR-2022, *above*, represented an advance. After serious problems with their earlier VCR, the VR-2020, they needed a strong contender. So eager were they to erase the mistakes, that an improved version of the VR-2020, the VR-2021, appeared in the interim.

The VR-2022 has all the trick frame facilities that were omitted from the older models, and these achieve the high quality of which Video 2000 is capable. The 'Search' and 'Store' functions still have to be operated by a point such as a pencil, a seemingly crude method which, however, prevents accidental operation.

Philips VR-2020 (Europe)

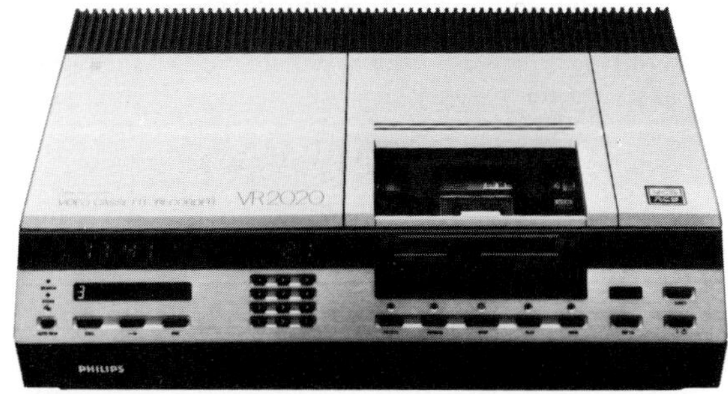

The Philips VR-2020, *above*, was the first Video 2000 VCR on the market, arriving in 1981. Although there were problems with this model in its early days, caused partly by the Dynamic Track Following (DTF) system which is Video 2000's great strength, it was seen even then to be a promising format. The troubles with the early VR-2020s may have been precipitated by Philips's desire to put Video 2000 on the market before Betamax and VHS gained an unbreakable hold.

The VR-2020, like Grundig's 2 × 4, is basically a mass-market VCR (if not actually budget) and lacks the still frame facility.

Technicolor Microvideo

At present there are three popular formats in home video, but a trend for specialization may be started by a fourth entirely portable format on sale in Europe and the USA.

The Technicolor 212, *below left*, weighs a mere 7 lb (3.2 kg) and its tuner, the Model 5112, only 4.2 lb (1.9 kg), making them the lightest in the world. As the parachutist, *below right*, shows, this sets a new level of portability.

The Technicolor 212 uses tape only ¼-in (6-mm) wide, and cassettes which can play for a maximum of one hour. Video enthusiasts will probably shoot on ¼-in (6-mm) tape and then dub and edit the footage across to ½-in (12.5-mm) tape. This portable format will complement a home system.

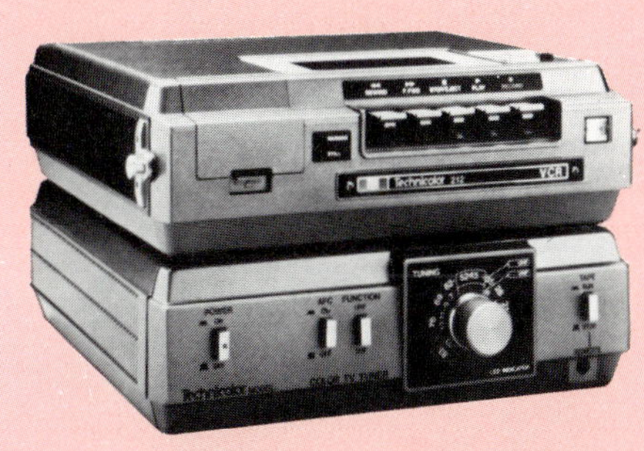

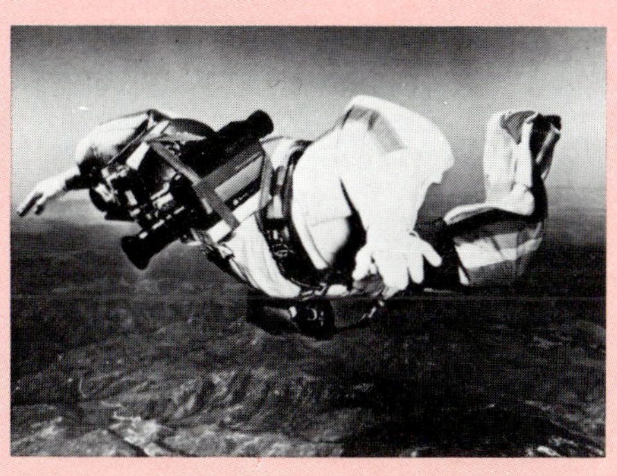

ITT VR-483 (Europe)

The ITT VR-483, *above*, is the Philips VR-2020, sold under ITT's label. Of the models they market, the VR-580 is the VR-2022, and the VR-482 and -483 are the earlier, less sophisticated and cheaper models, slightly restyled.

Another well-known electronics firm, Bang & Olufsen, best known for their futuristic hi-fi designs, are also joining the market with a repackaged Philips VCR. This one, however, looks very different from the Philips model.

The Pye 20VR20, *below*, is the Philips VR-2020 by another name. It is common for makers of electrical goods to market some models bought from other manufacturers, often with little or no alteration besides the change of colour and label. Pye also sell the VR-2022 under the name 20VR22, and a 20VR23.

Unlike Philips, Grundig is so far the sole marketer of its own VCRs.

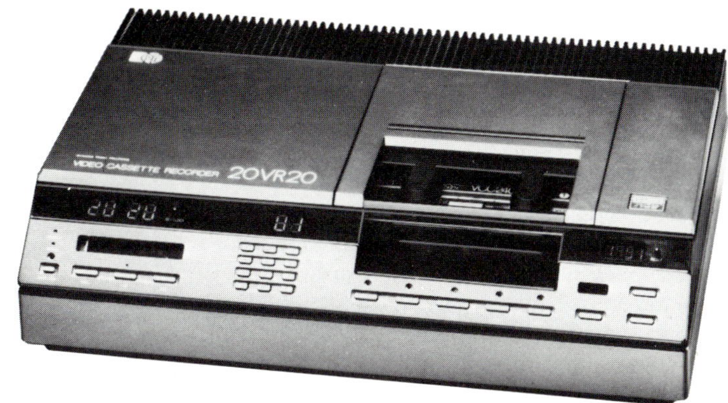

Pye 20VR20 (Europe)

The U-matic system

When it was introduced in the early seventies the U-matic format revolutionized video recording. Before U-Matic, all video recorders were reel-to-reel, winding from one open spool to another. They had to be threaded, at least partly, by hand, through a tricky tape path (although the tape used was rather tougher than the cassette tape now used). Cassette-loading is much less troublesome, and despite its comparatively narrow gauge of three-quarter inch (19 millimetre) as opposed to 2 inch (50.8 millimetre) for professional reel-to-reel tape, U-matic can give a quality acceptable for editing and broadcast.

As well as a wider tape, the tape speed, at 3.7 inches (95 millimetres) a second, and the writing speed of 26.3 feet (8.54 metres) a second, are much faster than for domestic systems. The cassettes, 8.6 × 5.4 × 1.1 inches (219 × 138 × 31 millimetres) and the machines, 25.4 × 18.2 × 10.4 inches (646 × 425 × 226 millimetres) are about twice the size of the biggest domestic versions. To carry the portapack for any length of time requires a backpack and a certain dedication.

U-matics are industrial machines. In professional terms they are small, mobile, cheap, convenient and easy to use. The range and flexibility of their recording facilities for both video and audio is well beyond that of domestic models, although the gap is narrowing as far as facilities are concerned. However, the sheer extra space for electronics inside a U-matic chassis will ensure that the differential in quality is maintained.

A whole new era in video may well be opened up by the arrival of industrial-quality VHS and Betamax machines, hybrids between the domestic formats and the U-matics. Approaching the quality of U-matic, these VCRs will play ordinary Betamax and VHS cassettes, offering new opportunities for interchange between the amateur and domestic, and the professional and educational user. For broadcast and near-broadcast quality, however, the wider-gauge tape will probably continue to be a necessity.

The main users of U-matics are colleges, industry and video recording studios who need cheap recorders giving professional-quality recordings. For the serious home video enthusiast they are the most ubiquitous of all VCRs, opening the way to full broadcast-quality recording.

JVC's CR-6060ET is a player/recorder VCR which comes in two versions, the standard machine, handling PAL and SECAM signals as a matter of course, and the triple-standard version, also able to replay on NTSC.

U-matics, unlike domestic VCRs, are issued as part of a series designed to work together. Instead of issuing a new model, the manufacturer will bring out a whole new series.

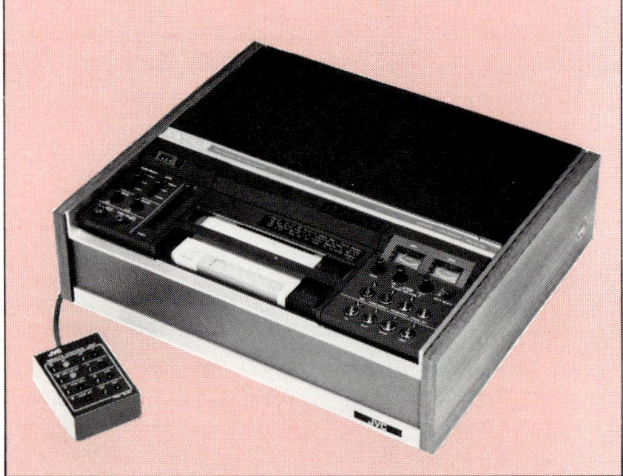

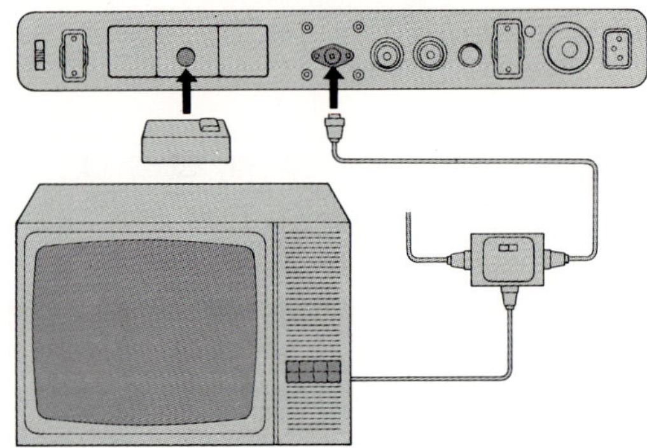

To play a tape on an ordinary TV receiver, the video signal from the VCR must be converted into an RF signal which the TV can receive.

An RF converter, plugged into the back of the CR-6060, must be designed for the same TV system (PAL, SECAM or NTSC) as the TV being used, and tuned to a channel used by the TV receiver.

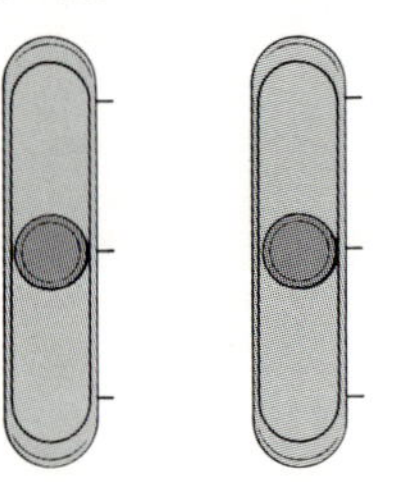

The simple selector, *far left,* switches two audio channels between separate and combined use.

The 'Auto' lever, *left,* will replay a tape continuously if set to 'Repeat'; on 'Search', it will stop when the counter reads '000'.

The 'ET' in the serial number of the CR-6060 stands for 'Europe/Triple-standard'. The same basic VCR is sold in the USA in its NTSC version. The PAL VCR can record and replay in PAL and SECAM automatically, but a switch, **1**, on the front panel, *bottom right*, must be operated for replay in NTSC; recording is not possible in NTSC.

The CR-6060 is a typical record and replay U-matic VCR. Some of the controls on the rear panel, *below*, will be unfamiliar to home video users.

The 'Playback mode' switch, **1**, has settings for normal playback ('Auto'), monochrome ('B/W'), or for dubbing a video recording on to another machine.

The TV receiver/monitor connector, **2**, is for transmitting video signals to a monitor for playback only, a feature not found on domestic formats, which are made to be permanently connected to a TV.

If recording is to be done from a TV, an RF decoder must be plugged into a cavity, **3**, in the back of the U-matic. The 'Video in'/'Video out' sockets, **4**, the professional VCR's normal connection to another machine, are becoming more common on home VCRs: the 'Audio in'/'Audio out' connection, **5**, is already common.

It is not as easy to alter the supply voltage as it is on most domestic VCRs; although the voltage selector, **6**, can be altered by the user, the manufacturers suggest consulting a dealer for advice.

The U-matic control panel, *right*, is a little more complex than that of a home VCR. There are two audio tracks, and one or the other can be brought into use with the audio select lever, **2**. U-matics have audio level controls, **8**, and meters for each channel, and the 'Audio Dub Ch. 1' switch, **7**, allows sound to be added on one audio channel of a recording.

A feature not often found on domestic formats, although likely in future, is continuous rewind and replay, operated on the CR-6060 by the 'Auto' control, **3**.

The 'Colour Lock' control, **4**, enables the operator to adjust aberrant colour effects on colour or monochrome pictures.

The 'Skew' control, **6**, allows the user to adjust picture distortions which would need an engineer on a home VCR.

The 'Timer Record' control, **5**, allows the CR-6060 to be used with an external timer, a facility which enables it to be used for time-shifting.

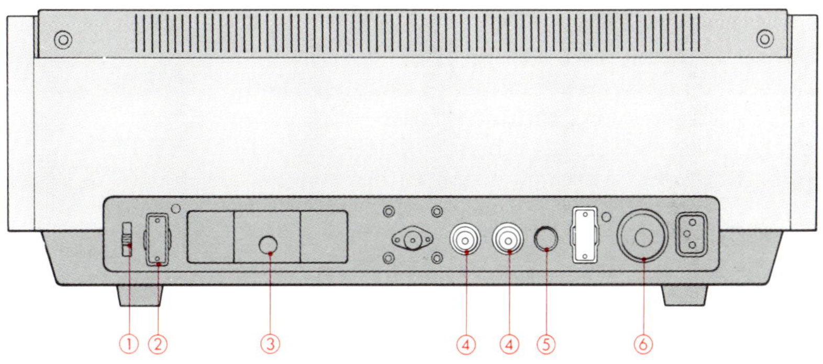

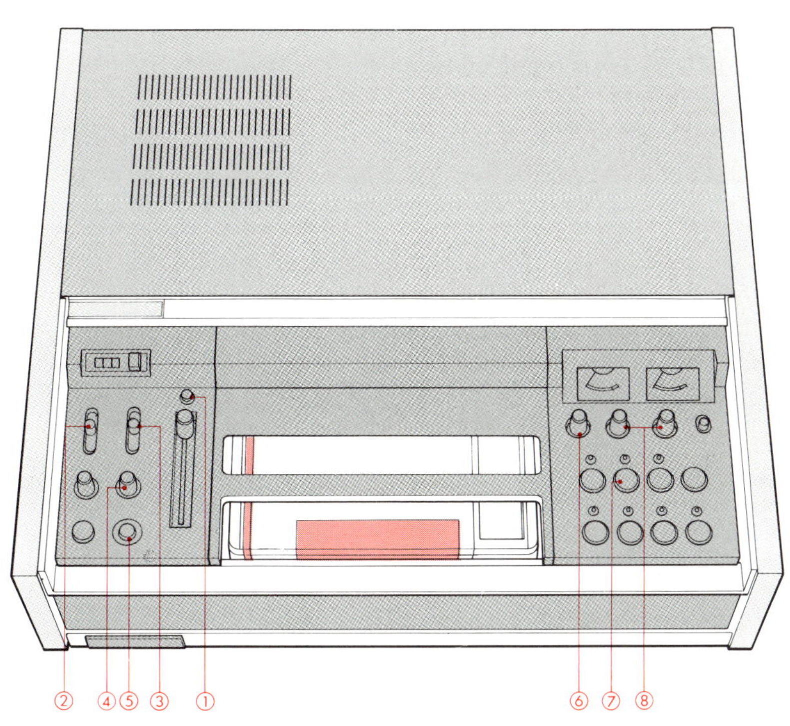

On the market/U-matic recorders

U-matic machines are designed for industrial and professional use; the choice of models is, therefore, arranged very differently from household VCRs. The model you buy will depend more on the application than the price or the facilities. Many of the features common on home VCRs do not exist on U-matics, as they are unimportant in the context in which these machines are used.

The basic U-matic, which is usually used for educational video in schools and colleges, is a playback-only VCR, with no recording mechanism to increase the price and endanger valuable pre-recorded tapes. It can be a triple-standard machine which, with a triple-standard monitor, will play back tapes in PAL, SECAM and NTSC, which is obviously useful for teaching. All U-matics can record and play back two audio channels.

The closest equivalent to domestic VCRs is the player/recorder, which will record from a camera or another machine but not usually from a television, unless the television has its own video/audio output, or a second RF modulator unit plugged in. U-matics normally have auto tape search and pause/freeze frame, but not on-screen picture search.

Dual-sound-track recording is a feature that will come to home VCRs but is now only available in Video 2000 and a few Betamax and VHS machines. You can have a commentary on one track and music on another; You can record in stereo; and the tracks can be played separately or in combination. Dual-sound-tracks are available on industrial VHS and (in the USA) Betamax VCRs.

The portapack is designed to use a small version of the U-matic cassette, with the same spool size and playing for twenty or thirty minutes, and will not take full-size cassettes. Unlike mains machines, portapacks tend to have on-screen picture search and give a very clean edit when shooting.

Many U-matic users want to link their VCRs to a full-function editing suite. This requires the most complex and expensive machines with the full logic needed to control them. These will fast-wind or give slow motion on screen, in order to select an edit point accurate to the nearest frame, and make perfectly noise-free edits. The editing VCR can be linked to a Time Base Corrector so that its signals can be perfectly synchronized with another VCR or camera.

Most U-matics on the market are designed as part of a series of machines from one manufacturer, and even where VCRs from one series are compatible with others, it is safest to stick to machines from one manufacturer for reliable results.

Edit controllers designed to match a particular series are usually compatible only with that system.

Panasonic NV-9200 (Europe)

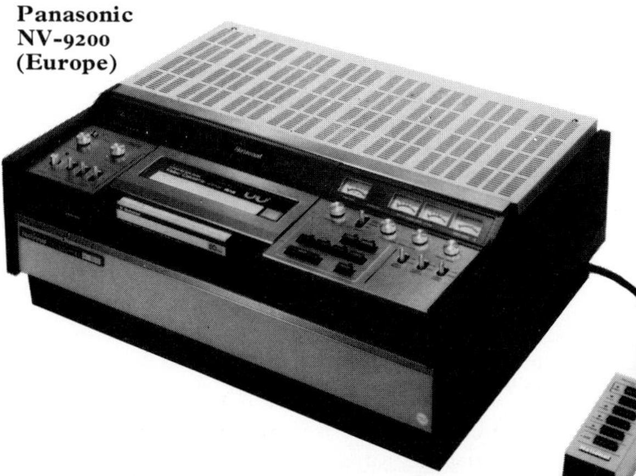

The Panasonic NV9200 is Europe's triple-standard version of a typical U-matic.

The rest of this series includes the basic NV-9210 (which can handle PAL and SECAM, but not NTSC); an editing VCR; an edit control unit; a portapack; and a special 'High-performance' VCR, the NV-9240, which has extra controls to make it a first-class input machine for an editing suite.

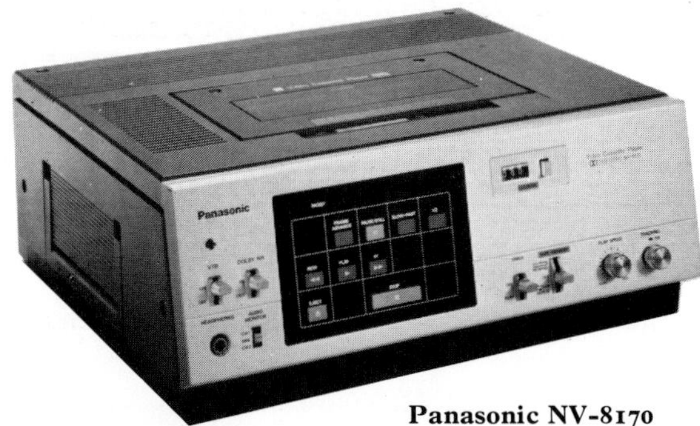

Panasonic NV-8170

The NV-8170, *above,* is a Panasonic VHS machine built to industrial standards. This is the player-only model; there is a record/replay VCR in the series (thê NV-8200), and while they are compatible with ordinary domestic VHS models, they can also be used with an edit controller.

In the USA there is already an industrial Betamax series, but this is not fully compatible with home Betamax VCRs: it uses a faster playing and recording speed.

JVC CP-5500

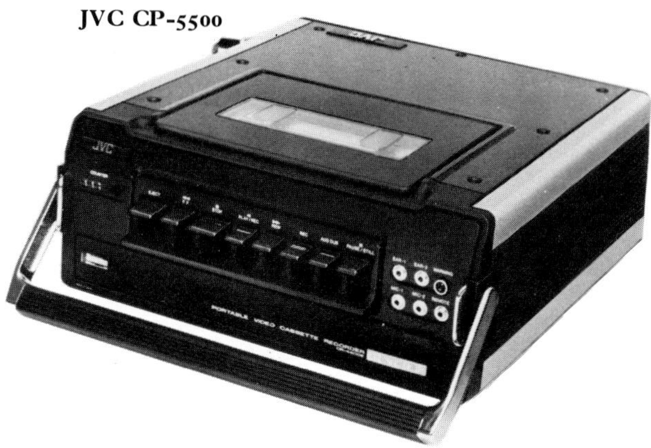

U-matic portapacks such as JVC's CR-4400, *left*, use smaller versions of mains cassettes, which have the same spool dimensions. Portable tapes can be used on home decks, but not vice-versa.

U-matic portapacks are especially rugged. This weighs over 24 lb (11 kg) with built-in rechargeable batteries. It can also run from a car battery.

Automatic assemble edit gives perfectly clean cuts between edits, and the two audio channels have balance controls and an audio peak level limiter.

JVC CR-4400

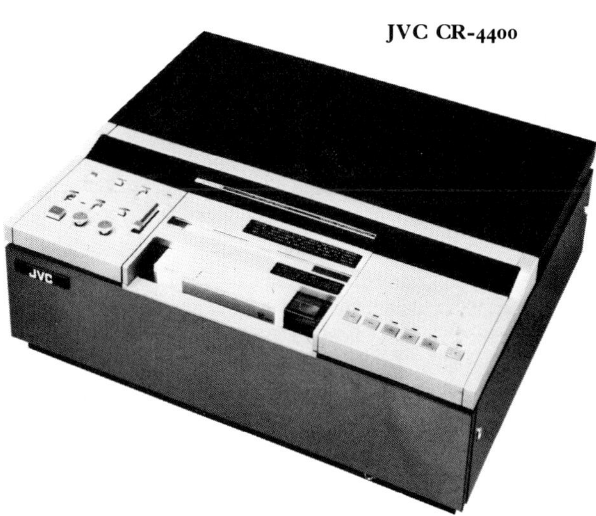

JVC's CP-5500, *right*, is a new-generation player-only VCR with some unusual features.

The CP-5500 and its recorder/player partner, the CP-6600, can be externally sychronized from a Time Base Corrector to give high-quality playback suitable even for broadcasting.

More unusually, these VCRs can take an FM (frequency-modulated) signal directly from the tape and record it on another machine with minimum loss of signal quality.

Usually, the signal has to be converted to video, and back to FM.

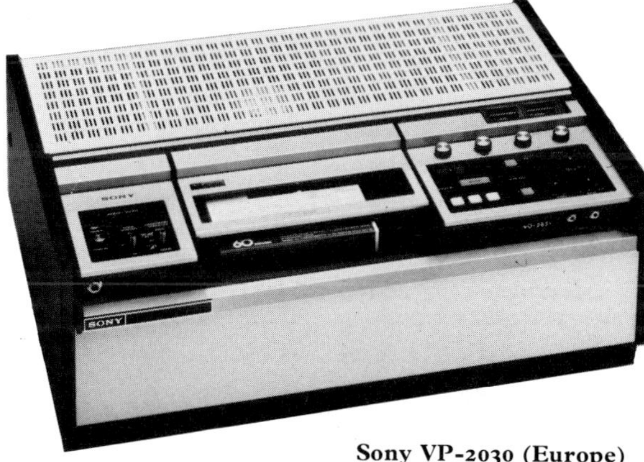

Sony VP-2030 (Europe)
VP-2000 (USA)

Sony's VP-2030, *left*, is a PAL machine; in the USA, the NTSC model is the VP-2000. It is a playback-only VCR. Its sister, the VO-2631 (VO-2611 in the USA), is a player/recorder, which can utilize Sony's Random Access Controller, the RX-353.

This device can record and memorize an audio tone on up to 63 segments of recorded tape. It can return to any of the segments on command, or play back any eight segments in a pre-set order. This facility, especially useful for educational video, is available for most U-matic systems.

The VO-5800, *right*, is a VCR belonging to Sony's new Type 5 series, released in the USA in 1982.

Type 5 is the first front-loading U-matic system, designed to fit a 19-in (48-cm) rack-mount. To make the controls more visible, the lower control panel tilts upward.

On-screen picture speeds can be varied continuously from 1/30th normal speed to five times normal by means of a rotary switch, and high-speed picture search is possible when it is linked to an edit controller.

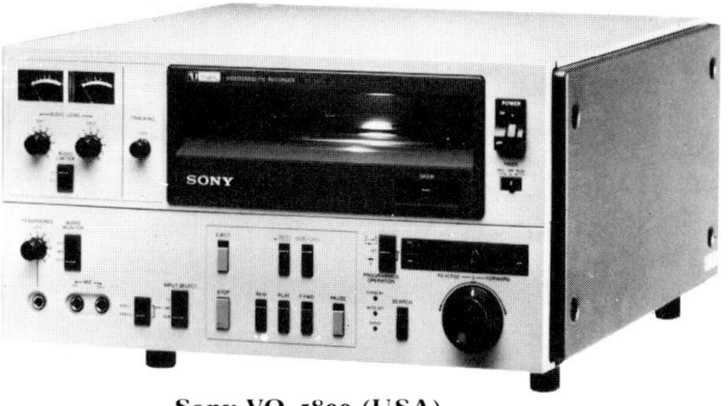

Sony VO-5800 (USA)

Choosing and using accessories

The accessories available to supply the needs of enthusiastic and demanding video users is growing daily. Yet the task of evaluating video accessories is almost impossible, for two reasons. First, the market is in a constant state of change. As improvements are made to VCRs, so some accessories are becoming redundant, while others are being superseded by the demand for more sophisticated, or more fashionable, alternatives. Second, one person's accessory may be another's necessity. Whether it be a device to improve the sound or picture of a recording, or cut out commercials, any video accessory's usefulness is always dictated by its owner.

Although no two video users will agree on the same order of priorities, some items are more or less vital to serious video recording. These include a tool kit for fault-finding and cable repair and, since most of the day-to-day problems involve cables, a set of spare cables. The connections and combinations will depend on the details of your system and the continent in which you live.

Adaptors, which convert one type of connector to another, can be a useful stopgap, especially if you frequently work with other peoples' units. An alternative is to buy a kit which allows you to build a set of interconnections between VCRs, television sets and peripheral hardware, such as games and computer units and video disk players. Some such systems, for example the VAK400 made by Total Video Supply for the American market, provide a set of interconnections to suit most simple needs.

More sophisticated kits, such as the RMS Electronics range, supply signal amplifiers, switchers, splitters and other components. As the ultimate in complexity you could switch all your systems through a box such as VideoMate or Video Commander, although, to date, both of these are still only available in the USA. Whatever your choice, the only rule to remember is to buy with your own system and needs in mind.

From splicing kits to extensive storage systems, dust covers to detailers, cleaning cassettes to crimping tools, the range is vast. If, even so, the precise accessory you want is not available, do not worry. The chances are that it will be on the market soon, or is, in fact, already on sale at a specialist shop – if not in the exact form you want then as an easily modified version of it.

Little black boxes
Many of the accessories most useful to the VCR owner have an unprepossessing box shape. Here is a list of some of the most useful:

1. Commercial cutter: records TV broadcasts but cuts out the advertisements.
2. Stereo simulator: splits the audio signal to give a 'mock' stereo effect.
3. Cable by-pass: useful for cable subscribers. It can record a broadcast on one channel while you are watching a cable-fed broadcast on another.
4. Image enhancer: improves the smoothness of scene changes in tapes edited at home or made jerky by pressing the 'Pause' button during recording to cut out the commercials. It can fade out the signal to black or fade in extra colour as required.
5. Detail enhancer: increases the sharpness of the video signal, so minimizing loss of quality during recording.

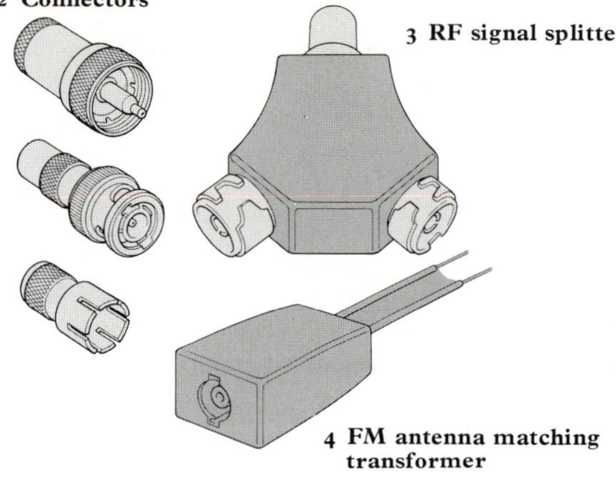

A useful selection of video accessories could well include the items illustrated here. Releasable cable ties, **1**, not only help keep your cables neat and undamaged, but are also re-usable. A set of emergency connectors, **2**, will probably be invaluable, especially for the camera owner, to replace snapped-off connections. An RF signal splitter, **3**, can split the signal you receive to serve two TVs or VCRs. An FM antenna-matching transformer, **4**, is the

To splice tape ends, lift the clamps, **1**, then apply recording tape, glossy side up, overlapping over the diagonal slit, **2**, and clamp into position. Holding the tape with a finger, cut across the diagonal slit, **3**, and remove surplus tape. Apply splicing tape, **4**, and carefully cut off any excess.

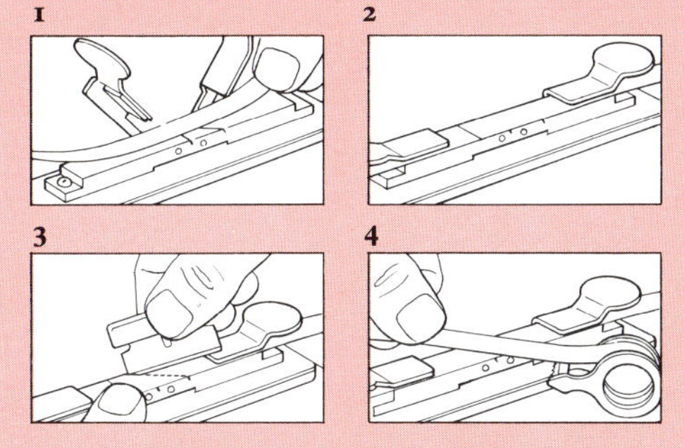

A crimp tool is often a useful alternative to a soldering iron. First trim the cable, **1**, then insert the central conductor, **2**, which acts to increase the energy supply across the connection. Twist the correct connection on the prepared cable, **3**, and, finally, use the crimp tool to seal the joint, **4**.

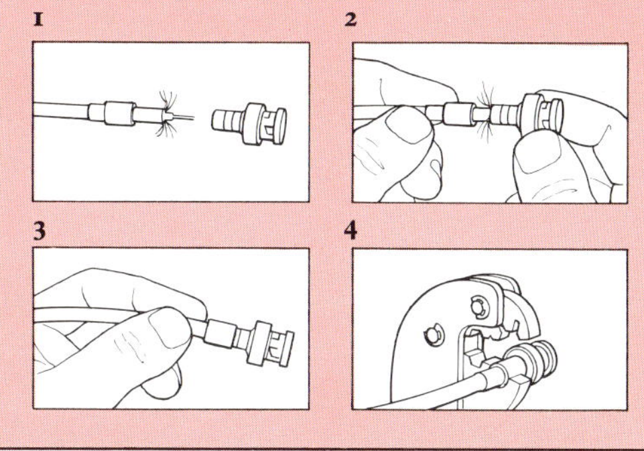

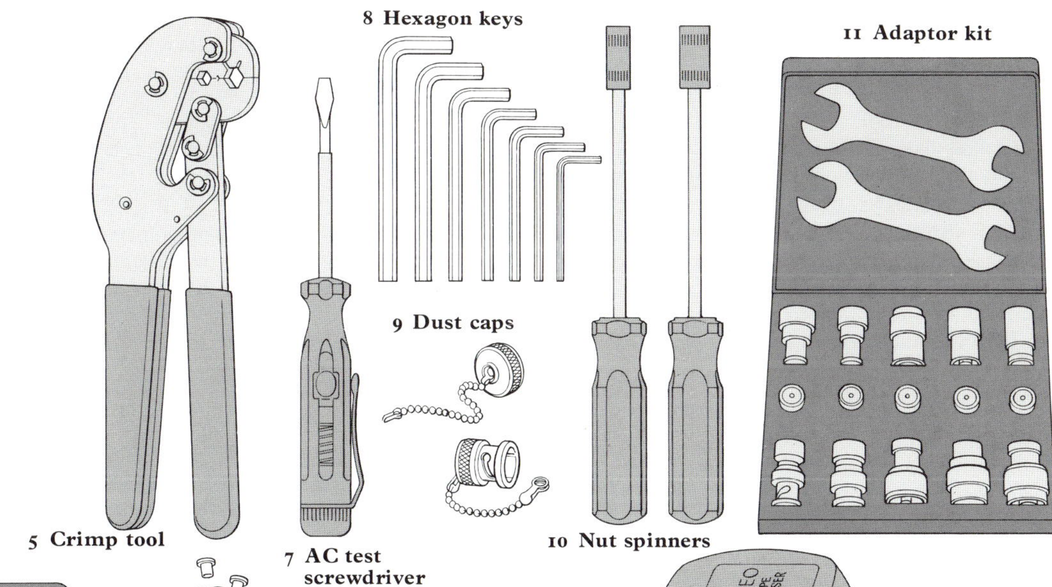

device you need to match the signal from your VHF antenna to an FM tuner. A crimp tool, **5**, with metal crimps, **6**, is the easy way to effect BNC and UHF repairs and reconnections with no fuss or mess. An AC test neon screwdriver, **7**, has a handle which lights up if an AC of 220 to 250V (110 to 125V, US) passes through it, so is handy for testing the integrity of circuits. A set of hexagon keys, **8**, available in metric and imperial sizes, will help you loosen and tighten hexagon sockets effectively. Use dust caps, **9**, over unused sockets and mountings. Nut spinners, **10**, give an improved grip for tackling a variety of electrical connections. An adaptor kit, **11**, comprising a selection of male and female connectors, and spanners, will improve the versatility of your equipment. To make sure that an old video recording does not 'haunt' a new one, use a hand-held bulk eraser, **12**.

Video disks

Video disks look rather like long-playing records. Their surfaces are pitted with microscopic indentations and both visual and audio information is impressed into them. The player is plugged into a television set on whose screen the pictures appear.

The first video disk player, conceived in the twenties by John Logie Baird, had a rotating drum, carrying a precisely arranged set of perforations. This scanned a picture and broke it down into a radio signal. The receiver had a similar drum, synchronized to the camera drum, which reassembled the picture into recognizable images. Baird used the existing technology of the gramophone record industry to produce the 78 r.p.m. video disk. The signals created when the disk was played on a record player were fed into a Baird receiver to give just a dozen stills.

When Baird's television system was killed by the arrival of the EMI electronic system in the late thirties, his video disks died with it. Only in the seventies was the notion revived by the European companies Telefunken and Decca. The resulting TELEDEC machine incorporated an adaptation of an even earlier recording idea – the undulating grooves of Edison's original audio cylinders. The reason for this move was to accommodate more information into narrower grooves on the finished disk. Even so, it had to rotate at 1500 r.p.m. (a speed used by many of today's systems) for acceptable picture quality, and the disk gave only ten minutes' playing time.

As a refinement of the TELEDEC system, Matsushita Electric of Japan produced an ultra-fine pressing that crammed an hour of picture material on a disk scanned by a pressure-sensitive head. This head sensed the undulations and converted them into the video signals fed into the television.

Only with the arrival of the precise, sophisticated technology of the eighties has the Matsushita system evolved into a viable product. First on the scene were RCA with their SelectaVision system. This works by means of a head, which senses billions of undulations in the disk surface and converts the information into a video signal, which varies with differences in the capacitance as the head moves up and down. The JVC Video High Density (VHD) system works in a similar way, but Philips have opted for a novel optical system, LaserVision, in which the disk is 'read' by a laser beam.

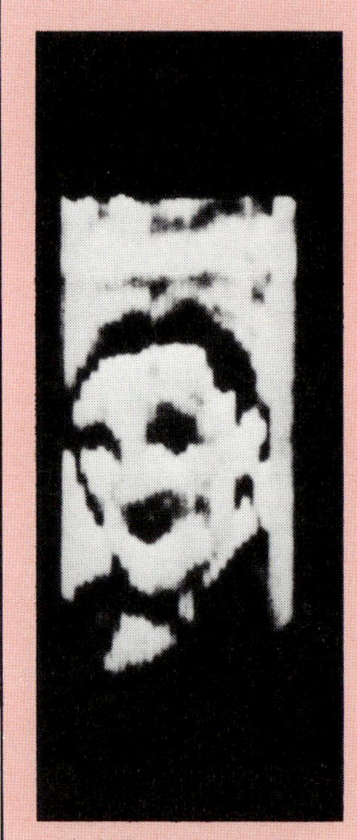

The first video disks appeared as long ago as 1928. Constructed from brittle plastic shellac, like the gramophone records of the day, they were played on a gramophone linked to a Baird TV receiver. The disks gave 12 still images of rather poor quality and were sold by Selfridges, the London department store, in the thirties.

In the SelectaVision system developed by RCA, information is etched on a disk as vertical ups and downs in a spiral, V-shaped groove, *below right*. When the disk is played back, a diamond stylus rides in the groove. As the many undulations pass under the stylus, a metal electrode creates the signal readout which is fed into the TV. Unlike an audio system, the stylus does not move in an arc but travels in a straight line across the disk radius, *below left*. This gives a clear, steady picture.

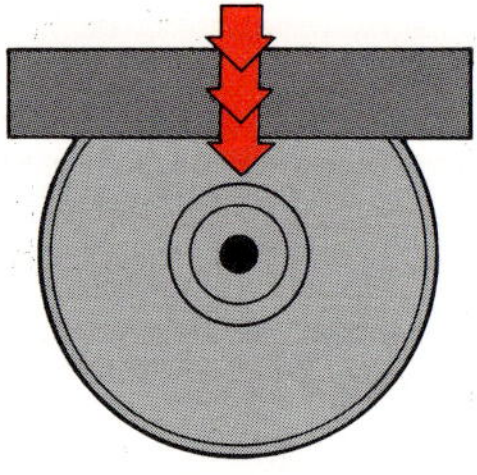

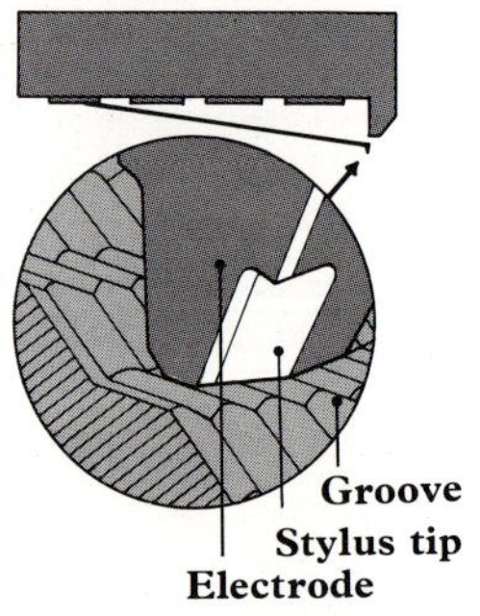

In JVC's VHD system, *right*, information is stored in the disk in 'micropits' of two kinds. Some pits contain audio and video information, the rest tracking data. The scan of the sapphire stylus over the disk (unlike the RCA system it is separated from the disk by a plastic sleeve) is directed by the tracking pits.

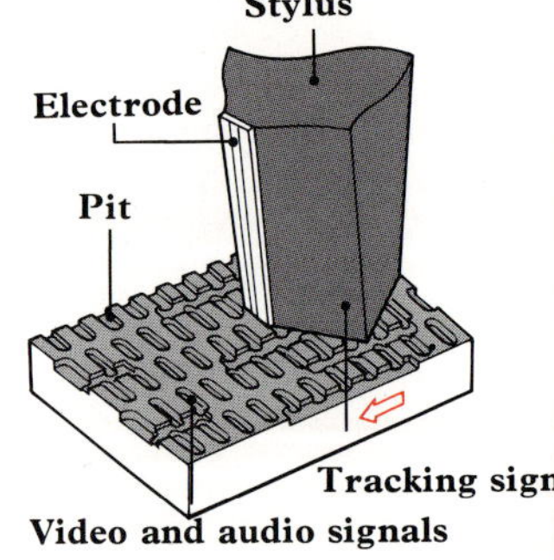

To re-create sound and moving pictures, all video disks have to store a huge amount of information. More than 10,000 pictures are needed for only an hour's visual entertainment. The grooves on the disk have, thus, to be fine, packed tightly together, and etched with great accuracy. All the modern disks are made from plastic. For the RCA and JVC systems, this has to be highly conductive. For the Philips system, it is coated with a reflective material.

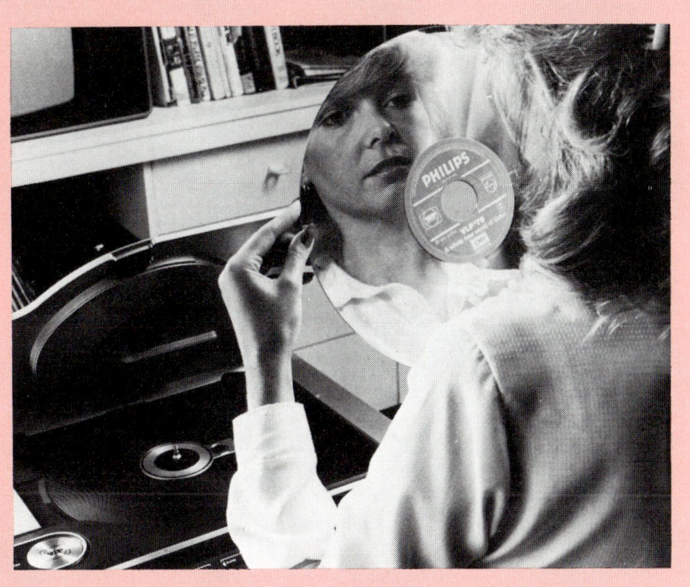

LaserVision, *below*, the brainchild of Philips, uses a completely different technology from the RCA and JVC systems. The great advance is that information, etched on the reflective material of the disk in a series of pits, is read off with a fine beam created by a helium neon laser. By means of a series of lenses, gratings, prisms and mirrors, the laser beam is directed to the disk underside where it is moved by a scanning lens. When it hits a pit in the silvery surface, the beam is reflected. The reflected beam then passes back to the photo diode which creates the signal output.

In a similar system, which is currently being worked on in France by Thompson CSF, the laser beam is not reflected from the surface of the disk, but actually passes through it. Variations in the quality of the beam, created by the information etched on the disk, are then used to construct the electrical signals needed to create a TV picture.

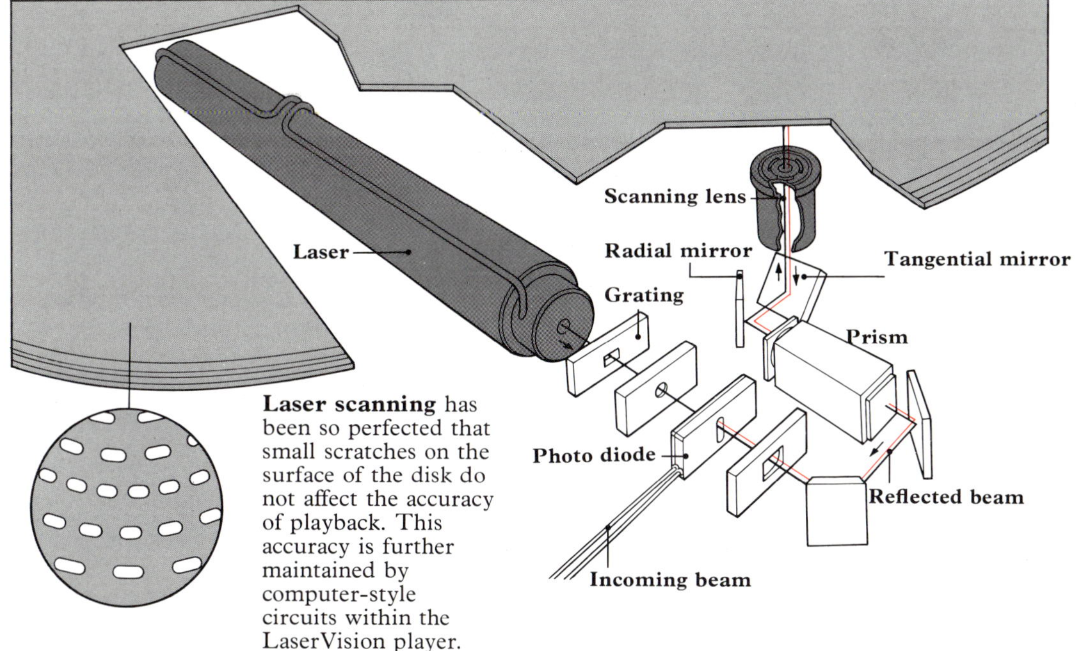

Laser scanning has been so perfected that small scratches on the surface of the disk do not affect the accuracy of playback. This accuracy is further maintained by computer-style circuits within the LaserVision player.

On the market/Video disk players

The video disk could turn out to be the vanguard of a new generation of video, or a complete commercial disaster. For the consumer, deciding whether to opt for a video disk system to plug into your television depends most on what you want video to do for you. Because of the technological requirements of their manufacture, the systems presently available are all players only: they cannot record television broadcasts, neither on nor off air, nor let you record your own material with a video camera.

On the credit side, supporters of the disk system claim that the picture quality is better than that of any tape, and that, in time, prices will be lower. At present it is difficult to evaluate such claims, but there is certainly a possibility that disks protected from physical wear could have a longer life, and be potentially more durable library material, than video tapes.

By the early eighties, the video disk market developed into a three-horse race, with RCA SelectaVision, priced at about $300, having a head start as the only one on sale. Like other systems of its kind, which are being manufactured in prototype by Sanyo, Hitachi, Toshiba, Zenith and Radio Shack, it is susceptible to interference from dust; and even minute scratches on the plastic disk can impair its playback performance.

The LaserVision system by Philips incorporates a machine which looks rather like a conventional record player, but is far from conventional in its laser-operated function. One of the great advantages of this system, available at first only in the USA for around $700, is that the disks are coated with a protective layer of plastic one sixteenth of an inch (1 millimetre) thick. Also, the absence of mechanical friction means that there is no wear on the disk. The Pioneer DiscoVision System, also launched in USA, is laser-operated and programmable.

The JVC Video High Density (VHD) system works, like the RCA system, by contact between stylus and the disk, but as in LaserVision the signal is stored in pits, and the stylus can move freely across the disk's surface in random access. Until all three systems are established and consumers have had sufficient time to use and compare them, there is no knowing which system, if any, will be the most successful.

The oldest form of video disk on the market is the one that has been used since the late sixties for providing action replays on TV sports programmes. The disk is magnetic, and works on the same principle as magnetic video tape. Signals are recorded and stored on the disk in concentric magnetic tracks, each track accommodating 30 seconds of action. The disk can be replayed at any speed, both forward and in reverse, and can freeze frame at the touch of a button.

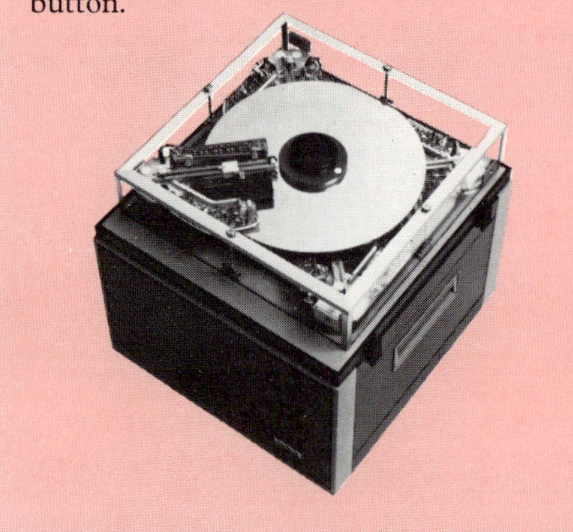

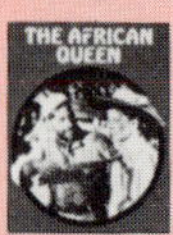

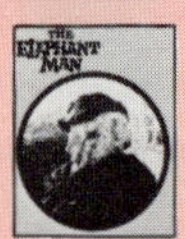

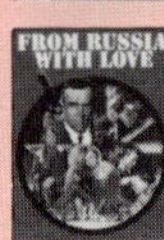

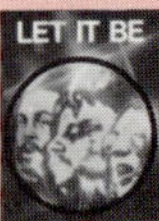

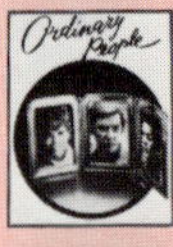

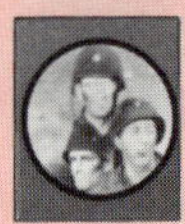

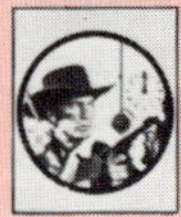

The range of video disks varies in detail from system to system, but offers an increasingly wide choice, with popular movies at one end of the scale and solely educational disks at the other. The disks, which cost just over a third of the price of pre-recorded video cassettes, play for about an hour on each side, but many of the broadcasts come packaged in multi-disk sets. Of the ranges announced in 1982, that of RCA SelectaVision offered the widest choice of popular entertainment and an increasing amount of educational material, but expansion of the list seems to be limited by the lack of a freeze-frame facility on the disk player. Philips LaserVision, with more sophisticated players, has marketed two sorts of disks – long play disks for straight play through, and active play disks with shorter playing time but with slow motion and freeze-frame potential built in. The JVC VHD system promises disks with the same possibilities as those of the LaserVision system, and a good range has been forecast.

To use the RCA video disk player, you slide the disk into the loading slot, **1**, remove the sleeve, and press the function lever, **7**. Lighted dials, **2** and **3** indicate which side is being played and how long it has been running. 'Rapid access' buttons, **4**, allow you to speed forward or back to a particular time segment, but during this there is no sound or picture. Forward and backward 'Visual search' buttons, **5**, play 16 times normal speed with perfect vision, but muted sound. If it is necessary to stop the disk during play, you press the 'Pause' button, **6**.

The JVC video disk player, incorporating a microprocessor, should give great flexibility. Disks are loaded into a slot, **1**, by pressing the 'Load' button, **11**, and sound selected, **2**, as stereo or bilingual. Lighted displays indicate the chapter, **3**, time, **4**, and the side playing. An array of controls allows such operations as 'Manual search', **5**, and 'Chapter search', **6**, programmed by the machine's chip. For chapter repeats you press the 'Memory' button, **7**. Frame playback in both directions is available, **8** and **9**, and a 'Pause' button, **10**, is a standard fitting on the player.

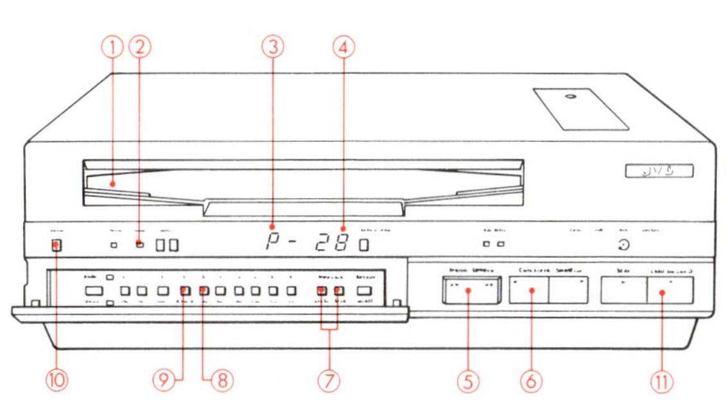

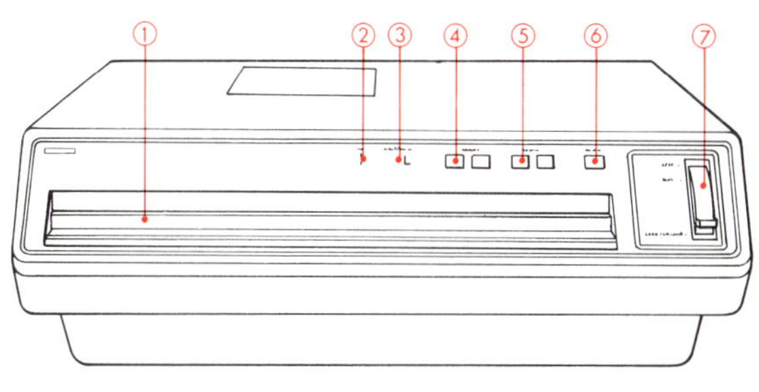

To play a Philips LaserVision video disk, you turn on the power, **1**, lift the lid, **4**, and insert the disk. Closing the lid automatically begins play. The in-play facilities that are offered by this system include still or freeze frame, **2**, which can be moved forward or backward a single frame at a time. Slow motion, **3**, is possible in both directions. For locating particular frames you press the 'Fast forward' button, **5**. Pressing the 'Fast search' button, **7**, plays the whole of one side of the disk, in about 25 seconds with its pictures displayed on the screen. The 'Audio' button, **6**, gives stereo or bilingual sound (in a variety of combinations) on selected disks.

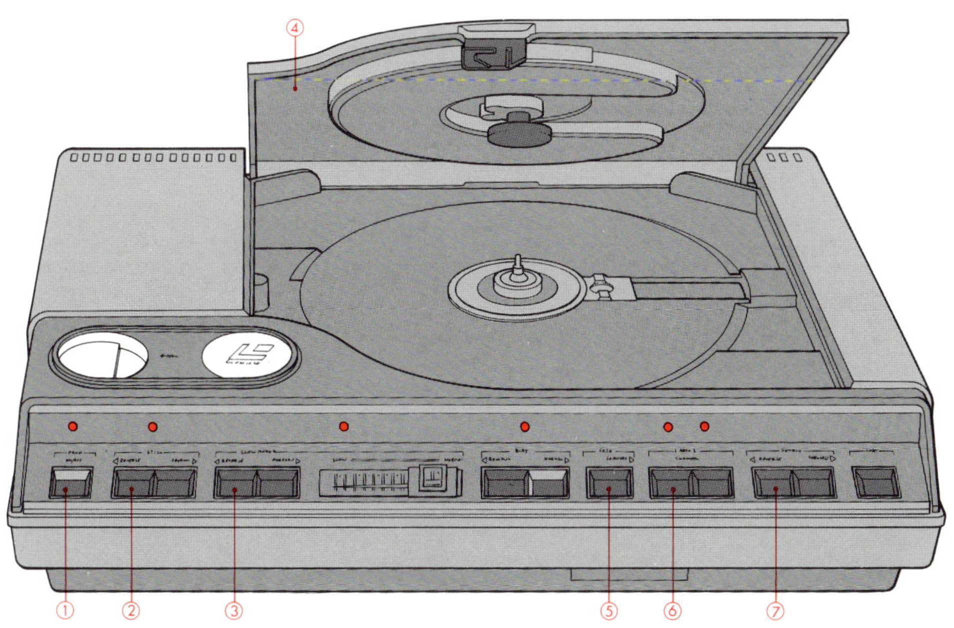

Interactive video systems

When video can be made to respond to the wishes and instructions of the operator, it is said to be interactive. Both video tapes and disks have a role to play in this new application of video, which promises to introduce new methods of learning.

The aim of interactive learning through video is to let the machine take over from the teacher. If it incorporates a programmed tape, a machine on which to play it, a monitor and a device, linked to a print-out, into which the student can punch responses, an interactive video system can check a student's progress and understanding step by step, and discover which areas of study demand the teacher's personal attention.

The next step in sophistication is to link the tape player of the system to a computer such as the Apple II. With the potential of a computer added to the inventory, the level of education can be made more flexible and more technical.

Whenever interactive programs are used with VCR systems, they always meet the same drawback: the access time. This is the time it takes to spin the tape from the end of one section to the start of the next. The solution is to use video disks instead of tapes. Because of the way in which they are constructed and played, video disks allow virtually immediate access to any other part of the disk. All the machine has to do is to move its playback head a few centimetres and the next section of the recording appears on the screen. The player incorporates a microprocessor, so a hand-held infra-red remote control pad can be used to interact with the disk. On advanced machines the sound can be set for stereo, bilingual or a choice of commentaries at two levels of difficulty.

If you already have a television monitor, you could acquire an interactive video disk system for around £1,200 ($600). The limitation of the system, compared with the tape systems so far available, is the lack of facility for teachers and parents to make their own recordings. Instead, they have to rely on material made by the disk companies. Despite this, interactive video disks are already proving useful commercially in the marketing of all types of commodities, from holidays to baby clothes. In the USA, parents can buy a children's disk offering a wide variety of education and entertaiment.

Linked to a computer, interactive video disks are proving to have some fascinating applications. The American Heart Foundation, for example, has a disk on mouth-to-mouth resuscitation, in which the computer is linked to a dummy. As the first-aider goes through the course, the program flashes back comments on progress such as 'first breath too hard, third too weak'.

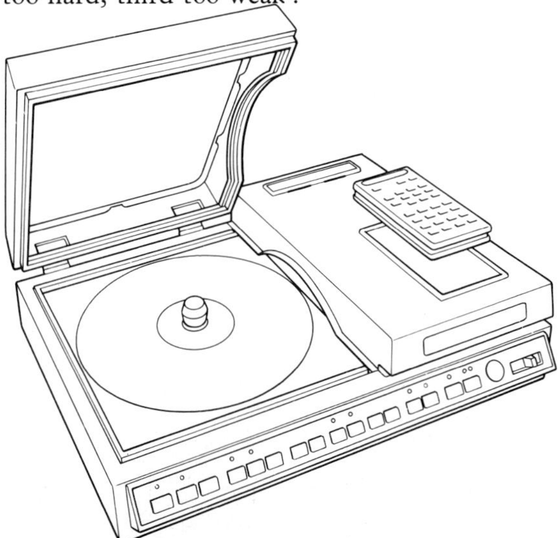

Attached to a TV monitor, the Pioneer DiscoVision player becomes an interactive video system. Sound and picture are both recorded on the video disk, which is played at 1,800 r.p.m. and read by a medium neon laser beam.

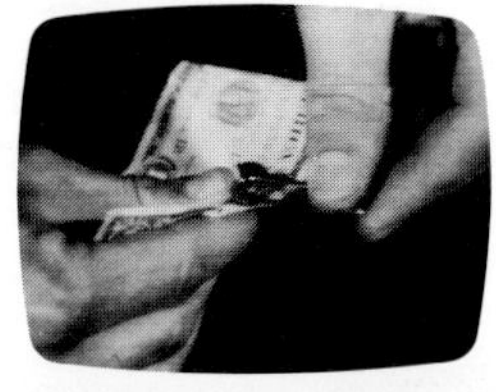

One side of a video disk is composed of 54,000 still frames, any one of which, at the touch of a button can be frozen on the TV, for a long period.

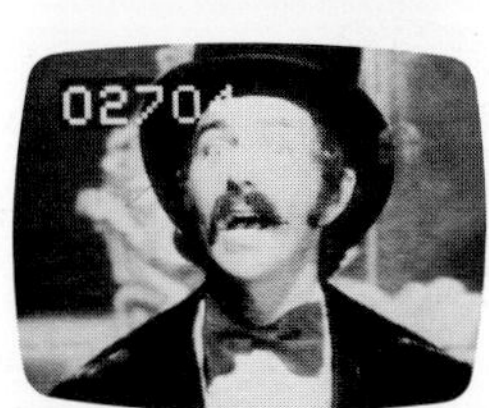

For easy reference, each frame on the disk is numbered, but these numbers can be made to disappear during play to prevent them from interfering with the image.

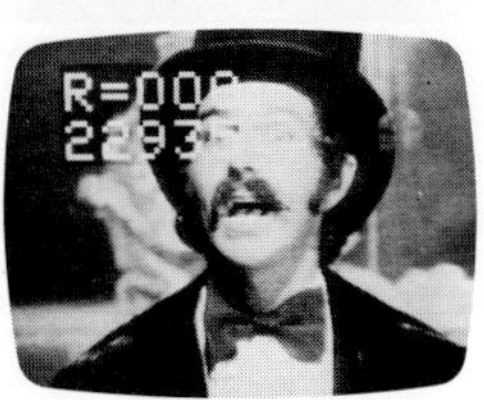

To switch to another part of the disk, you simply code-in the appropriate reference number on the control pad, then press the 'Search' button.

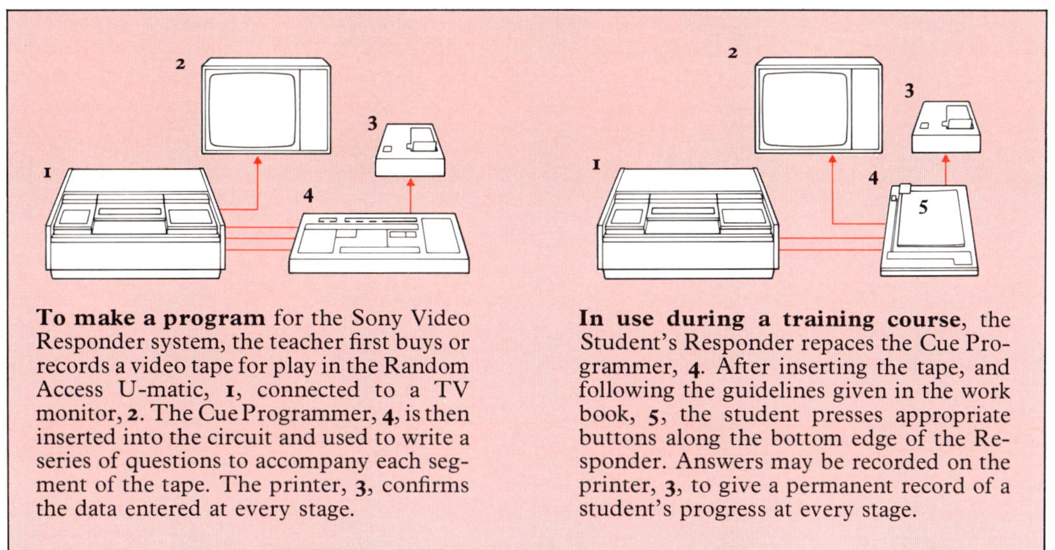

To make a program for the Sony Video Responder system, the teacher first buys or records a video tape for play in the Random Access U-matic, 1, connected to a TV monitor, 2. The Cue Programmer, 4, is then inserted into the circuit and used to write a series of questions to accompany each segment of the tape. The printer, 3, confirms the data entered at every stage.

In use during a training course, the Student's Responder repaces the Cue Programmer, 4. After inserting the tape, and following the guidelines given in the work book, 5, the student presses appropriate buttons along the bottom edge of the Responder. Answers may be recorded on the printer, 3, to give a permanent record of a student's progress at every stage.

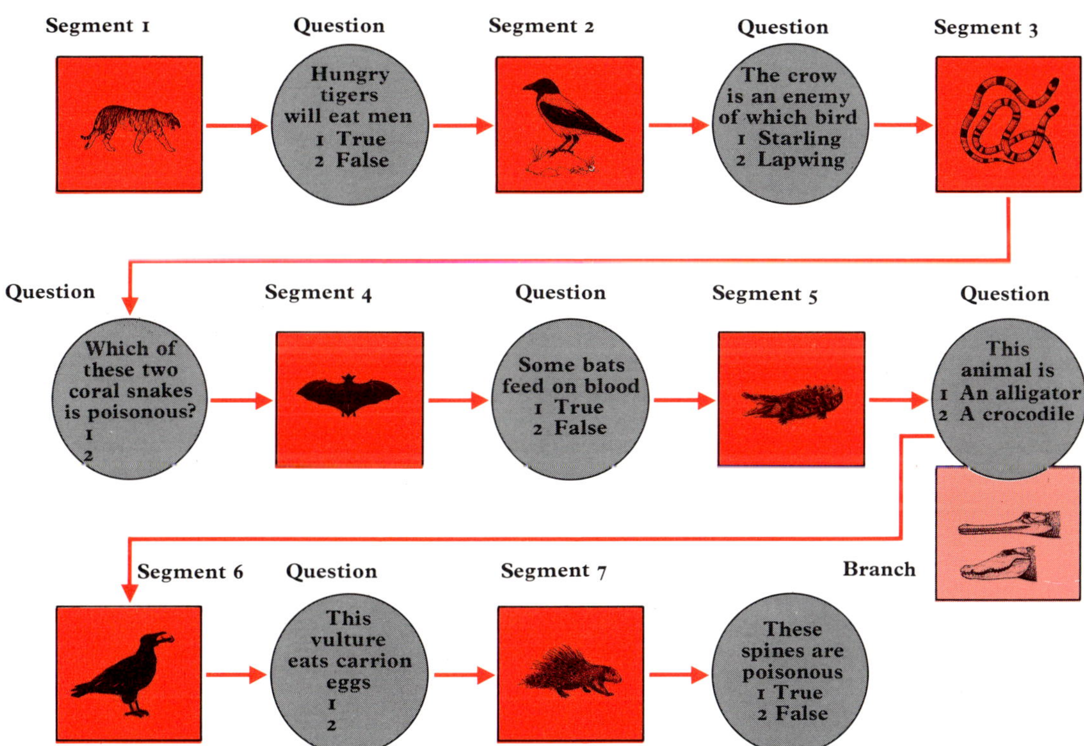

The Sony Video Responder is designed to allow students to progress at their own pace. The tape that is inserted into the U-matic part of the system is programmed in a series of up to seven segments. At the end of each film segment, the student is given a series of questions to answer relating to what has been shown.

In writing the questions to accompany the video tape, the teacher may specify up to three attempts for each question, and make provision for branching into more detailed treatment of any topic. A typical series of questions, and the images on the preceding segment of tape to which they relate, are shown in the program traced out above through seven segments.

The future of video

Video is progressing at such a fast and furious pace that all we can be sure of is that most of today's most outrageous ideas will have become a reality long before the magical date of 2001. Leading the way toward this bright future will be improvements in the versatility, quality and scope of existing systems. On to these will be welded novel technologies which will make video a near-perfect communications system.

As video becomes cheaper and more accessible to all, more and more VCR owners will suddenly become video movie directors. This will be made possible by means of narrower, 8 mm tape and the CCD, or charge coupled device, already being incorporated into the new generation of video cameras. The CCD is a microchip which effectively replaces the camera tube, reducing the camera's weight. Although prototypes do not yet produce recordings of professional quality, this new camera responds to much lower light levels than a conventional video camera and is immune from burning by over-bright light sources.

Today, even the best of video is limited in quality by the coarse screens on which it is displayed. Researchers in Japan are already working toward the 2,000-line television picture which will, for the first time, compete with 35 mm film on grounds of quality. When pictures of such clarity and detail can be scanned by a CCD device and reassembled on large-screen systems, the way should be clear for another step forward, namely increased speed and ease of copying of video recordings. At present, copying involves inevitable loss of sharpness and quality because the copying machine is trying to draw an accurate picture of a fluctuating signal. If, however, the signal is recorded as a set of precise digits, each specifying a particular element of brightness or colour, the recording becomes a rigid set of instructions which incur no loss of quality.

By the end of the century, video will be everywhere: in home, office, school and factory, in supermarket and sports stadium. Video, linked to computers, will spearhead the new industrial revolution. To aid and to entertain, video will change from being a single entity into a whole system, offering the consumer a diverse range of facilities as offshoots from the basic camera, tape and recorder.

The video cassette has a novel future as an integral part of the truly portable video movie camera now being developed by Sony. Replacing the conventional camera tube is the CCD, or charge coupled device, a microchip which splits up the picture into thousands of fragments. These are scanned by logic circuits to create a video signal, which is transferred to 8 mm tape ready for instant playback.

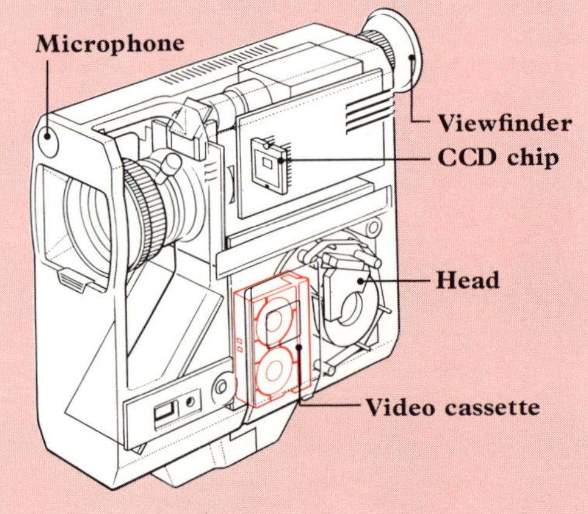

In the Longitudinal Video Recorder (LVR), developed by Toshiba, a high-speed graphite-lubricated tape is divided up into a series of no fewer than 316 separate parallel tracks and so can store a vast amount of information. When played in a specially adapted VCR, the tape shuttles backward and forward so that about 25 seconds of material is played from each track in succession.

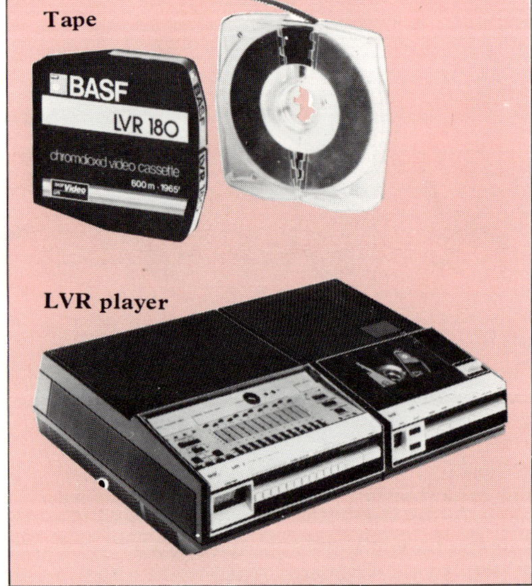

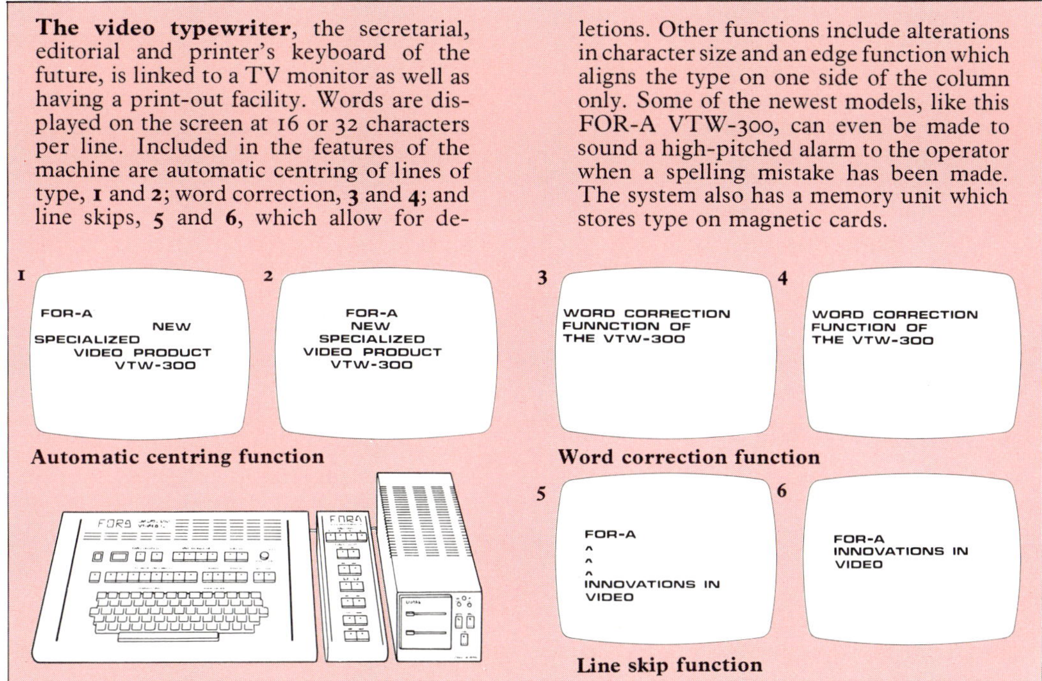

The video typewriter, the secretarial, editorial and printer's keyboard of the future, is linked to a TV monitor as well as having a print-out facility. Words are displayed on the screen at 16 or 32 characters per line. Included in the features of the machine are automatic centring of lines of type, **1** and **2**; word correction, **3** and **4**; and line skips, **5** and **6**, which allow for deletions. Other functions include alterations in character size and an edge function which aligns the type on one side of the column only. Some of the newest models, like this FOR-A VTW-300, can even be made to sound a high-pitched alarm to the operator when a spelling mistake has been made. The system also has a memory unit which stores type on magnetic cards.

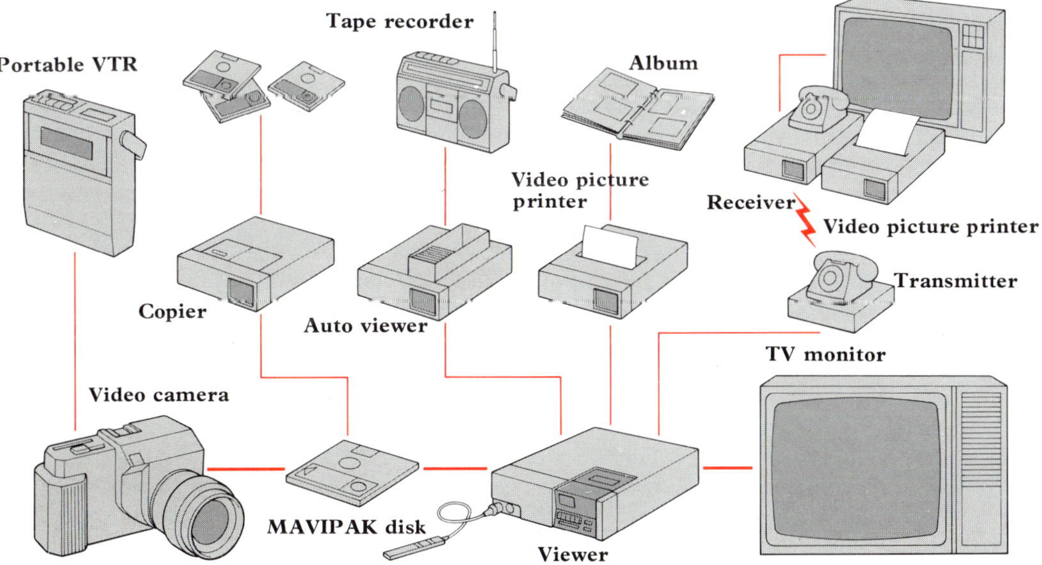

The Sony MAVICA system plans to offer consumers complete video. The basic elements are a small portable video camera containing a CCD to record signals on a small magnetic disk, the MAVIPAK. With the help of a special playback system, the pictures can be seen at once on a TV monitor.

To these basics can be added a whole set of extras. These include the portable VTR; a copier for duplicating the disks; an autoviewer and tape recorder for separating off the video and audio components of the disk and a video picture printer, which can convert stills into prints.

By means of a transmission facility, the video owner of the future will be able to send signals along the telephone cables to transmit all or part of a video recording anywhere in the world, to be seen by the recipient as a print-out or as a series of images on a TV monitor.

The TV connection

Video grows ever more versatile, but however it is used, there is always a common denominator, namely the television screen on which the visual information is displayed. Without television, video would not have been possible; so a basic understanding of television is vital to any study of video.

Television is arguably the most powerful communications system ever developed, and because it produces its pictures electronically, not chemically like film, it has an immediacy that film could never match. When television first began, broadcasts were always live, with events seen on the television screen a fraction of a microsecond after they had occurred in the studio.

With the advent of colour, and of video recordings, television has changed enormously since those early days, but the basic principles of the transmission of signals and their display on a fluorescent screen remain the same. Advances in television technology, such as sophisticated satellite relay systems, have given live broadcasting a new fillip. Through the medium of television, live events, from battles to baseball, and pageants to pop concerts, can be seen simultaneously in hundreds of milllions of homes all over the world. Never before has it been possible for people all over the world to receive the same message at the same moment. Never before has it been possible for Earth's inhabitants to witness events far out in space, on the moon, or even beyond the planets.

The biggest problem with transmitting television signals is the enormous amount of information that has to be broadcast. A typical television picture transmission is completed in just 64 millionths of a second. In this time, the transmission has to convey signals indicating the thousands of separate differences in light intensity which go to make up every single line on the screen. If the signals were carried in radio space they would occupy more than the entire band taken up by medium-wave radio transmissions. This is why television transmissions have been pushed into the ultra-high frequency (UHF) ranges, and in the USA the very high frequency (VHF) ranges. Here the frequencies needed to transmit from a whole series of television stations can be accommodated without interfering with one another, or with the signals of any other transmission.

The greatest disadvantage of the UHF/VHF transmission systems is that they tend to limit good television reception to areas relatively near the transmitter. Also, high buildings, hills or even low-flying aircraft can interfere with the signals. These are just two of the reasons why it is essential to have the correct antenna for receiving UHF signals, and why it must be properly positioned to obtain the best reception.

The television signals picked up by your antenna are translated into pictures by your television set or receiver. The image that you see is built up from a series of minute dots, which the human eye reassembles into a detailed picture. This image has enormous potential for the viewer. It can be used, for example, to display constantly updated information. This may be transmitted piggy back with regular television broadcasts as the teletext system. Or the screen can be used to provide you with viewdata information fed, at your push-button command, from a computer into the television via the telephone cables. With the addition of a keyboard and mini-computer, your television can be used as an educational aid, a games machine and personal accounts clerk all in one.

The greater the possibilities of television, the more difficult it becomes for the buyer to choose exactly the right television for his needs. So what sort of a set should the VCR owner choose? The answer depends largely on what you want your television to do.

For the VCR owner, or prospective owner, the most important point to look out for is whether the set is a television monitor/receiver. Most ordinary television sets are receivers only, designed to take UHF signals from an antenna. Video signals, however, are low frequency, and when connected to a UHF set, the VCR has to convert the UHF signals to low frequency to record, then convert them back to UHF again to produce an image on the screen.

This two-way switching inevitably results in some loss of quality. The monitor part of the combination takes care of the problem by taking a direct low frequency signal. The monitor cannot act as a receiver: it merely displays what is fed into it; but it is the ideal connection to the VCR because it operates at low frequency.

A television monitor/receiver has a separate channel on the tuning buttons which

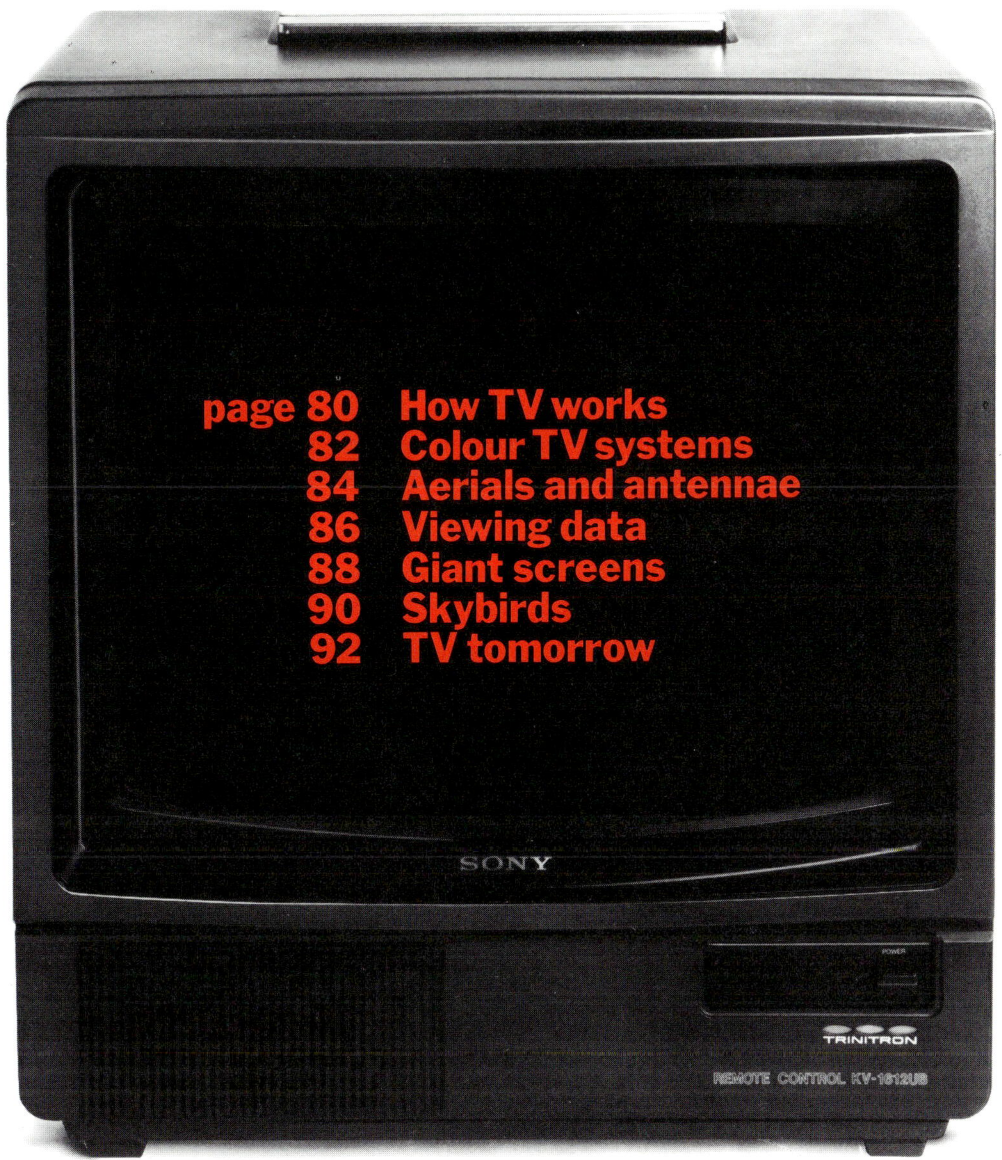

selects direct video input. At the back or side of the set there will be separate cabling for attaching to a discrete connection on the VCR. It is possible to convert an existing set into a monitor/receiver.

The diversity in size, shape and complexity of today's televisions reflects both the video and computer revolutions. The versatility of television is being extended way beyond the dreams of its pioneers.

The next novelty will be to have a colour television small enough to put in your pocket, flat enough to hang on the wall like a picture or with the capacity to create a high-definition picture large enough to fill your living room wall. The day of component television, with knowledgeable enthusiasts mixing and matching different parts of television and video systems to produce exactly the results they want will soon dawn.

Along with diversification and development of television will come a vast increase in the amount and variety of television material. Television of the future will cater for minorities as well as the majority. With the development of cable and satellite television, the choice of viewing – and video recording – will become international.

How TV works

Television has become such a commonplace part of our lives that we barely pause to wonder how it works. The form in which this technological revolution first spread across the world was as black and white television, and the basic principle on which all television, even today's complex colour systems, works is the same. Essentially, the picture seen by the camera is broken down into a series of dots of varying brightness which are then scanned in order, and the information passed as a signal to the television receiver. This reassembles the picture on a screen.

Like cine film, television does not consist of moving pictures. Instead it produces a series of still pictures at the rate of 25 to 30 every second (compared with 24 per second for cine film) which deceives the eye into thinking it is seeing smooth, steady motion, since it cannot distinguish the change from one split-second image to the next.

The black and white television picture that appears on your screen starts its life within the television camera in the studio or out on location. The light from the subject being televised falls on the camera lens, and is focused on a target plate whose surface is covered with a mosaic made up of spots of light-sensitive material. Each of these dots corresponds to one of the dots that will make up the finished picture. The light falling on the target plate will vary according to the areas of light and shade in the picture, so that the plate will, in effect, carry a map corresponding to the picture seen by the lens, but made up of electrical charges. The brighter the area, the higher the positive charge carried by the corresponding dot.

The electrical information on the target plate now has to be scanned, first to convert it into a signal for transmission and second to clear the plate ready to receive the next signal. This scanning is performed within the camera tube. A beam of electrons (negatively charged particles) is emitted from a plate (the cathode) and is focused by magnetic coils so that it strikes the target plate in a fine point. Under the direction of the coils, the beam sweeps backward and forward across the target plate in a series of lines, covering the whole surface 50 to 60 times a second. This is double the 'picture scanning rate' because, to prevent flicker, each picture is scanned twice on alternate lines.

The electrons in the scanning beam strike the target plate at a constant rate. As any one dot is hit, the positive charge on the dot is discharged by the negative charges of the electrons hitting it. The greater the charge on any dot, the more electrons are absorbed by it. The electrons which pass on through the target plate fall on a signal plate which lies behind it. Here they are used to generate a signal whose intensity varies directly with the intensity of light in the original picture, and it is this which is transmitted.

When, a fraction of a second after transmission, the signal reaches your receiver, it is used to produce an electron beam which scans the screen. The number of electrons hitting a particular dot on the screen corresponds exactly to the intensity of the original, and the more electrons hit a dot, the brighter it will glow.

The fluorescent screen of the TV tube is coated with dots or particles which glow when struck by the high-speed electrons of the TV signal. The beam of electrons is focused by magnetic coils which cause it to scan to and fro over the screen, making the dots glow in the forward movement only. The brightness of each dot depends on the number of electrons hitting it. When the scan reaches the bottom corner of the screen, it is deflected diagonally back to the top corner.

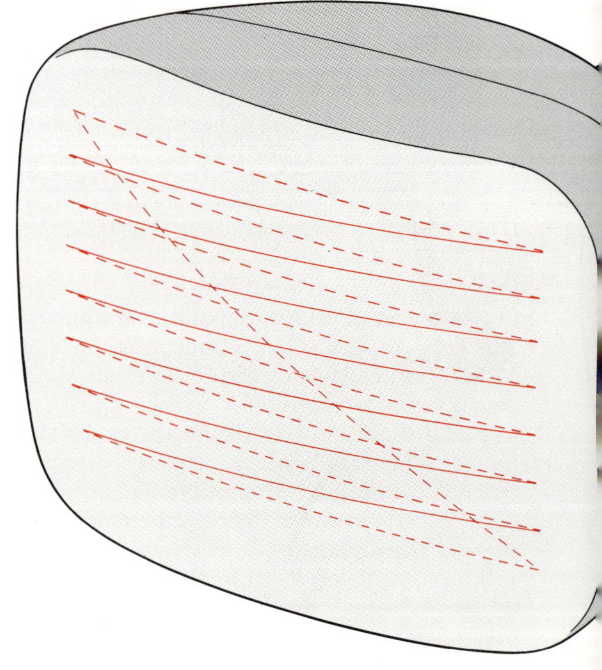

A TV monitor, unlike a TV receiver, does not pick up UHF/VHF signals. It displays a picture fed into it directly from a TV camera or VCR. For the best VCR reproduction, you need a TV monitor and receiver combined, the receiver for picking up UHF/VHF signals, the monitor for high-quality video fed in at low frequency.

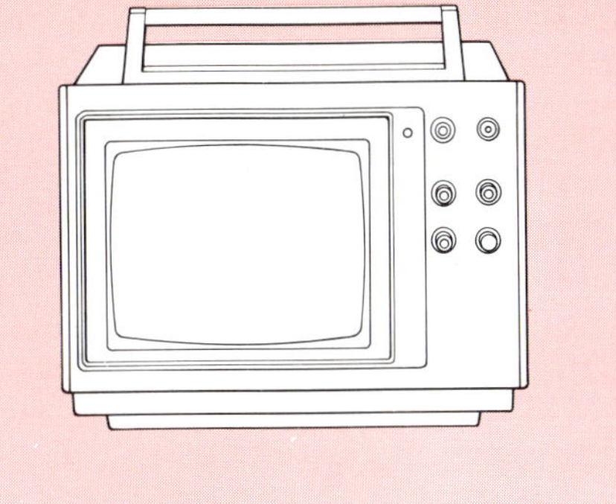

The signal which is transmitted from a TV station, *below*, is first generated as a steady high-frequency wave signal known as a carrier wave, **1** (black). Superimposed on this is the lower-frequency video signal (red). When the video signal is added on in this way, the carrier wave becomes modulated, **2**, so that the height, or amplitude, of the carrier wave (black) is altered to adapt to the video signal it is carrying. When this modulated vision carrier signal is picked up by the TV antenna, it is fed into the TV receiver. Here, the video signal, **3**, is first separated from the carrier wave by the tuner. The video signal is amplified and, after various modifications, fed into the TV picture tube, where it is used to generate a beam of electrons which scans the screen and varies with the intensity of the original signal.

In the TV, visual signals are used to create an electron beam, audio signals to create the sounds emerging from the loudspeaker.

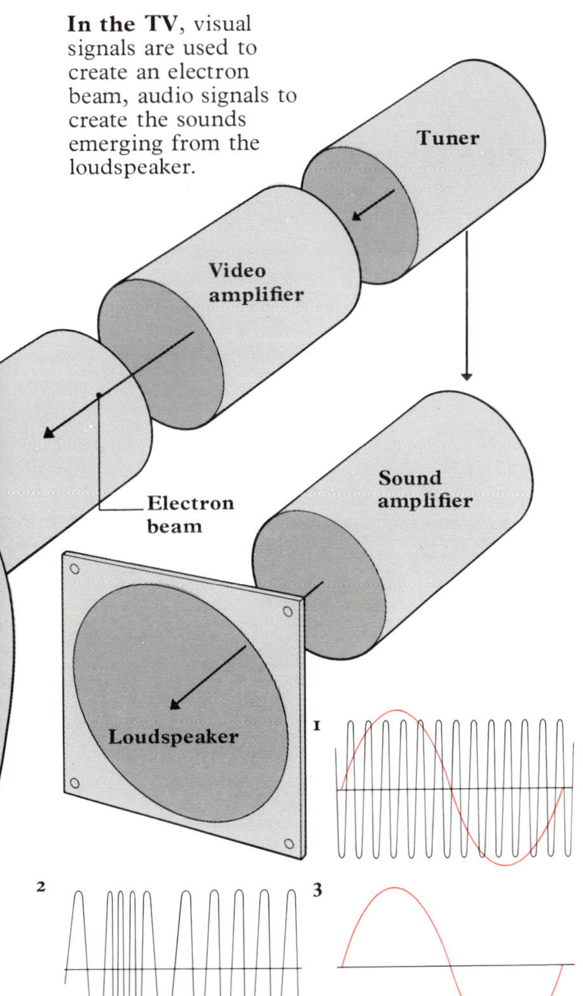

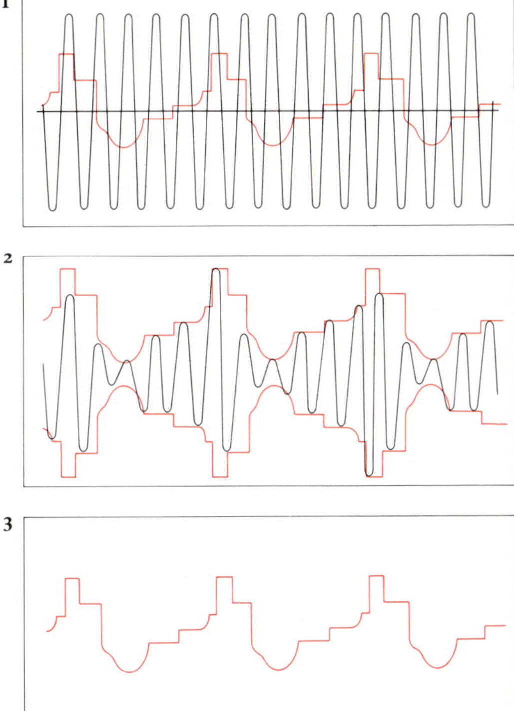

Sound signals, *left*, reach the TV in a similar way to video signals. First, a sound carrier signal is produced, **1**, which is modulated, **2**, by having a sound signal superimposed on it. This sound signal is carried at a frequency just above that of the video signal. In the TV, the tuner removes the sound signal from its sound carrier, **3**, and then amplifies it before feeding it to the loudspeaker.

Colour TV systems

Black and white television was a brilliant achievement, but adding colour was even more spectacular, for colour television involves the transmission of a vast amount of information. Moreover, when transmissions began, the new colour system had to enable people with black and white sets to continue receiving acceptable pictures from colour transmissions, and those with colour receivers to watch in black and white.

Just as the human eye produces a view of the world in two separate components, one a detailed picture in black and white, the other information about colour, so colour television works on a similar principle. One signal, the luminance signal, carries the information needed to build up a detailed black and white picture in exactly the same way as the existing black and white television system, so it can be received by black and white sets in the normal way. The second signal, the chrominance signal, is then added, and directs the television receiver to 'paint' on colours in the correct hues and intensities.

To produce colour television, the colours seen by the television camera are split up into the three primary colours, red, blue and green, by means of an arrangment of colour-separating mirrors or prisms. Each colour is focused into a different camera tube to produce a signal directly related to the areas and intensities of that colour.

Before transmission, the signals are first combined in the proportion of 30 per cent red, 59 per cent green and 11 per cent blue. This produces a completely black, white and grey picture which is transmitted as the luminance signal.

Next, the chrominance signals must be created in the following manner. The luminance signal already carries some of the colour information, since it is made up of the three primary colours in fixed proportions. Transmitter circuits now take the red signal and subtract it from the luminance signal, to produce a signal of green plus blue. Similarly, subtracting the blue signal gives a signal of red plus green. From these two so-called colour difference signals, a set of circuits in the receiver can break the information down to re-create the original red, blue and green signals and produce a picture.

Colour television has considerably complicated the variables between different television systems. The world's first commercial colour system was NTSC (National Television Systems Committee) introduced into the USA in 1954 and later adopted by Canada, Mexico and Japan. The main drawback of the NTSC system is that even slight errors in the phase between the colour difference signals produce errors at the decoding stage so that the set applies too much of one colour; hence the mnemonic for remembering the name: Never The Same Colour Twice. NTSC receivers have a hue control.

The PAL (Phase Alternating Line) system, used in Britain, Australia and most of Europe, produces better colour by means of more complicated receivers. SECAM (Système couleurs à memoire), adopted by France, Hungary, East Germany, Algeria and the USSR is simpler, but does not give such a good picture.

In the NTSC system, *below*, the two colour difference signals, **2** and **3**, created by the subtraction of red and blue from the total signal, **1**, are not transmitted together, but with one a quarter of a cycle behind the other. The signals are then added together to form a single chrominance signal, **4**. When it reaches the receiver, the decoding circuits, **5**, inside the TV break down the chrominance signal and separate it from its carrier wave. The two signals are fed into a matrix, which then combines them with the luminance signal to re-create the three original colour signals. These then create beams in three electron guns.

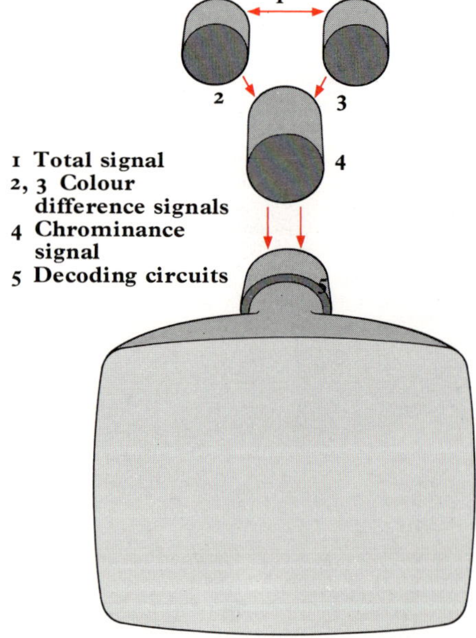

The screen of a colour TV is composed of minute phospor dots, which glow red, blue or green when struck by streams of electrons. Behind the screen is a shadow mask, a plate studded with thousands of holes. The three electron guns of the TV are arranged so that the beam from each gun is directed through the mask on to a dot of the correct colour. The picture you see is created by the addition of the three primary colours.

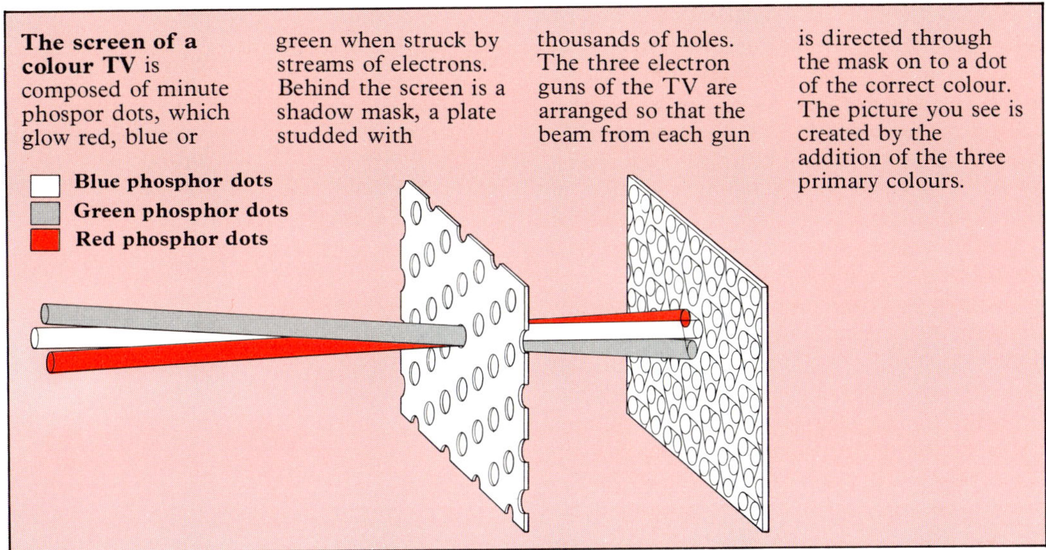

The PAL system, *below*, aims to improve the colour picture. The signals are transmitted in the same way, but the receiver delays the information on every line by means of an ultrasonic device, **6**, for the exact time needed to compare it with the signal for the next line. If, for example, a certain line of the picture, as received, contains too strong a red signal, the system ensures the next line will be too low on red by reversing the polarity of alternate lines. The final information passed to the picture tube is the average of the delayed first line and the corrected second one, so cancelling out the error. No hue control is necessary.

The SECAM system, *below*, has a different mode of transmission. The colour difference signals, **2** and **3**, are not arranged a quarter of a cycle apart, but are kept separate by transmitting them on alternate lines of the picture. Delay lines, **4**, inside the receiver then hold up one set of signals so that they can be recombined to build a picture from alternate lines of signals from the original scan within the camera. The SECAM system does not, however, give good pictures on a black and white receiver because it is difficult to separate the signals from their carrier.

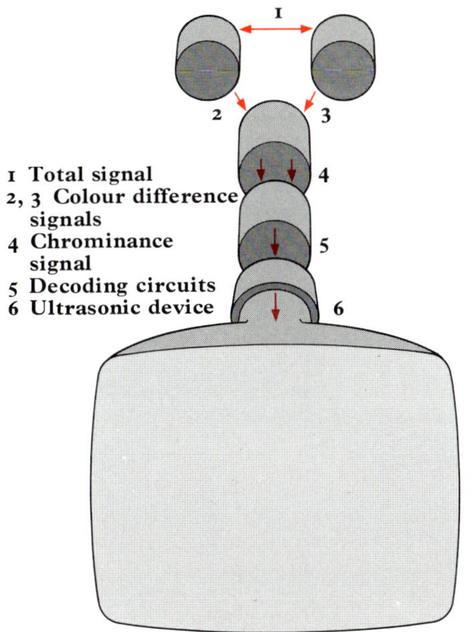

1 Total signal
2, 3 Colour difference signals
4 Chrominance signal
5 Decoding circuits
6 Ultrasonic device

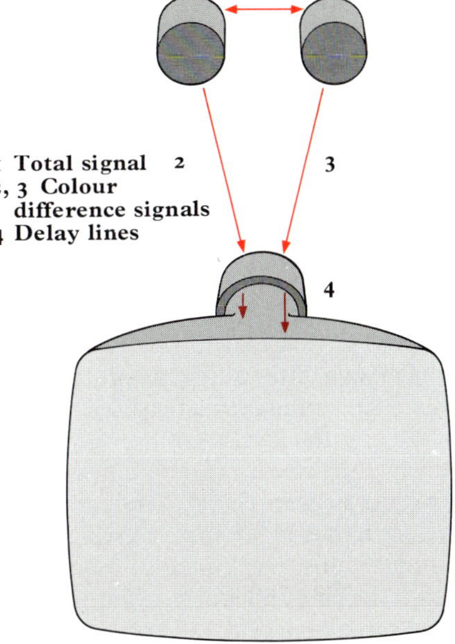

1 Total signal
2, 3 Colour difference signals
4 Delay lines

Antennae

Today's television sets and VCRs can produce pictures of remarkable quality. Nevertheless, however sophisticated your equipment, the biggest limiting factor to picture quality is the strength of the original broadcast signal relayed from your antenna to the input system of your terminal. If that is deficient, nothing you can do to improve or update the electronics of your video system can make the slightest difference. Yet any improvement you can make to the incoming signal will pay instant dividends in terms of better-quality recordings and, as an added bonus, improved television reception.

There are two reasons for a poor signal. Either the broadcast signal is poor, or you are the victim of a geographical accident. If you are a long way from the transmitter, or are overshadowed by hills or tall buildings, then you may have to come to terms with a poor-quality signal. To assess your situation, check with your neighbours; if everyone else seems to enjoy substantially better reception than you do, then there is probably room for improvement.

Whatever the standard of signal you can receive, an antenna adequate for the job is a must. In an area with good signal strength an indoor antenna in the attic, or V-shaped, loop or other antenna mounted on top of the set, may be enough. If these seem inadequate, you may need a roof antenna. These come in many sizes and degrees of elaboration. They may be highly directional (sensitive only to signals coming from one particular direction) which is ideal if your signal suffers from interference from other stations, or from echo causing ghosting. If you are lucky enough to receive reasonably good signals from several transmitters, you may need an antenna with less directional sensitivity to provide the best choice of viewing.

A useful extra to give increased flexibility is an RF signal splitter. This is a plug-in device which can split up the signals into two or more inputs so that either of two television sets can be switched to the VCR, or direct to the RF out for receiving broadcast television, without moving the VCR.

Whatever your needs, the best person to ask for advice is a specialist, who can take accurate test readings of your signal, check the existing antenna, assess the losses that occur down the cabling, and suggest ways to improve a weak or distorted signal.

A UHF amplifier is a useful device for improving the picture on your TV screen. Connected between the antenna lead and the VCR or TV, it can triple signal strength in areas where reception is generally poor.

A distribution amplifier is useful if you have several TVs or VCRs but only one antenna. It obviates the need to change the connection from each VCR as you use it. It can increase the signal strength to each by 75 per cent.

The area over which the receiver, or dipole, of an antenna receives signals can be drawn out as a simple circle, **1**, *below*. The addition of a reflector, **2**, cuts down the amount of signal picked up from the rear; a director, **3**, elongates the shape. Lack of rear pick-up eliminates inter-channel interference when you receive signals from only one transmitter, but you may need a less directional RF out to pick up signals from several transmitters.

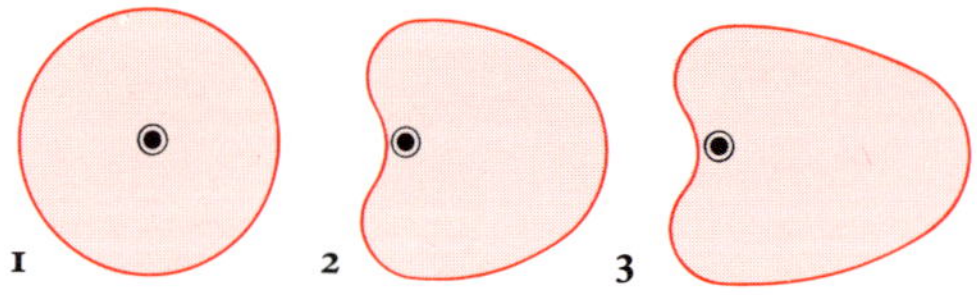

An indoor, set-top antenna, *below*, is most effective if you obtain good signal strength from the transmitter, since a signal weakens as it penetrates a building. Such an antenna can also be an asset if severe power losses are caused by long cable connections to your TV or VCR.

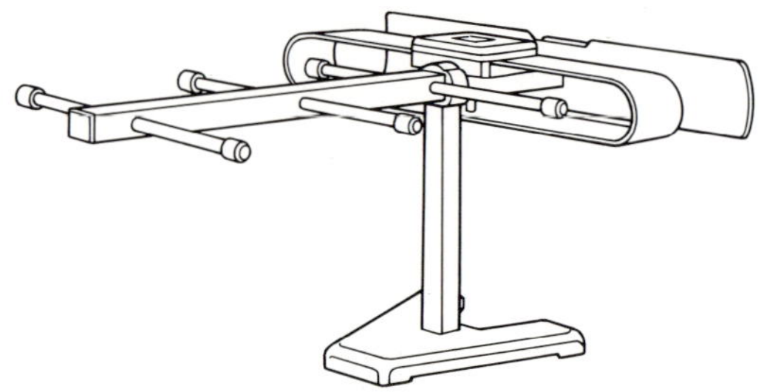

Exorcising ghosts
Poor reception is a problem when the area between the TV transmitter and a house fitted with a UHF antenna is obstructed by tall buildings or hills. In such situations, the most common reception problem is ghosting: the appearance of a shadowy second image to the right of the main image on the screen. What happens is that the antenna first picks up the strong, direct or primary signal from the transmitter, then, a fraction of a second later, picks up one (or sometimes more) weak or secondary signals which arrive by reflection from hill or housetop. Since the strong and weak signals arrive from slightly different directions, the best way to overcome the problem of ghosting is to fit a high-gain antenna, *below right*, on your roof.

If reception from your nearest transmitter is unacceptably poor, a specialist might suggest you tune in to another transmitter.

Twin-stacked UHF antennae positioned for best possible picture

Signal bent (diffracted) by hilltop

Strong signal

Transmitter

Weak signal

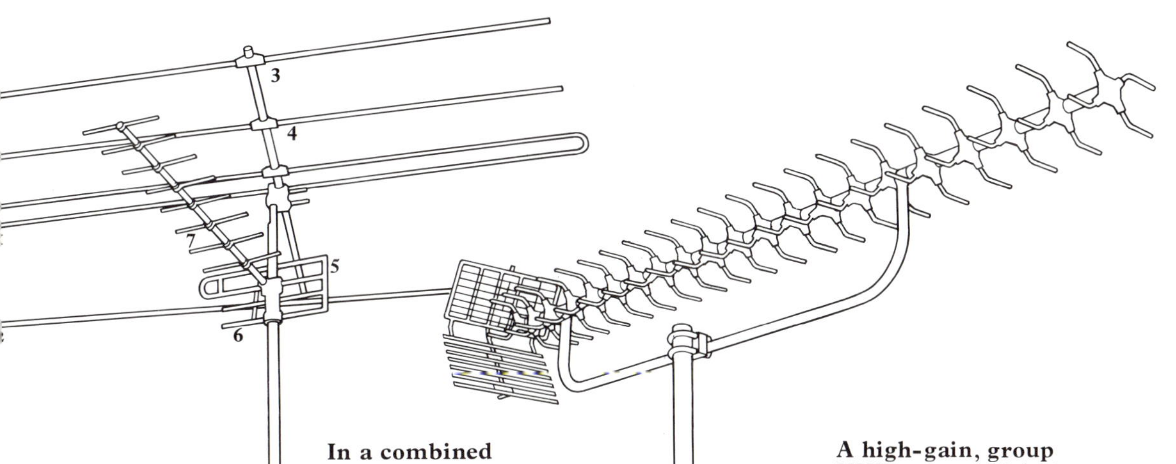

In a combined VHF/UHF antenna, *left*, the VHF part consists of a receiving element or dipole, **1**, on to which signals are directed by means of a reflector, **2**. Two directors, **3** and **4**, are positioned on the transmitter side of the dipole, and help to increase the power of the signal that reaches the dipole. The UHF part of the system has a gate-shaped reflector, **5**, compact dipole, **6**, and many directors, **7**.

A high-gain, group UHF antenna, *right*, is essential to obtain a good TV picture in areas where the signal strength from the transmitter is poor to moderate. Having 21 or more curved directors, the antenna works by picking up signals coming from one particular direction only. By rotating the antenna you may also be able to get rid of ghosts.

Viewing data

Silent but lively, videotext is invading our television screens with a dazzling array of computerized information. It was born of an idea conceived only a decade ago by engineers working for the British Broadcasting Corporation, who began sending internal messages along some of the 16 or so unused lines at the top of the television signal. Experiments in transmitting digital information along these lines brought home to the BBC the commercial potential of the discovery and as a result, videotext made its world début on British television in 1976.

Today, videotext comprises two types of service, teletext being the older and simpler. It is a one-way system: it piggy-backs digital data on to the unused portion of the broadcast television signal (it can be superimposed on a broadcast image) and in its unconverted state it resembles a twinkly band of Morse code just above the screen image, visible only when the vertical hold has slipped.

The teletext signal converts directly to a display of words or graphics on your television screen, but in order to receive any videotext system your set must be fitted with a decoder to transform the signal into the display. You can buy adaptors for older sets.

Simply by keying coded numbers on a remote control unit, you can conjure on to the screen news flashes, weather forecasts, sports results, even jokes and puzzles and your horoscope. The information is provided free by the broadcasting company, but only during transmission hours.

Viewdata originated with the British Post Office. It is a two-way system which exchanges information (chiefly by telephone link, but also by two-way cable or satellite) between your television set and data bases filled by commercial organizations such as airlines, financial institutions and retailers.

To receive viewdata you telephone a data bank, identify yourself as a subscriber, and use your remote control unit to call up the pages you want on your television screen. The data, stored in digital form, is converted first to audio signals for transmission by telephone, then back to digital form by the decoder in the television set, before finally being displayed on the screen.

The future of videotext is impossible to predict at so early a stage, but it bestows upon television the potential to become an instrument of information retrieval.

The numbers, letters and graphics that appear on the teletext or viewdata screen are built up as a pattern of dots, *below*. The UK authorities allow 24 lines of 40 characters; the USA broadcasts 20 lines of 32 characters. Both systems have the same visual format, with 96 characters and symbols, **1**, upper and lower case, as well as double size type and 32 control codes for flashing or boxing items, and so on. White, blue, green, red, magenta, cyan, yellow and black are the colours available. The dot matrix is so versatile that it is easily adaptable to display other types of written language such as Arabic, **2**.

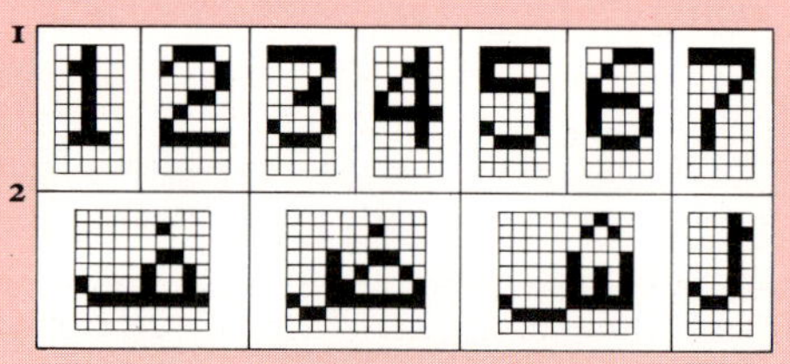

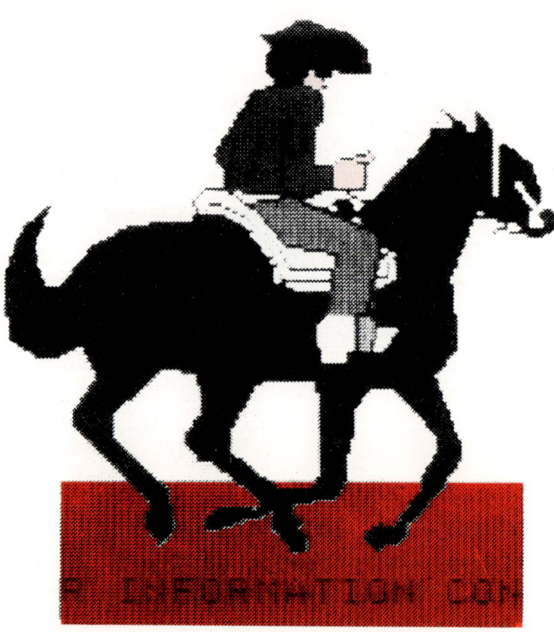

The viewdata system with the most realistic graphics is based on the Canadian Telidon system. This uses an alphageometric technique which gives a more curved display, *above*, than the alphamosaic system developed in the UK; however, it is much more expensive. It has been adopted by the Americans, who argue that the improved graphics will attract advertisers.

Selecting data
A keypad, which usually also acts as the audio-visual remote control and channel switcher for the TV set, is needed to call up teletext or viewdata information. Each teletext system publishes its own book of code numbers, telling you which digits to dial on the keypad for specific information or services, from commodity market prices to subtitles for the deaf.

With a keypad you can freeze a teletext page display for as long as you want it on the screen, or enlarge part of any page to double its size. You can superimpose a page on any TV show you are watching, or have newsflashes and updates, or the latest sports results, flashed up on the screen.

On the viewdata, which is transferred from computer to TV set along the telephone wires, it is possible to use the keyboard to answer back. Punching certain digits on the keypad thus allows you to book theatre tickets, make a hotel reservation and order food for dinner (and pay for all three with your credit card) without getting up from your armchair. The next step in sophistication is to replace the keypad with a typewriter-style attachment by which the viewdata system can be used as an intelligent mini-computer terminal.

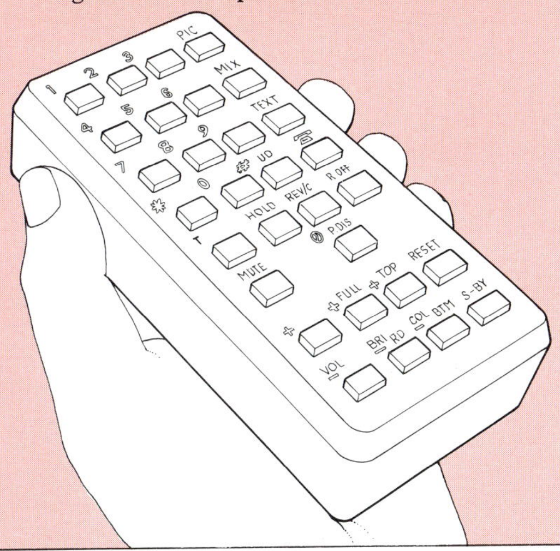

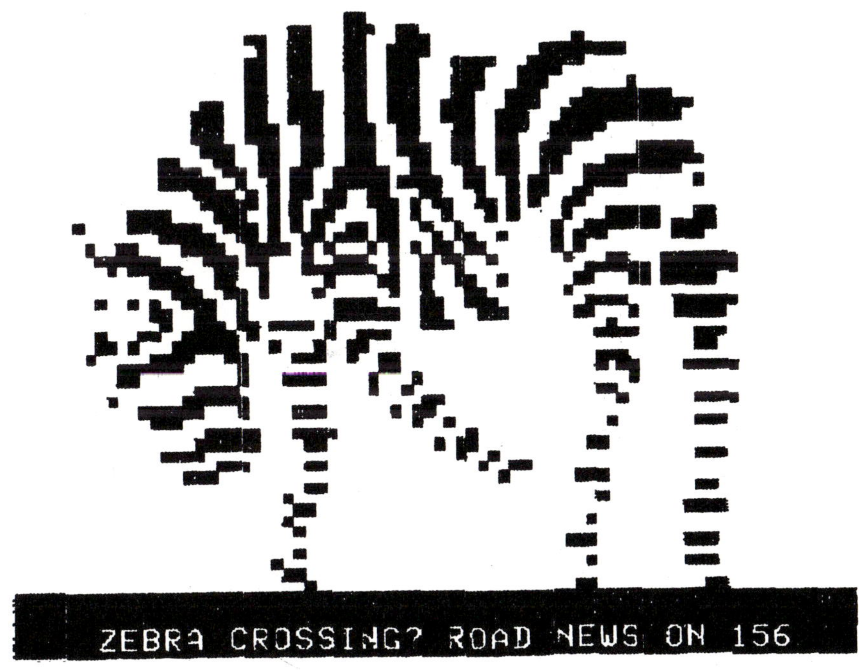

Teletext pages, *above*, are accessed cyclically, like slides in a carousel; the most-used ones are repeated so they may be retrieved in fewer seconds. The viewdata system of supply on request is immediate. the teletext page repertoire is a fraction that of viewdata, but free. With viewdata you pay for the telephone call to the computer, a connection charge and a variable charge per page retrieved. Both systems are constantly updated, but while teletext relies on the broadcaster to act as editor, viewdata is updated by direct input into the computer from many suppliers.

Teletext systems in current use include Ceefax, Orbit and Oracle in the UK. Canadian Telidon and French Antiope are teletext systems which can be extended to accommodate viewdata.

Giant screens

The homely, small-screen television is perfectly adequate for most viewing. But when your video tapes begin to seem poor imitations of their wide-screen cinema originals, or when you hanker after re-creating all the atmosphere of a live sporting occasion, if you have the space and the funds, then could be the time to consider acquiring a large-screen television.

A modern large-screen home system consists of three items which may or may not be built into the same unit: a television receiver, an enlarger using one or three lenses for optical magnification, and the viewing screen. The least expensive systems have the screen separate from the projection unit so that the television stands back from the screen rather like a movie projector. This system uses a modified, conventional television set with a huge lens fitted over the screen to project the image on the viewing screen. However, whereas a film projector employs a bright light to boost the illumination on the screen to an acceptable level, the brilliance of a normal television picture cannot be increased, making the vast picture rather dim.

The more elaborate systems use three tubes, each projecting one of the primary colours: blue, red, or green and three lenses which increase the brilliance of the picture and which, with the screen, are contained in one package. A built-in system of mirrors and back projection reduces set size to manageable proportions.

Other techniques of achieving large-screen television include the do-it-yourself conversion of an existing set using a screen, lens and manual supplied by the kit manufacturer. Alternatively, a plastic fresnel lens may be erected in front of the normal television set. The use of most such conversion kits demands considerable technical competance, so they are for experts only.

Apart from home conversions, the most daunting aspect of large-screen television is its cost, which starts at around £800 in the UK and $1,000 in the USA, and rises to as much as £2,500 and $4,000 respectively. But for anyone who relishes the prospect of big-screen viewing at home, for video recordings or ordinary television viewing, then it is certainly worth considering. Projection television systems also have practical possibilities in the leisure industry, in business and in education.

Purchasing check list

1. Make sure you have enough room to accommodate the equipment and to obtain a good view of the picture.
2. Make sure the fixed furniture in your room will not obstruct your view.
3. Try to view the device in conditions similar to those in your own home.
4. Buy only from an established and reputable specialist video dealer who will be able to offer maintenance services when necessary.
5. Survey all the machines on the market and choose only a well-known brand.
6. Ask for a week's home trial before committing yourself to a large sum.
7. Ask if the unit has infra-red remote control. If not, will you be satisfied?
8. Make sure that the dealer will install your chosen system and will guarantee to maintain it for 5 years.
9. Choose the most up-to-date and sophisticated model you can afford.
10. Ask whether the system will allow you to play video tapes recorded on all the world's TV systems (PAL, SECAM and NTSC).
11. If you have any queries that your prospective retailer cannot answer, always contact the manufacturer direct.

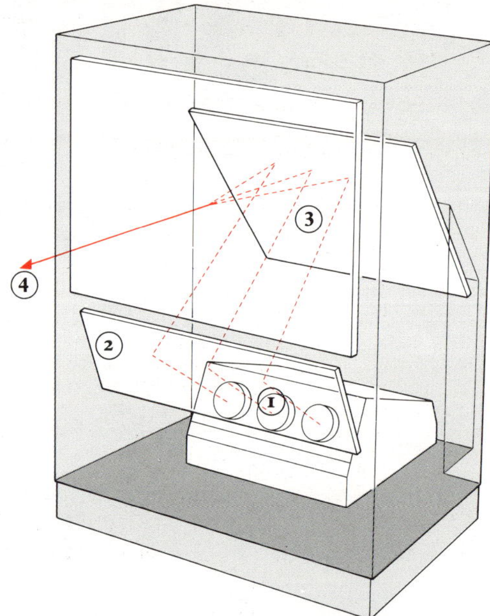

Big-screen TV is made possible by back projection. In this system 3 tubes each project one of the primary TV colours via lenses, **1**, on a mirror, **2**, which reflects them on to a flat screen, **3**. This screen bounces the colours off at such an angle that they are combined, **4**, to form a picture. Sound is also bounced off.

Ample viewing space is essential for watching big-screen TV. Most projection TV systems take up an area of floor space at least 3 ft by 2 ft (1 m by 60 cm) and are designed to be viewed at a distance of 5 to 20 ft (1.5 to 5 m). The primary, and best viewing zone, **1**, is contained within a wide angle, but the best viewing position is well back from the screen. The secondary viewing zone, **2**, provides a less than perfect view of the screen and should be avoided unless the screen is completely flat. To make the most impact, large-screen TV is best viewed in a dimly lit, cinema-like atmosphere with the lights down low.

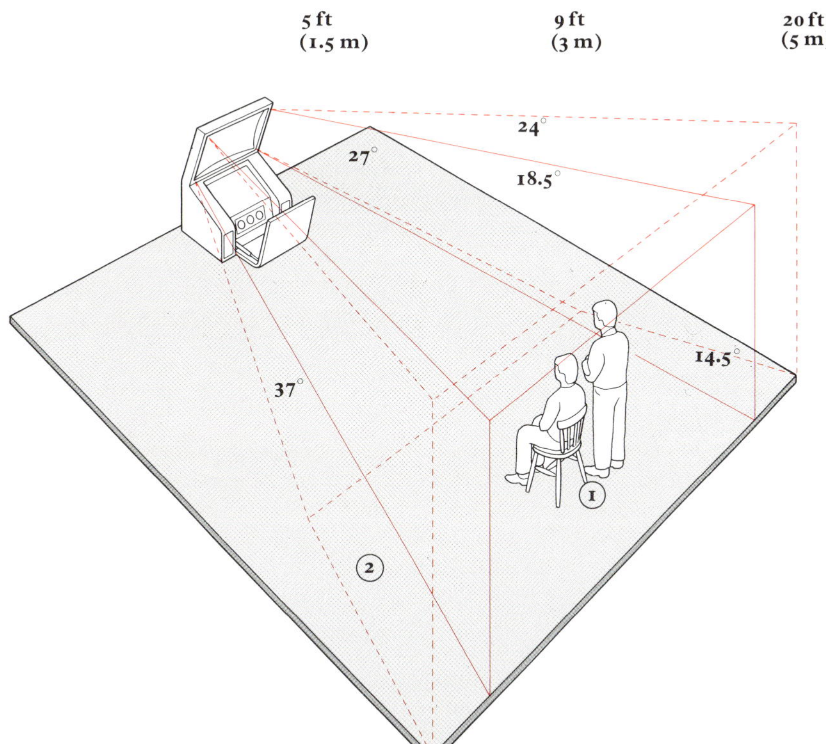

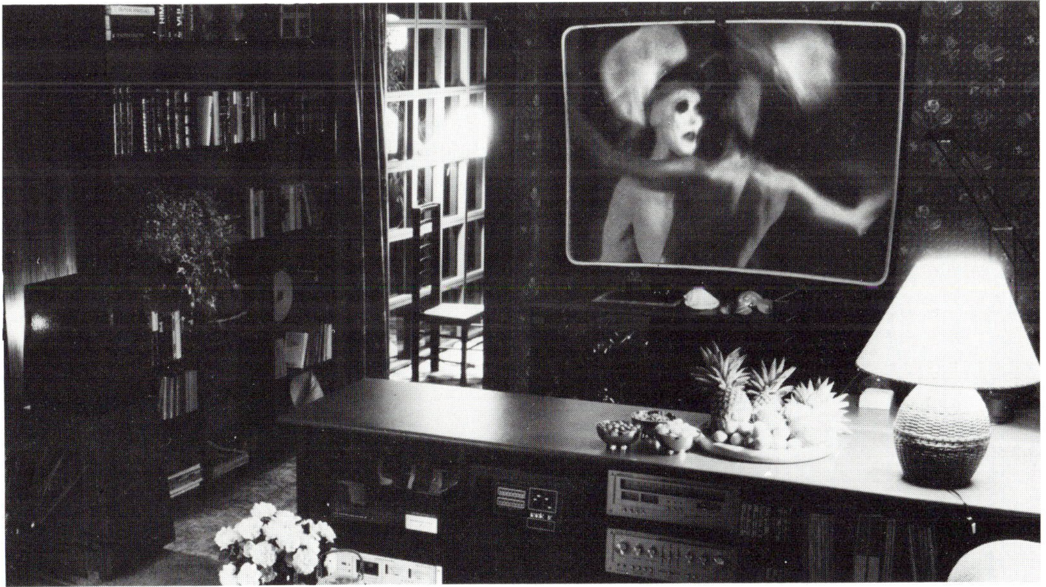

Picture quality on a large screen, with diagonal measurements of 50, 60 or even 80 in (125, 160 or 200 cm), is only as good as the source from which that TV picture originates. Any TV projection system will, by its very nature, magnify any shortcomings in a picture which might well pass muster on a normal TV. This is especially true of snowy, weak or ghosted broadcast TV reception and poorly recorded video cassettes. A good TV signal from your antenna or cable TV system is, thus, the first requirement. Another must for a perfect picture is a screen with a highly reflective beaded glass surface.

Skybirds

For the television of tomorrow, the future will involve both cable and satellite broadcasting. Neither system is novel, but what is new is their potential impact on viewing.

Cable television dates from the earliest days of broadcasting. In these times the coverage of the transmitters was much less than perfect, with several areas, obstructed by tall building or hills, where the signal was either weak or non-existant. The answer was to set up antennae at sites where the signal was acceptable and to pipe it along coaxial cables for distribution to any viewer prepared to pay the subscription fee. The engineering standards employed by these networks were of variable quality, and in many cases the technical system demanded the use of special or modified television receivers.

Modern cable television has developed in many ways, with cable companies sending out specialized broadcasts, often to target audiences. However, the economics of cable television mean that it is viable only in densely populated areas.

Where cable television is too costly, or where isolated communities are cut off from information sources, the answer to improved communications may lie in satellites. Since the launch of Telstar in 1962 satellites have been used to link continents, but satellite broadcasting to the general public is only just becoming an economic reality. With an appropriate dish-shaped antenna and a down-converter to make the signals usable, satellite television beamed from an orbiting 'bird' will soon be available to all consumers for the same cost as a video recorder. In the near future, cable television networks will probably receive and redistribute some of the satellite channels; only serious viewers will need to own a receiving dish.

While the technology of satellite television has been worked out, the politics certainly have not. Hitherto, governments have been able, if they wished, to control the pattern and content of broadcasting, but now transmitters can beam a particular brand of uncensored entertainment, cultural or commercial, into another country.

Cable and satellite television have a linked future and together form an enormously powerful medium for reaching the mass of viewers. While neither is likely to replace conventional broadcasting, they provide a means of catering for special interests.

Cable TV world-wide
Around the world, cable TV has developed in different ways. In the UK the original networks were developed to carry the three broadcast channels and even some of the newer systems have little or no capacity for additional services. In 12 localities, however, feature films are available for a premium subscription

On the continent of Europe, cable TV is more established, with most major cities wired for it. In Belgium, 75% of all homes are wired for cable TV and, apart from the four national programmes, viewers can receive broadcasts from France, Germany, Luxembourg, the Netherlands and the UK.

In the USA, about 45% of homes (a total of more than 20 million) are wired up to cable TV. The majority are in metropolitan areas where 30 or more different channels may be available, providing a wide range of services, including updated displays of news, weather and financial reports. There are cultural, community and ethnic channels, and channels devoted to films, sport and children's entertainment.

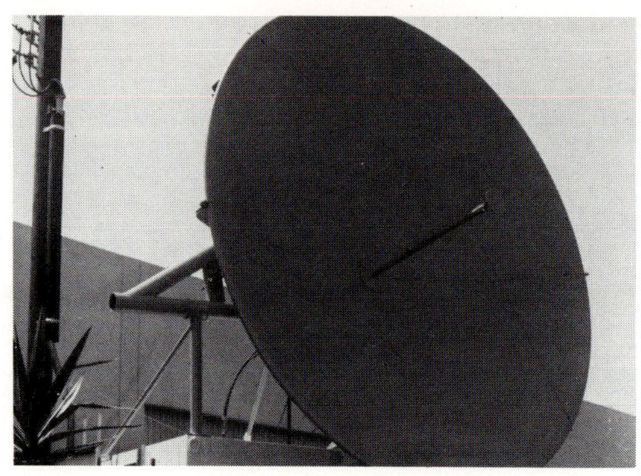

A dish on the roof or in the garden makes satellite TV available at home. Yet even in the USA, where there are over 50 satellite TV channels, and where such dishes are commonplace, the satellites are really intended only to supplement cable TV systems and do not generate their own broadcasts.

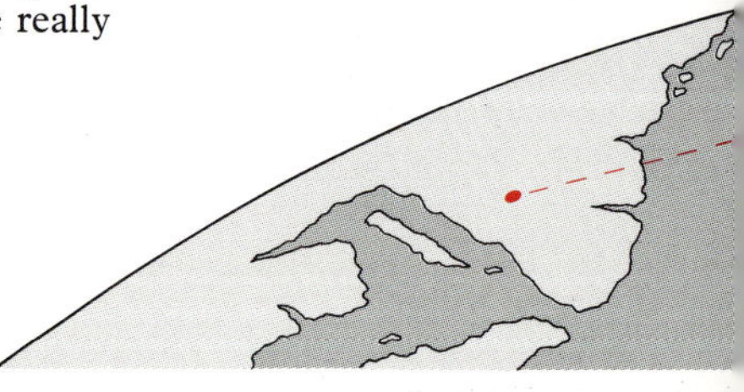

Communications of the future will almost certainly involve the use of optical fibres. Compared with copper coaxial cables now in use, optical fibres carry more data much faster. In this system, electrical signals are converted into a binary code of on-off signals which is then transformed into flashes of light. These are transmitted along transparent glass fibres and, when they reach their destination, are turned back into electrical signals. Some fibres smooth out the overlap between light rays, **1**, but the best, **2**, give higher quality by allowing one ray to pass at a time.

Advances in fibre optics promise wide repercussions for cable TV. In the Japanese town of Higashi-Ikoma, 160 homes are linked by two-way fibre optic cable TV to a central studio and eight other points round the town, including the emergency services. Every home TV set also has a computer terminal equipped with a mike so that residents can call up the various services, see the personnel, and have access to central computer banks. But because the central studio also has a TV window into private homes, the experiment has been considered a violation of privacy.

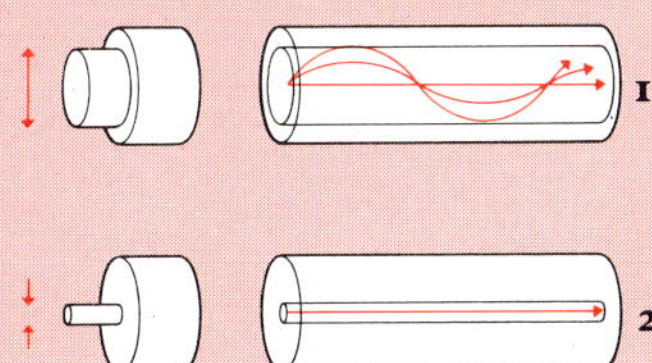

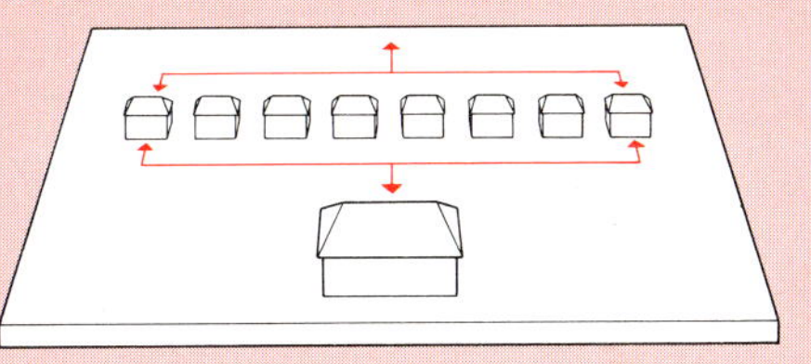

Like a giant bird, a Direct Broadcasting Satellite orbits the earth. In the future, such satellites will beam down signals powerful enough to cover whole countries and allow reception with a dish only a few feet across. The signals will be transmitted at much higher frequencies than our regular UHF transmission of 500 to 800 MHz, so special downconverters will also be needed. With a big dish, say 8 ft (2.4 m) across, it may be possible to pick up signals from satellites sent up by other countries, but the dream of Europeans watching US TV will not be possible until the Americans beam it down on them deliberately.

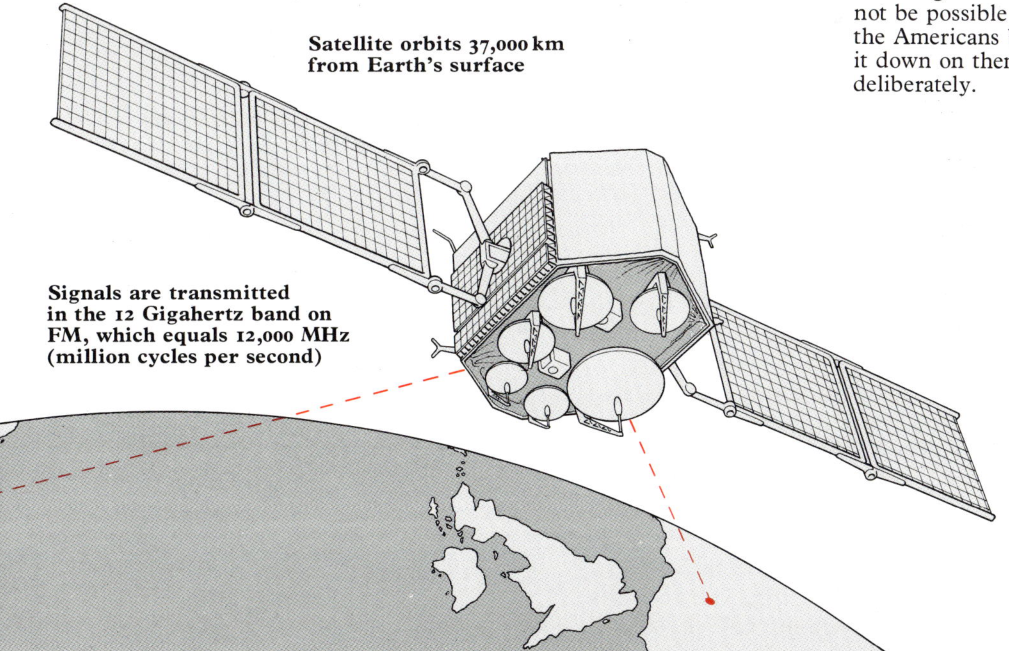

Satellite orbits 37,000 km from Earth's surface

Signals are transmitted in the 12 Gigahertz band on FM, which equals 12,000 MHz (million cycles per second)

TV tomorrow

Television, now inextricably bound to video, has a rosy future. So what does that future hold for us, the viewers? First and foremost, conventional television will quietly improve.

Already American manufacturers are using comb filters to improve picture reproduction, and manufacturers the world over are acheiving new standards of reliability. On the audio side, stereo broadcasting, already a reality in Germany and Japan, will become commonplace.

Soon, we can expect our televisions to be 'user friendly'. By 1988 a television will have, as standard, remote control responding to spoken commands, teledata and cable television facilities, and easy connections for video games, VCR, home computer and reception of satellite broadcasts.

Television will change in shape and size, so that we shall soon become used to large-screen video 'cinemas', and the progressive miniaturization of the screen will be limited only by the abilities of the human eye.

The picture will be given added brightness, either by means of liquid crystal displays, similar to those already used in pocket calaculators, or by electroluminescence, created by light-emitting diodes (LEDs) similar to those used today as indicators on a variety of electronic gadgets. Liquid crystal LED displays will also have a role to play in making home computers, linked to televisions, smaller and cheaper. In addition, picture tubes using LEDs, now being developed in Japan and the USA, are proving to be safer because, unlike today's television tubes, they do not run hot and do not use a potentially dangerous combination of glass and a high-pressure vacuum.

High-definition and three-dimensional television are sure to have arrived by the nineties, but are unlikely to be commonplace. The former promises to be most advantageous to the video owner, and could give definition as good as 35 mm film.

With component television, the viewer will have the ultimate in choice. In Japan the Sony Corporation has already introduced the 'Profeel' range for which the consumer selects monitor, receiver and tuner. To this can be added a hi-fi system and VCR.

So (if you can afford it) the future offers a feast of new television. Add to this the promised proliferation of broadcasts, and that future promises to be rich indeed.

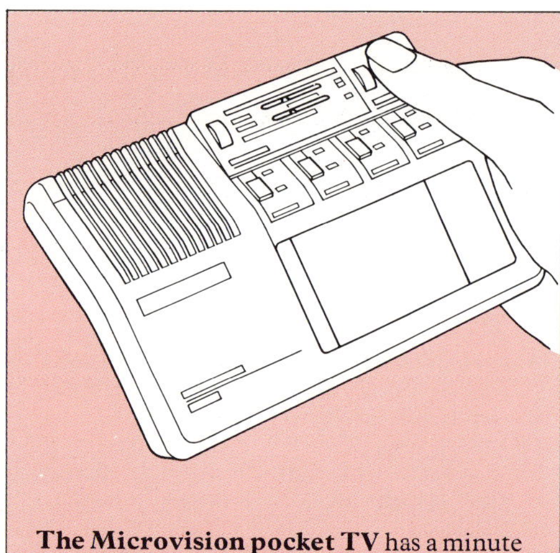

The Microvision pocket TV has a minute flat screen only 3 in (7.5 cm) in diameter, but is able to receive transmissions almost anywhere in the world (France is a notable exception). To do this, the set needs a telescopic antenna, and this, together with the fact that it will not operate in aircraft, trains or automobiles, also limits its usefulness. Despite these drawbacks, miniature TV is likely to be commonplace in the near future, and the day of the wrist-model TV cannot be far off.

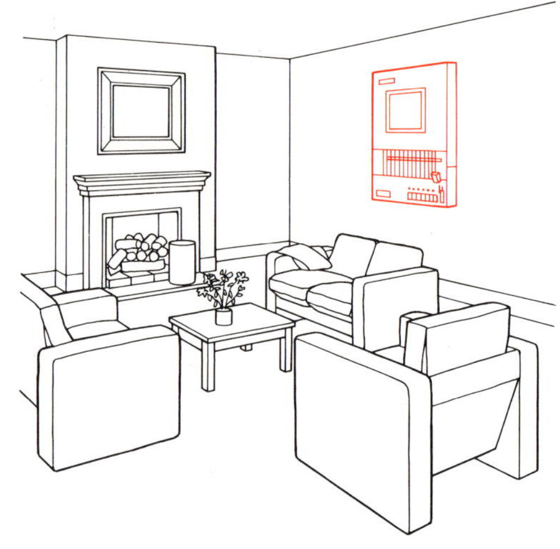

Flat-screen TV, hung on the wall like a picture, will have the advantage of taking up less space than today's TV sets. The Japanese are currently working on two possible models. In one, the screen is made up of a matrix of liquid crystal cells; in the other it contains a myriad of minute gas-discharging devices. These sets should also have the added asset of using less electricity in their operation than conventional sets.

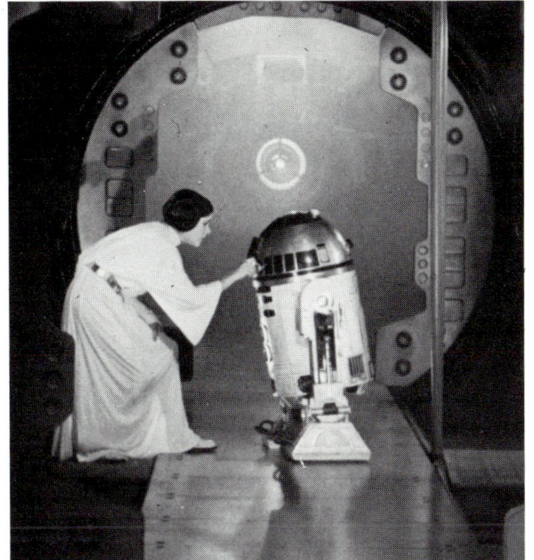

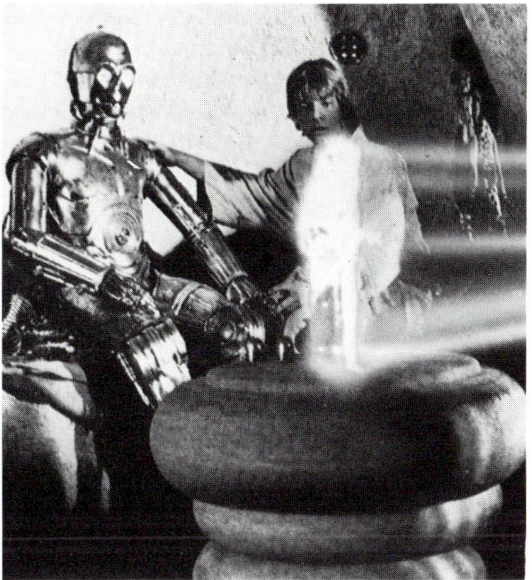

In the movie 'Star Wars', Princess Leia asking for help from Ben Kenobi in the Rebellion: it is a 3D holographic image of herself. Such 3D imagery (in realistic colour, not a weird, shimmering blue) is the TV technician's dream. The key to success could well lie with the hologram, a 3D image produced with the help of lasers. To create a hologram, a laser beam is split in two. Half is directed at an object, half at a holographic plate. Where the two beams overlap they set up interference waves which reproduce the shape and depth of the object. Adapting this technology to create truly 3D TV is the exciting prospect for tommorow.

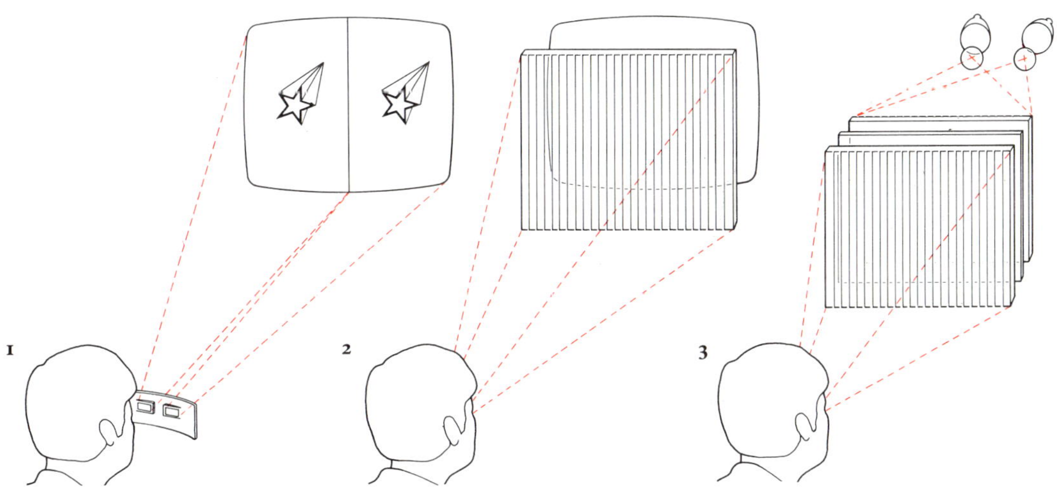

TV in 3D will be on the screen soon. Of the schemes devised to date, one of the simplest, **1**, is to create two identical images, one to be seen by the left eye, the other by the right. The TV is shrouded with a viewing hood containing lenses, to bend the sight-lines outward. Although cheap, this system gives an unacceptably upright image. The direction-selective 3D TV, **2**, has a picture made up of vertical lines with alternate lines designed to be viewed by left and right eyes. In front of the screen is a grooved lens, which translates the visual information and directs it into the appropriate eye. Unfortunately, the picture definition is poor, and even slight sideways movements of the head distort the image. To help solve this problem, **3**, two images from separate tubes are projected on a double-grooved lens. This gives improved definition and restricts viewer movement less.

All about cameras

A video recorder can do more than you realize. In addition to recording broadcast television shows and playing back pre-recorded cassettes, a VCR linked to a video camera is all you need to make your own recordings.

Video cameras are no longer out of the price range of the home-producer. Although a decade ago the only portable colour cameras were professional broadcast models costing thousands apiece, now a compact colour video camera costs little more than a good-quality Super 8 home movie camera. Every year sees further improvements designed to make cameras for home-producers more sophisticated, yet easier to use.

Since portable colour cameras were introduced during the sixties, black and white cameras have dropped in price so dramatically that older ones cost as little as one tenth of the price of a new colour model. These are ideal for beginners to try out. Once you graduate to colour, the monochrome camera will still be useful for adding titles and graphics to your colour movies.

As a result of the revolution in video technology, making your home movies with a video camera is easier than with a Super 8 home movie camera, although it has some disadvantages. One of the most important of these is that the video camera is not yet self-sufficient.

Unlike still and moving film cameras, you cannot load a video camera with film. All it can do is translate the picture framed by its lens into an electrical signal. This has to be passed down a cable to the recorder before it can be recorded on the cassette tape. This means that if your camera is connected to a home deck you can venture no farther than the garden.

For location work, you need a battery-operated portable recorder, which will give you freedom, but at a price several times that of a basic cine film outfit, although film cameras are more versatile. As a compensation, however, the cost of video tape is far less than that of film. Minute for minute, it is roughly one-tenth the price of Super 8 film.

Video has many advantages over film. You can record a shot, then play it back instantly, simply by pressing the 'Rewind' and 'Playback' buttons on the VCR. You will never have to wait, nor pay, for video tape to be processed. If you are not happy with the shot you can wind the tape back and record the next shot over it because tape, unlike film, can be used more than once.
Since your home deck will be connected to a television set, you can try out a shot, watching it on the screen without using the tape at all, to see if it is worth recording. Sound is automatically recorded alongside the picture information on the tape, so you always have a sound-track which is perfectly synchronized with the pictures on the screen. You can play back your movies over the television, without setting up a projector and screen, or darkening the room.

Nevertheless, video cameras are still heavy to operate for more than a few minutes at a time, especially if you are carrying a porta-pack over your shoulder, and film cameras are, by comparison, light. In addition, they can do things video cameras can never do. They can shoot frame by frame, and so create the illusion of movement by showing 24 frames, or separate pictures, every second. The eye's persistence of vision smooths out this succession of separate images into smooth, unbroken movement, a facility which is exploited in cartoons, or in time-lapse effects such as the speeded-up movement of a flower opening.

Luckily you will not have to make a definite choice between video recording and cine filming. Because Super 8 pictures can be copied easily and cheaply on video tape, there are many ways of combining the two media to take advantage of the best of both.

If you have a library of home movies you can transfer them to tape for longer life and ease of playback. You can use a Super 8 camera to shoot film in places where carrying video equipment would be difficult or impossible, and re-record the pictures on tape afterwards.

Finally, because professional editing on video tape can be expensive, limiting most home-producers to recording material in the order in which it is to appear in the finished production, you can shoot complicated sequences on film, edit them by cutting and splicing, and copy the entire sequence into the appropriate place on the tape.

Until you have used a video camera, this sounds complicated, but because tape is so cheap, and because you can see the results of your experiments immediately, you can learn from experience more quickly with a video

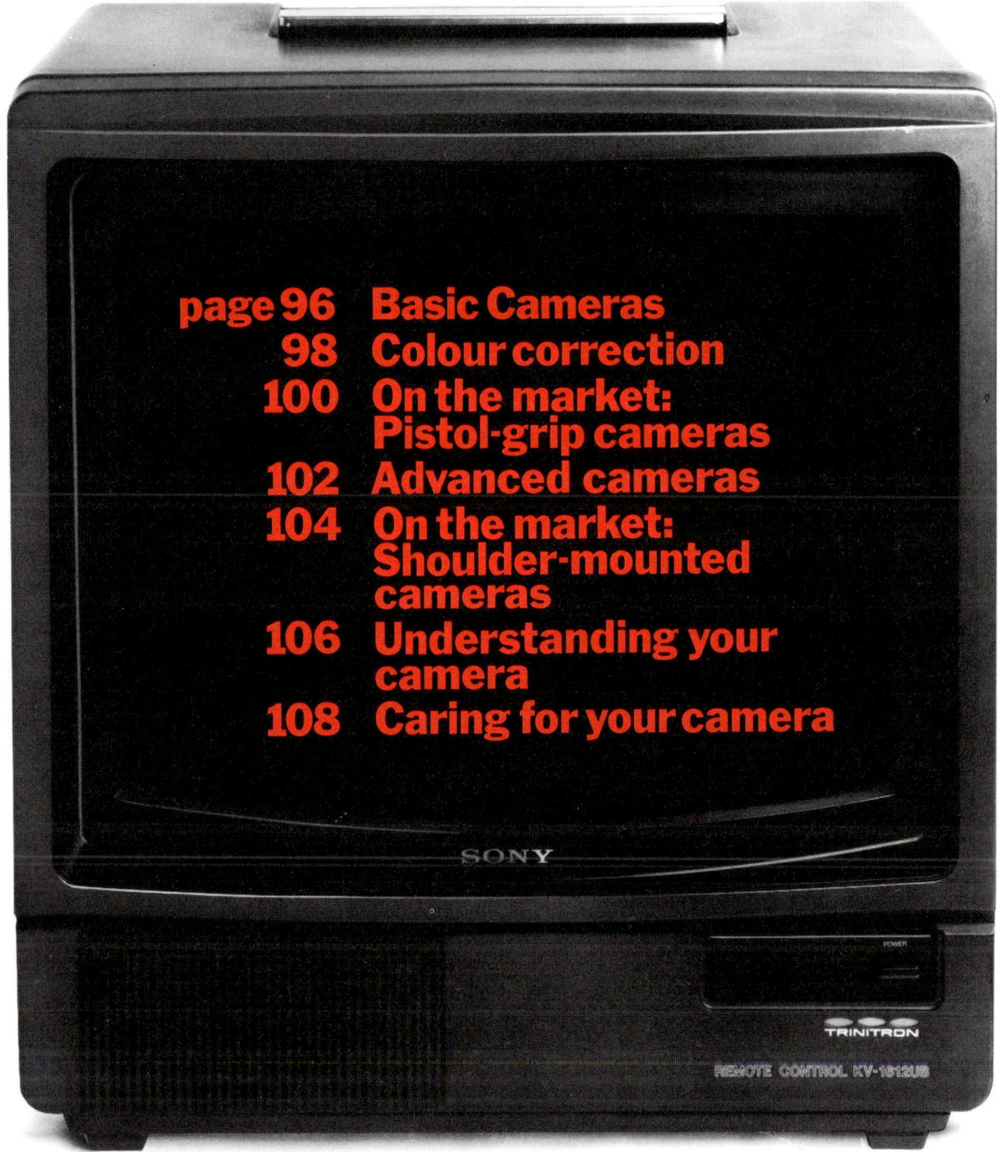

camera than would be possible with film.

The next few pages show how modern video cameras are removing the guesswork, and making it much easier to produce professional-quality results from the beginning. They progress from the simplest hand-held cameras to the most advanced semi-professional shoulder-mounted models, whose automatic features, adapted from professional broadcasting cameras, do almost all the work for you.

The portable colour video camera has now reached an advanced stage of development, and is unlikely to change a great deal in the near future. The appearance of Microvideo, a new quarter-inch cassette tape format aimed exclusively at the portable market, heralds the birth of a new generation of lightweight portapacks – but these have a very limited playing time.

The innovations of the distant future are, however, just in sight. Sub-miniature tape formats, using microcassettes similar to those you load into dictating machines, will eventually result in a combined deck-and-camera system – a video camera into which you load the tape. Inevitably, too, the microchip will invade the video camera.

Basic cameras

Mastering the technique of operating a video camera is easy enough, but the various types and models are so differently designed that whenever you buy a new camera always read through the manual before operating it.

Using the manual, identify the camera's components and put them together carefully. Most of the simpler cameras arrive ready to use, but more sophisticated models may have a separate electronic viewfinder and a microphone, each of which has to be fitted into place and its cables plugged into a special socket. If the lens has to be screwed into its mounting on the camera body take great care not to cross-thread it; the damage may be expensive to rectify. Put the lens cap on to protect the camera tube.

A video camera is designed for use with a portable VCR, but can be used with a home deck, if you connect both to a power adaptor. The various manuals will give clear wiring diagrams. If the camera fails to work the first time you press the stop/start switch, recheck the connections first; then make sure that the camera, the power adaptor *and* the VCR are all switched on. The camera may not produce a picture until two or three of its switches are in the correct position.

Pick the camera up carefully and point it at a well-lit, but not very bright, subject. In a dark room do not point it at a window; the contrast between dark interior and bright surroundings can cause flare if the iris is manually operated, and close down an auto iris so that the interior turns black. If the camera has an optical viewfinder, wait for it to warm up. If it has an electronic viewfinder the picture will appear on its television screen in the eyepiece, but in black and white, not in colour.

Remove the lens cap and zoom in to a subject, holding it in close-up. The picture will probably be out of focus and, if you switch on your television set to see the picture in colour, you may find the colours are unrealistic. Adjust the colour temperature controls, guided by the manual, and then focus the image to give as sharp a picture as possible. It should then remain sharp through the whole zooming range of the lens.

Treat the camera carefully – but at the same time try it out, put it through its paces, experiment with it, find out what it can and cannot do.

Many of the simpler video cameras have been designed along similar lines to Super 8 cine cameras, with pistol-grip handles. However, video cameras are heavier – this Hitachi camera weighs just over 4 lb (nearly 2 kg) – and can be tiring to use and difficult to hold steady. For this reason many pistol-grip cameras are designed to rest on the shoulder, and the angle of the grip is adjustable.

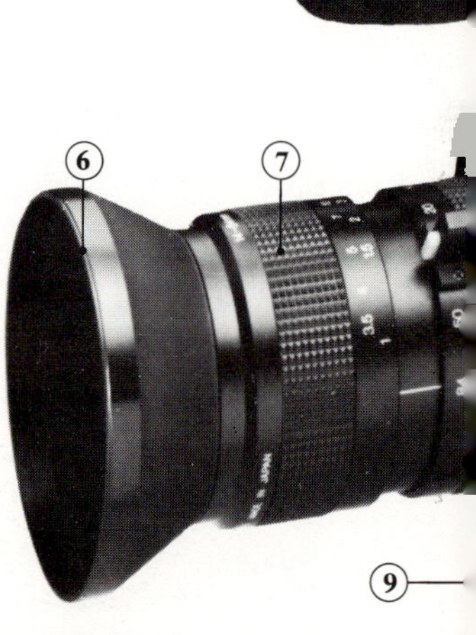

The lens, **6**, has three adjustments: aperture or iris (which controls the amount of light reaching the camera tube); focus (which controls the sharpness of the image formed on the camera tube), and the zoom. This allows the focal length of the lens to be varied so that the picture can be framed at any point between a shot of a whole scene and a magnified close-up of its central area. On this camera the iris is fully automatic, the focus, **7**, is manually controlled, and the zoom has a power switch, **8**, *and* manual control, **9**.

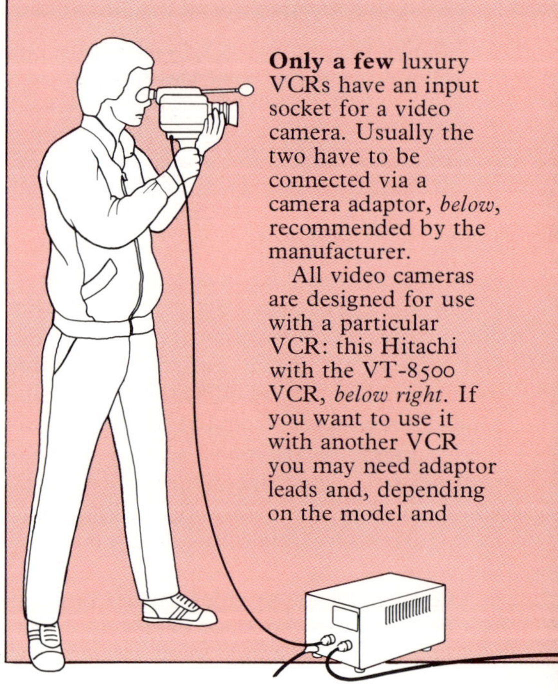

Only a few luxury VCRs have an input socket for a video camera. Usually the two have to be connected via a camera adaptor, *below*, recommended by the manufacturer.

All video cameras are designed for use with a particular VCR: this Hitachi with the VT-8500 VCR, *below right*. If you want to use it with another VCR you may need adaptor leads and, depending on the model and

Most cameras have built-in omni-directional mikes. These pick up sound from all round the camera – but that may include the sound of the auto iris and even the operator's breathing. This camera has a uni-directional boom mike, **1**, which picks up the sounds ahead.

You may need a more sensitive extension mike, for which there are special sockets in both camera and VCR. By plugging an empty jack into an external mike socket you can cut the sound.

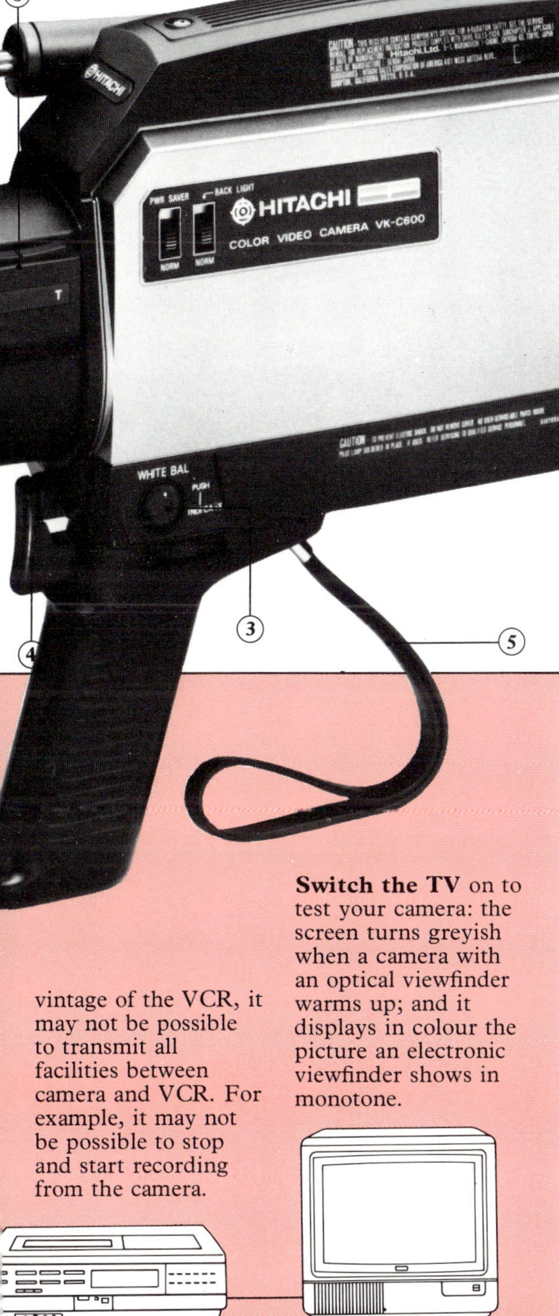

To operate the camera there is usually a stop/start trigger switch near the hand grip, but it may be hard to identify. On this camera it is marked 'Remote control', **4**.

A camera, unlike the human eye, cannot adapt to the colour of light. A white object looks white indoors or out, but through a lens it looks reddish or bluish. The white balance control, **3**, corrects this.

Simpler cameras may have one of four types of viewfinder: an optical viewfinder similar to that on a stills camera; a more sophisticated version which tracks with the zoom, permitting accurate focusing, and gives an idea of the lens's field of view; a TTL (through the lens) device, so called because it gives a picture of the scene framed by the lens; or an electronic viewfinder, **2**. This is a miniature TV monitor mounted in the eyepiece, which displays the picture in black and white.

If the camera has a handle strap, **5**, loop it round your wrist. You will be unconscious of it as you operate the camera, but it will stop it falling if you lose your balance, or if someone bumps into you.

vintage of the VCR, it may not be possible to transmit all facilities between camera and VCR. For example, it may not be possible to stop and start recording from the camera.

Switch the TV on to test your camera: the screen turns greyish when a camera with an optical viewfinder warms up; and it displays in colour the picture an electronic viewfinder shows in monotone.

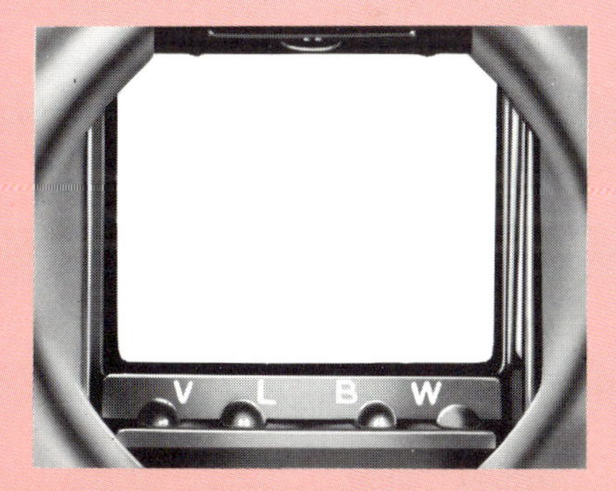

When you look into the eyepiece of an electronic viewfinder you usually see a number of lines, lights and motifs below or down the side of the screen and sometimes superimposed upon the picture. These are warning lights. In the Hitachi, *above*, 'V' lights up when the camera is recording; 'L' when the object on which the camera is focused is inadequately lit; 'B' when the batteries are running down and 'W' when the white balance is incorrect. For on-the-spot playback just rewind the tape.

Colour correction

An object looks coloured only because it reflects light of a particular colour to the eye. It can do this only if the light falling on it contains the particular colour it reflects. In fact, objects appear in what we see as their true colours only under white light, which is made up of light waves of all colours. Light which is not white, which has one or more constituent colours missing, may distort the colours of the objects upon which it falls. To take an extreme example, an object which looks blue-green under white light would look black under pure red light, since it reflects no red light at all.

This is important in film and video shooting simply because pure white light is surprisingly rare. What our eyes see as white light is usually tinged with blue, yellow or red. In fact, the colour of the light under which the camera is recording tends to vary with the location, the time of day, the weather and the type of lighting being used; and while our eyes automatically compensate for this changing colour balance, so that we always see white objects as white, the camera cannot.

This means that a video camera with its colour responses correctly set up for daylight, which has a distinctly blue cast, will produce pictures with an obviously yellow quality under indoor lighting; and a camera that is adjusted to reproduce the colours in a living room lit by tungsten light (which has a strong yellow tone) in the evening, will 'see' outdoor scenes during the day as much bluer than they seem to the eye.

Film makers solve this problem by using colour correction filters. Whatever the colour of the light falling on the outside of the lens, using the right type of filter ensures that the light reaching the film is always correctly colour-balanced. Similarly, outdoor working presents no problems if a filter which tones down the blue content of the light is fitted to the camera, leaving the light emerging from it perfectly balanced for the film. Many video cameras use filters in exactly the same way for indoor/outdoor adjustment, though they can also produce the same effects electronically. Often a good camera will use both methods: filters for basic colour compensation and electronic correction control for fine adjustment to suit the prevailing light conditions and so will have two sets of colour controls.

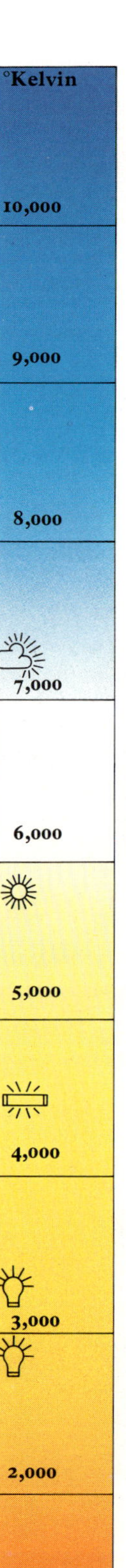

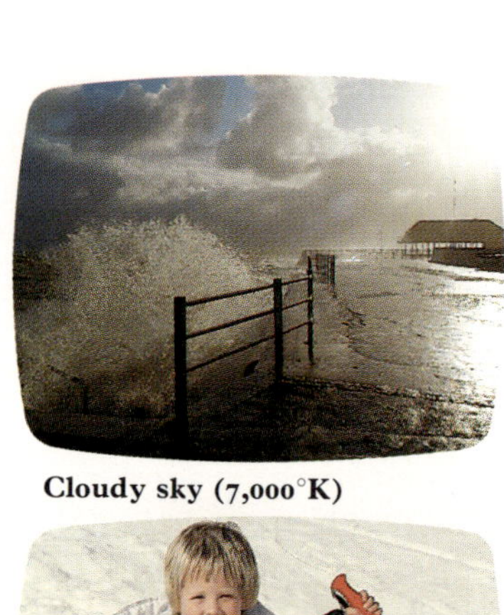

Cloudy sky (7,000°K)

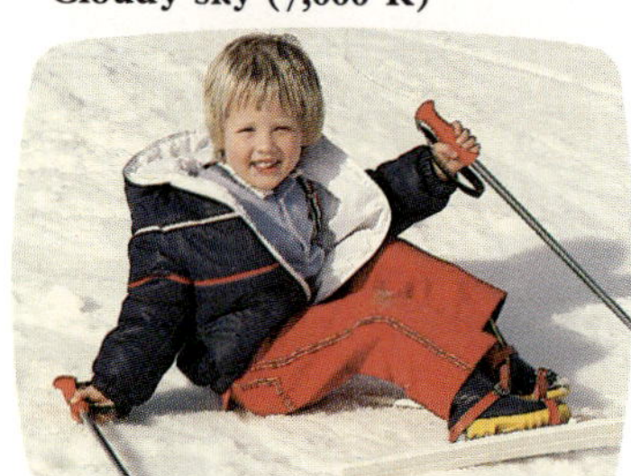

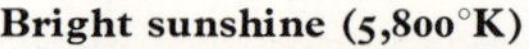

Bright sunshine (5,800°K)

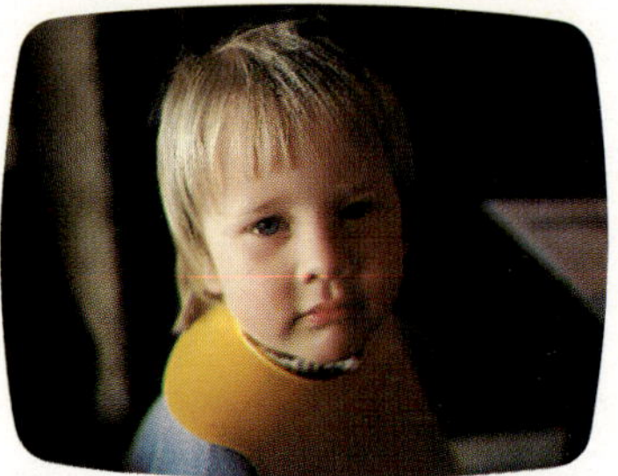

Average daylight (5,000°K)

The different colour balances of different kinds of light can be measured by relating them to the temperatures to which most materials must be heated to produce all the different kinds of light. For example, the reddish light from a candle measures around 1,000° Kelvin, so that is its colour temperature, *left*. Most materials heated to this temperature will produce the same reddish light. Photofloods produce a whiter light at 3,000°K, while flashbulbs, quartz halogen and fluorescent lights range up to around 5,000°K. Sunshine, usually thought of as pure white light, is about 6,000°K, while overcast skies become progressively bluer, at colour temperatures ranging from 7,000°K to 10,000°K. This is not because the sky is hot, but because the scattering of the light produces a bluish cast.

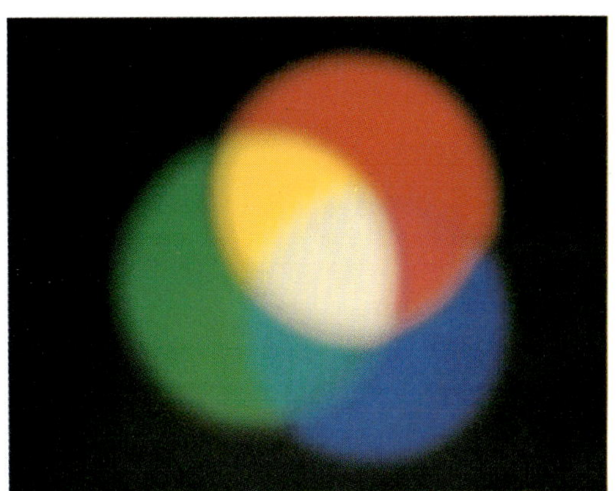

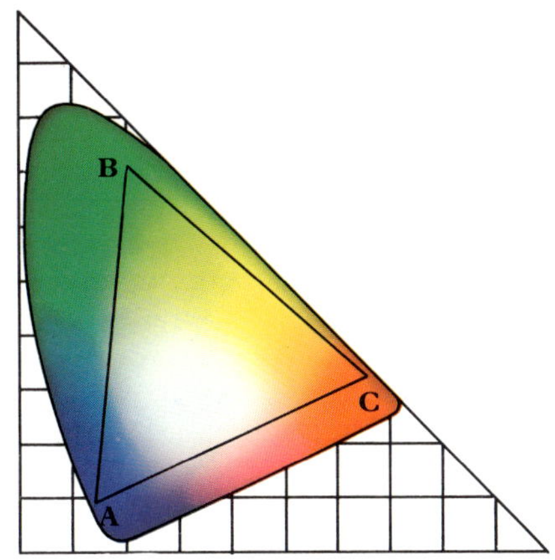

Video uses light of three primary colours: red, blue and green, which mix in the eye to re-create all the other colours you see on the screen. Light beams of these three primaries will mix into a single white beam, *above*, if their intensities are equal. Yet when they are mixed in different intensities, other colours result. The red and green beams, *above*, mix to yellow, the exact shade depending on the proportions of red and green: more red and less green will produce orange; the reverse ratio gives khaki. Moreover, any two colours can be tinted by different proportions of a third colour, and any colour mixture can be paler or darker, widening the colour range.

The chromaticity chart, *above*, shows this colour range. A, B and C mark the primaries. The shades in the inner triangle result from mixtures of these primaries and represent all the colours you see on a TV screen.

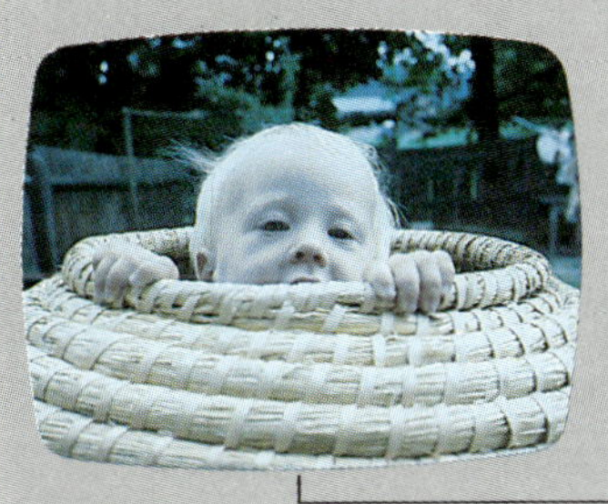

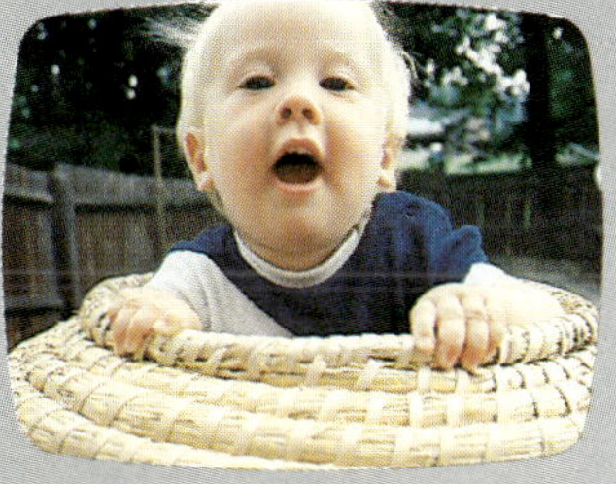

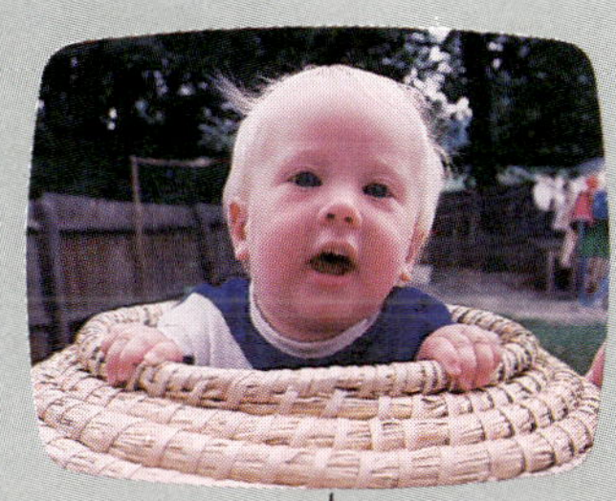

The camera's colour temperature adjustment compensates for the type of light being used for the recording. It may consist of a set of colour correction filters moved, often by a 'Daylight Filter' switch, between the lens and the tube, or a switch with up to four colour temperature settings, *far right*. The photographs, *below*, show how the colour of the prevailing light looks natural only when shot under the correct setting.

Instead of using an optical filter to remove excess colour from a scene, turning down the sensitivity of the blue circuits inside the camera has the same effect. The Tint, or the White Balance control, *above left*, varies the balance between red and blue. Turning it left makes the picture bluer, and right, redder, *top*.

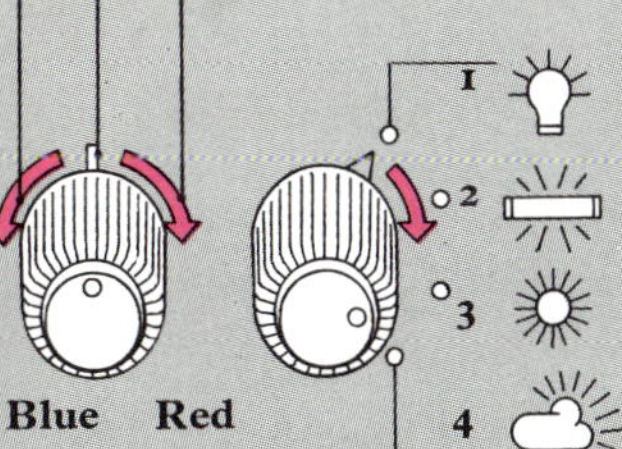

Iodine lamps (3,200°K)

Fluorescent lamps (4,500°K)

Bright outdoors, sunshine (5,800°K)

Cloudy, rainy (7,000°K)

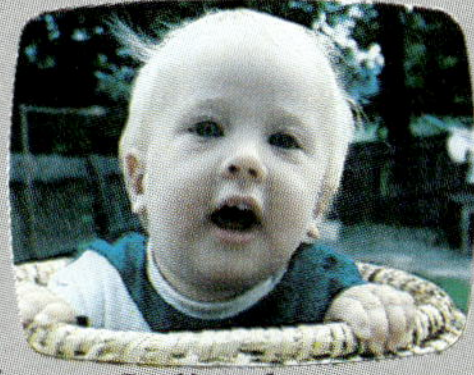

1 Iodine lamps

2 Fluorescent lamps

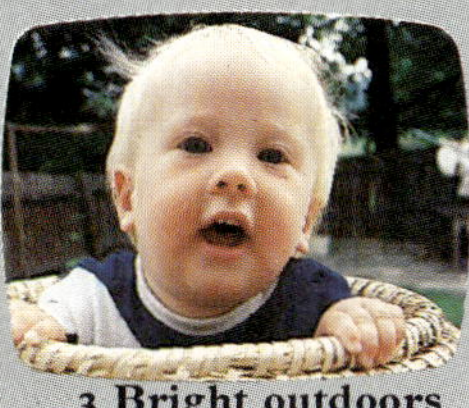

3 Bright outdoors

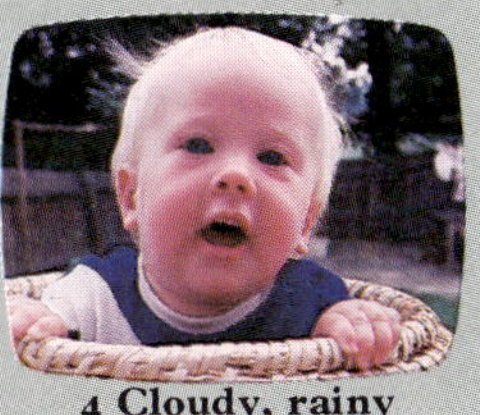

4 Cloudy, rainy

On the market/Basic cameras

Information and test reports on all the new cameras on their way in to the market can be found in all the major video magazines. However, the most important feature of any camera, and one that is not covered directly in the most detailed of specifications, is the quality of the picture the camera produces. This is a combination of sharpness of image, faithfulness of colour reproduction, and ability to capture fast-moving subjects and record at low light levels. The only way to be sure a camera meets all these requirements is to test it.

Take no notice of the photographs, apparently showing what the camera can do, which appear in manufacturers' brochures: they are invariably shot on film, not videotape. The picture a video camera produces is an electronic image and looks quite different. It cannot, however, look as good as the pictures you are used to seeing broadcast on television. These are shot under ideal conditions, using cameras costing thousands.

Begin by comparing the pictures produced by several different video cameras to assess current standards. The manufacturer's showroom or, better still, a video exhibition, may give you the facilities for doing this. Then test two or three systematically.

In a shop, insist on handling the camera yourself. Ideally you should try it with several different tape recorders and through more than one television set, especially if you spot a fault: it may not be in the camera. Point the camera at as many different kinds of subjects as possible. Notice how it copes with well-lit objects, dark areas, high contrasts. Pan the camera quickly across the scene to see whether the picture seems to lag behind the movement of the camera. Many single-tube cameras are faulty on the red colour balance, so notice whether red objects are tinted rust or maroon, and whether faces look too flushed, or pallid.

Most shops are on the dark side, so try to take the camera out into daylight. Record pans, zooms, fades and a variety of still and moving objects on a tape cassette, which you can then take home and compare on your own machine, through your own television, with tapes recorded on other cameras.

Really you need to try the camera out on your own ground, on subjects of your own choosing. You may be able to arrange a short rental or extended trial with the shop.

Hitachi VK-C750

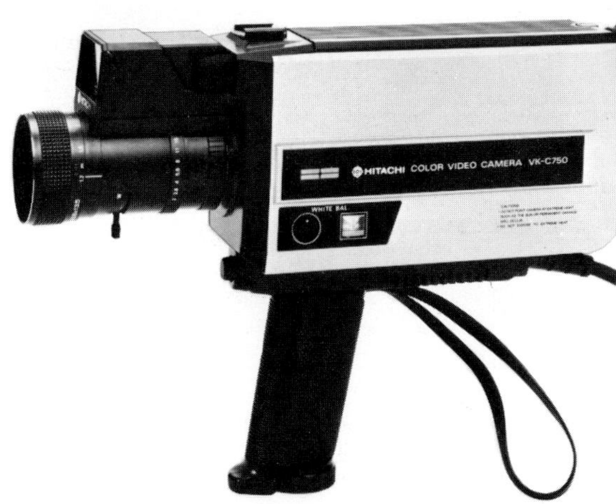

Hitachi produces a large range of single-tube colour cameras. In the earlier models the pictures are sometimes less crisp, particularly in long-shot adjustment, than in those produced by competitive cameras from other manufacturers. The new generation cameras are better in this respect. The VK-C750 is a lightweight model weighing only 4 lb (about 2 kg) and is one of the cheapest colour cameras on the

JVC GX-33

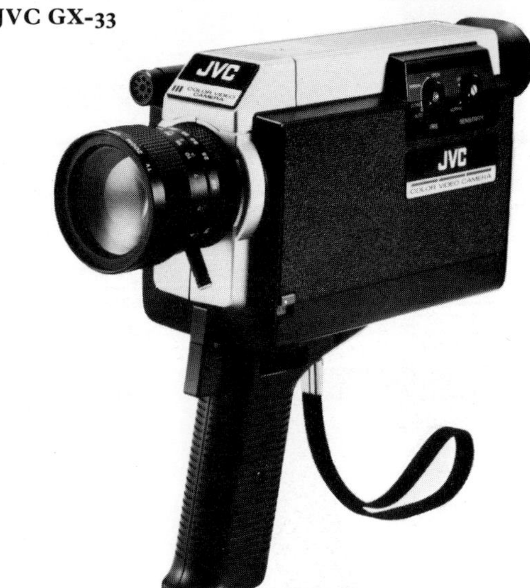

JVC produces a range of colour cameras so wide that it embraces both the cheapest and the most expensive cameras on the market. Their basic model is the GX-33, currently one of the lowest-priced designs in all markets. It has a through-the-lens optical viewfinder and a 3:1 manual zoom. This is an ultra-lightweight camera, weighing only 3 lb (1.4 kg). The

Hitachi GP-41D

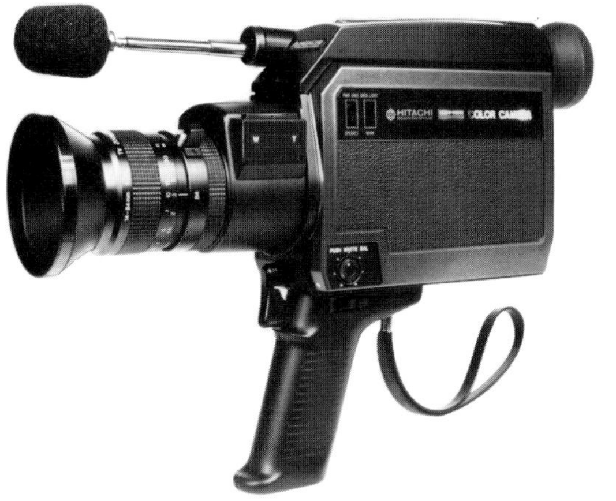

Panasonic WV-3000

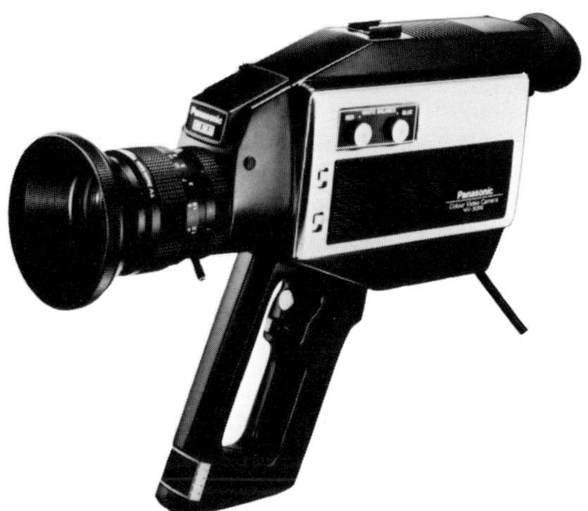

market. It has an automatic iris, a 2.8:1 zoom lens (the lens allows a 2.8 magnification of the picture in full close-up) with manual control and an optical viewfinder.

The GP-41D has many features the more ambitious enthusiast will want, such as an electronic viewfinder and a 6:1 power zoom lens.

Panasonic was one of the first companies to introduce a reliable colour video camera at a price the amateur could afford. Simple to use, but with a sharp and clear picture, our Panasonic camera performed reliably in locations ranging from the Irish Sea to the Himalayas. The WV-3000 is an up-to-date equivalent, with a 3:1 manual zoom lens and an electronic viewfinder.

JVC GX-88

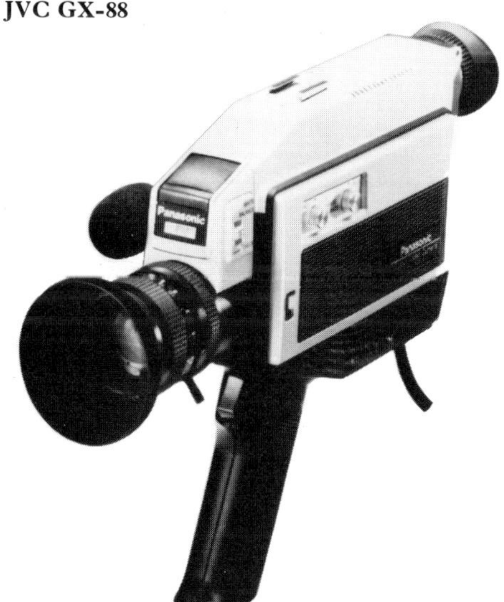

Panasonic
WV-3030 (Europe)
WV-3110 (USA)

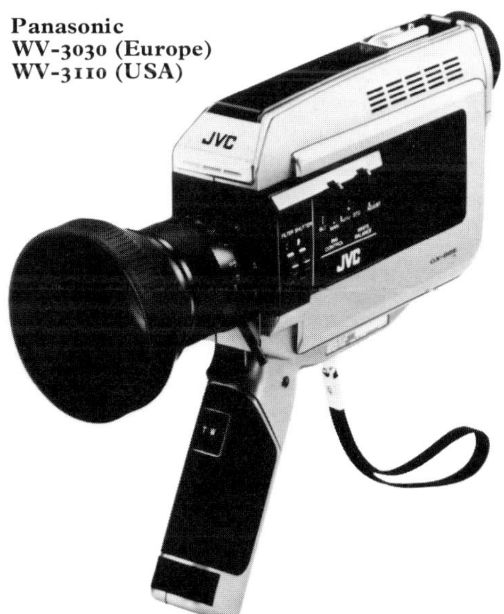

electronic viewfinder assembly, offered as an optional extra, would add to both weight and price.

The GX-33 comes in PAL and NTSC versions. The GX-88 has a SECAM variant, electronic viewfinder and a power-operated 6:1 zoom lens. It costs a lot more. It also features the super close-up facility which allows you to focus right in to the outer face of the lens.

The more sophisticated the features offered on a camera the more, as a rule, it will cost. Also, automation accounts for much of the price of the more expensive cameras, such as the Panasonic WV-3030. It has an electronic viewfinder, a 6:1 zoom lens with macro close-up facility and power zoom control, an automatic iris and automatic low-light-level compensation.

Advanced cameras

Video cameras vary in price and complexity from the basic home-movie type to sophisticated professional models. Drawing a clear line between the two types is difficult because the specifications for the cheaper cameras are improving year by year, and features such as power zoom lenses and semi-automatic colour-balancing, once limited to professional cameras, are beginning to appear on some of the cheaper models.

The biggest single difference between the two is in the way the camera is designed. All semi-professional and professional cameras have electronic viewfinders which, since they need no optical machinery but only a simple cable connection to the camera, can be located at the front of the camera body. This means that the operator can carry the camera on the shoulder while looking into the eyepiece, a much more comfortable arrangement than the pistol grip, and one which enables the camera to be held steady for longer periods.

Essentially, more complex versions of standard fittings make expensive cameras easier to use, more efficient in marginal conditions such as low light levels and more versatile. For instance, the relocated viewfinder on the shoulder-mounted camera can usually be pivoted in all directions, allowing the operator to aim it from the hip, or above the head, and even hold the camera on its side or upside down, and still see into the eyepiece.

Difficult and unusual shots often demand the ability to override automatic systems. As an example, in most situations an automatic iris is a sophisticated device, enabling the camera to react to changing light levels and so relieving you from constant rechecking and readjustment. What happens, though, when you want to zoom out from a close-up of a fairly dark subject to a brighter scene? As the light entering the camera increases, the lens aperture automatically closes down, and this can turn the dark subject into a silhouette. Some cameras have an aperture lock to allow you to guard against this. With the camera set in close-up you lock the iris so that it remains at the same setting as you zoom out. You then reset it to automatic. A backlighter compensator control is a useful alternative for situations where the subject is standing indoors in front of a window, or is outlined against a bright sky. This sets the lens open, usually by one and a half f-stops, so the exposure should be correct for the subject although the background will then be over-bright.

Most shoulder-mounted cameras, such as the Sony camera illustrated here, have a 'macro' position on the manual zoom control. Selecting it causes the components of the lens system to be rearranged into a different configuration, enabling it to focus on objects as close as the outer face of the lens for really detailed close-ups. However, you cannot zoom at the same time, and providing adequate lighting can be difficult as the camera casts its own shadow over the subject.

A red light, 10, called the VTR Indicator, Tape Run Lamp or Battery Lamp, lights up to indicate to the person being filmed that the tape is running. It also lights up during playback and blinks if the battery is about to run out.

The lens cap, 9, protects the lens from damage and prevents bright light from entering the camera during pauses in recording. On most shoulder-mounted cameras a white plastic lens cap is used, instead of a white subject, to adjust white balance.

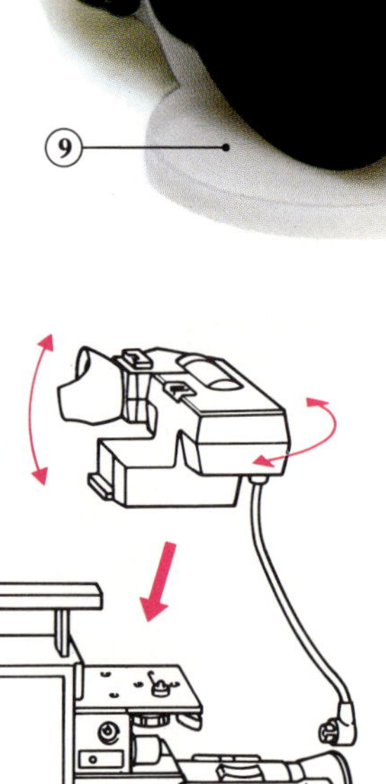

The electronic viewfinder, *right,* plugs into a socket just above the external microphone connection, and is fully adjustable. The viewfinder swivels and the eyepiece tilts up and down, giving scope for shots from unusual and difficult angles.

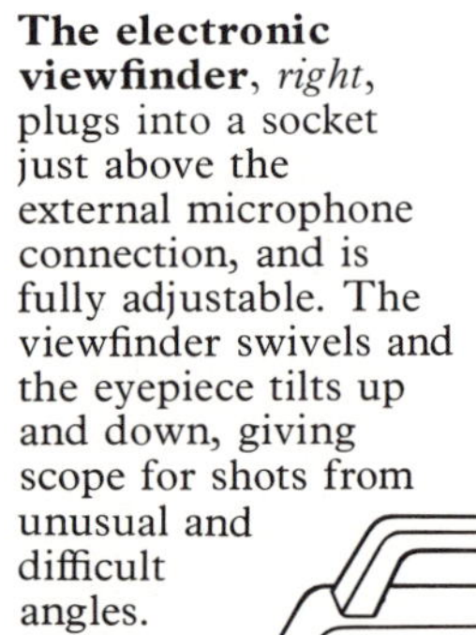

An external boom mike, or a video lamp, can be fitted into this receptacle, **1**. Since the built-in omni-directional mike is useful only to pick up general background sounds, a uni-directional boom mike is often needed. A light fitted to the camera is not so useful. It illuminates only subjects near the camera, and its beam may reflect off a shiny surface into the lens.

Low light conditions may make shooting impossible unless your camera is fitted with a control marked 'Low Light' or 'Sensitivity', **2**. This amplifies the video signal to create a brighter picture. Increased noise makes the picture appear grainier and less sharp, but it does allow the video camera to record an image where this might otherwise be impossible.

The iris control, **3**, has dual functions on this camera: push it in for auto and pull it out for manual control.

To fade the picture out to a blank screen, you usually shut down the lens aperture to the closed position, and you open it up to fade in. To overcome the difficulties of steadying the camera while manually opening and closing the aperture, some of the newer models have a fade control which does the job smoothly and easily. This camera has a pre-set auto-fade switch located on the back panel, **4**.

The camera body is shaped at the bottom to sit comfortably on the shoulder, **5**, but some manufacturers offer an optional, shaped pad that bolts on underneath the camera body and fits snugly around the operator's shoulder.

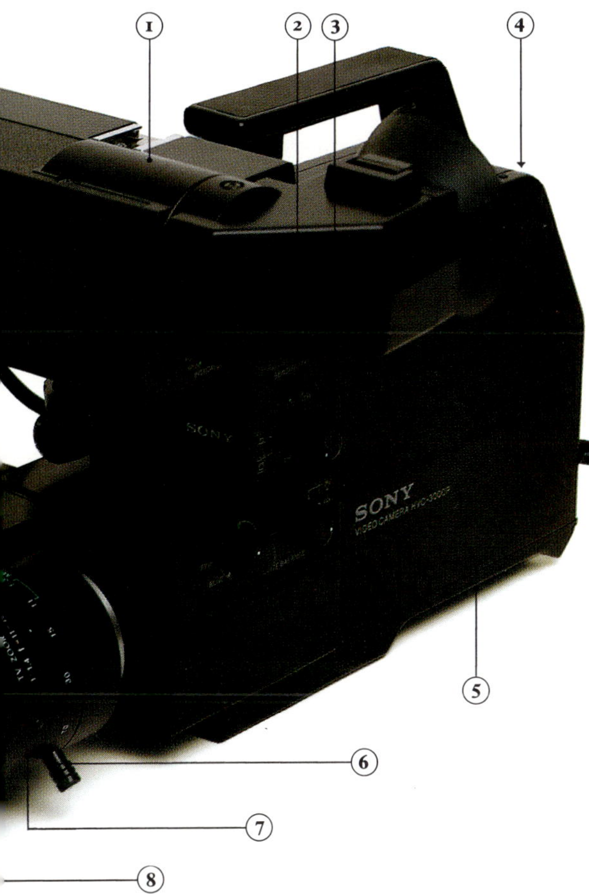

Every shoulder-mounted camera has a strong hand grip through which the right hand supports the camera, leaving the fingers free to manipulate the power zoom switch above it. The thumb operates the stop/start switch (called 'Tape Run' on this camera) located just behind the grip. The left hand moves the macro lever, the manual zoom and the focusing ring (**6**, **7** and **8**, *above*).

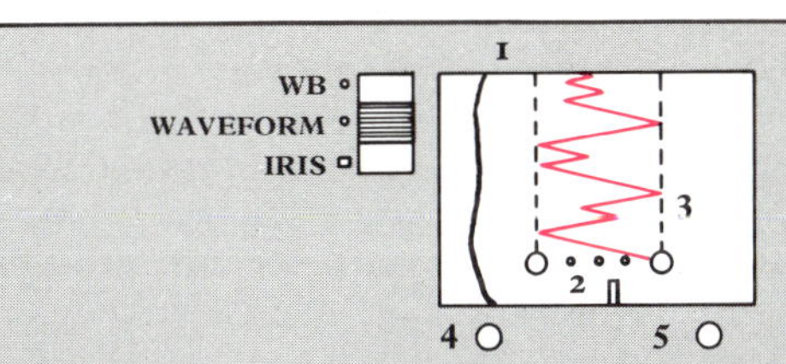

The brightness and sharpness of the picture in the viewfinder can be adjusted by two controls located beneath the eyepiece. With a third switch you select indicators to appear on the screen, *above*. The white balance (WB) indicator causes a white line, **1**, to appear; it moves to the left as the colour balance is adjusted. If the iris indicator is selected the white indicator, **2**, moves left or right as the iris is closed or opened. The waveform indicator causes the waveform, **3**, to peak at the lines only when there is enough light to record an image. Beneath the screen a yellow lamp, **4**, lights up when the light is insufficient and a red lamp, **5**, lights up when the tape is running and during playback.

On the market/Advanced cameras

The more complex shoulder-held cameras have developed directly from the vastly more expensive ENG (Electronic News Gathering) cameras now used by broadcasting professionals. With every new model launched on to the market, features originally introduced to meet television broadcasting requirements are coming into wider use. This makes it easier to predict the changes and improvements that are likely to appear on the next generation of video cameras.

Zoom lenses are now standard fittings on almost all video cameras, but the range of the zoom is being extended. Once a 4:1 zoom lens (allowing a 4:1 magnification of the picture in full close-up) was a popular fitting. Now most zoom lenses in this market are 6:1, with 10:1 and even 14:1 lenses beginning to appear on the costlier cameras.

Most cameras in this part of the market have power-driven zooms, but these work only at a single pre-set rate. The next step is an additional switch to select a slow or a fast zoom or, eventually, a pressure-sensitive zooming control. This allows you to vary the speed of the zoom to suit the shot, or even to vary the speed in a single shot by starting slowly, speeding up and then slowing down to a gentle finish, all using a single control.

Setting up the camera is also being made easier. Many semi-professional cameras have filters which can be slid into place behind the lens to compensate for colour temperature. White balance can be set simply by pressing a button with the camera focused on a test card, or a white object, or with a special white lens cap fitted on to it. Another useful feature is a memory circuit with its own long-life miniature battery. This maintains the control settings on location if the camera has to be switched off in between successive recordings in identical conditions.

Finally, more and more information will be displayed in the viewfinder, where the camera operator can see it most easily. Most cameras have some form of warning light to indicate when the recorder is running. Many have another to warn of light levels below normal operating limits. Other light displays can alert the operator to the fact that battery life is running out, or that the tape in the recorder is nearing its end.

Cameras like these are costly, and produced for the semi-professional operator rather than the home video enthusiast.

Sony HVC-2000 (Europe)
HVC-2200 (USA)

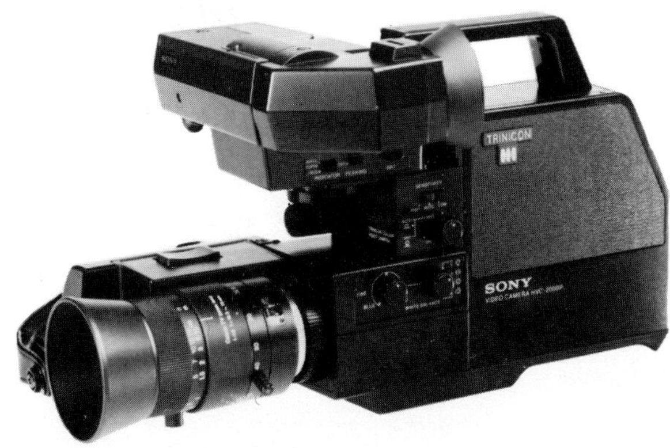

Shoulder-mounted cameras are now being fitted with the newer and more efficient Saticon tube. The immediate benefits are sharper pictures and better performance in marginal conditions. Current low-priced cameras such as Sony's HVC-2000, 2200 in the USA, still have the well-established vidicon tube. However, buyers willing to pay the higher price for the 3000 version will get all the features of the 2000 plus better low-light performance.

Philips V200

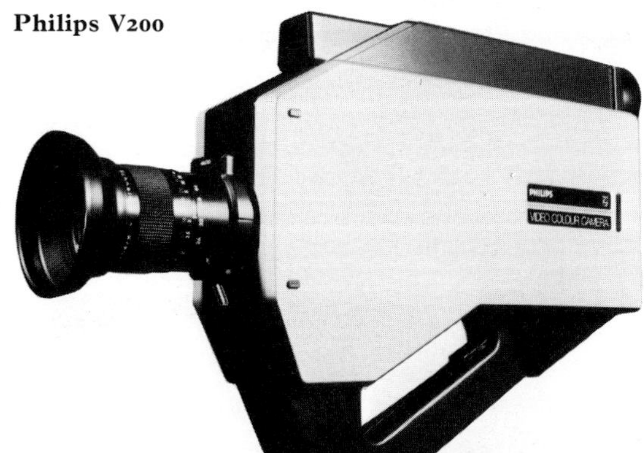

The Philips V200, a European model, is the exception to the rule that all low-priced cameras have single-tube systems. For the same price as the more elaborate single-tube models it has three vidicon tubes. In theory, a 3-tube camera should give a sharper colour picture – each tube receives a different primary colour – but 3-tube cameras have recurrent, though readily correctable, registration problems as a result of the tubes becoming misaligned.

Sony DXC-1800

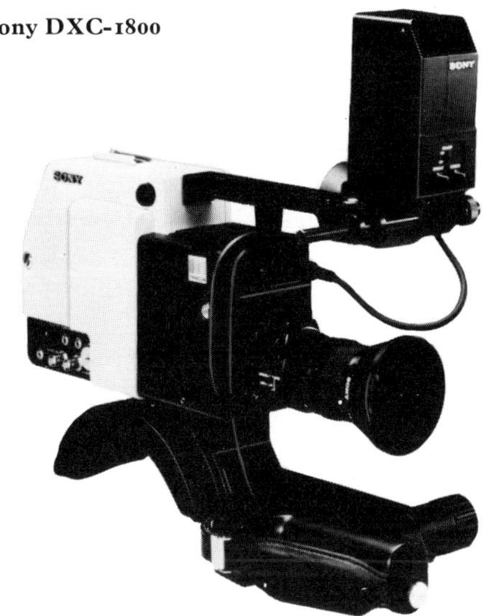

The Sony DXC-1800 costs 3 to 4 times the price of the pistol-grip cameras – but it has most of the features of professional ENG models: a Saticon tube, a 6:1, two-speed, power-operated zoom lens; an automatic iris with a pre-set auto-fade facility; automatic white balance with internal battery circuits to maintain the settings when the camera is switched off, and genlock capability.

JVC KY-1900

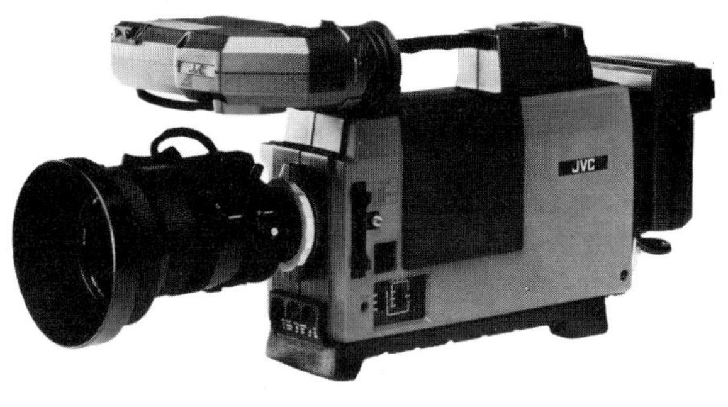

Although higher in price than the Sony DXC-1800, JVC's KY-1900 still offers remarkable value. It has three Saticon tubes, a 10:1 power zoom lens (with an optional 14:1), auto iris and automatic white balance adjustment. It also has genlock capability, which allows it to be linked via a special cable into a synchronized, multi-camera system, controlled through a vision mixer or special effects generator.

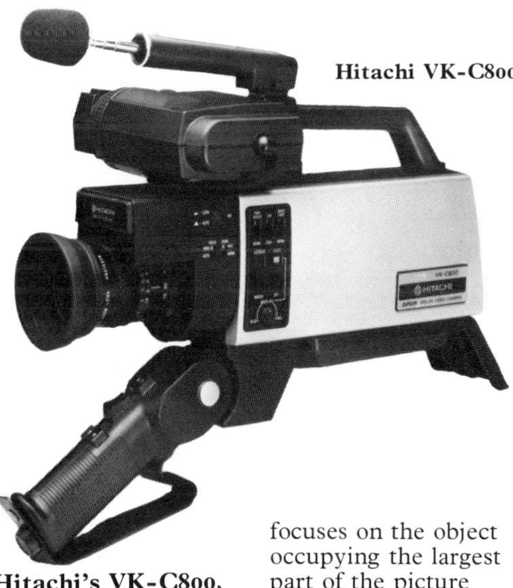

Hitachi VK-C800

Hitachi's VK-C800, has an unusual feature: automatic focusing. In theory this takes yet another chore off the camera operator's shoulders, but the manual over-ride may often be needed. The system focuses on the object occupying the largest part of the picture area, but there could be confusion if a zoom-out shot includes objects much closer to the operator; or if you want to shoot your subject past or through a closer object.

JVC S-100

JVC's S-100 is expensive – almost twice the price of many of the cameras designed for amateur use – but it has a 10:1 power zoom lens for really spectacular shots, such as zooming slowly in on a football crowd to pick out a small group, or even a single face, in close-up. It has an auto-iris, macro focusing and automatic white balance controls, and a genlock facility, enabling it to be linked into a studio system with other cameras.

Understanding your camera

A video camera works like a television receiver, but in reverse. It breaks down the picture framed by its lens into a coded video signal; the television receiver reassembles it into a complete picture. The heart of this process is the modern camera tube.

Fifty years ago, there was only the cumbersome television camera invented by John Logie Baird, which used a rotating drum and a light-sensitive cell to produce a crude, 30-line picture. Within 20 years cameras had electronic picture tubes called orthicons, which gave better pictures, but which were so large they needed huge lenses to focus the picture. Their average life was only 100 hours, so they often failed during live recordings, and their reaction to red was so poor that performers had to wear black lipstick.

The vidicon tube appeared in the early fifties. It was a fraction of the size, ten per cent of the price, and had 40 times the life of the orthicon. This is the tube that made possible the first colour television pictures. At first, picture quality was too poor for broadcast cameras, but refinements made it so successful that it is still in widespread use: Sony's Trinicon tube, used on the HVC-3000, illustrated here, is an advanced vidicon.

However, slow to respond to changes in illumination, especially at low light levels, the vidicon is gradually being replaced. The Plumbicon tubes of the sixties use more sensitive photoconductive materials, and

The light that enters the camera is focused by the lens, **1**, through the glass faceplate, **2**, at the end of the camera tube, and passes through a splitting device, **3**. This separates it into its three primary colours: red, blue and green.

The split beam passes through an electrode, **4**, and travels to the target plate, **5**, which, coated with photoconductive material, gives off electrons (negatively charged particles). These travel backward toward the electrode.

Each electron emitted by the target leaves behind an equal, but positive, charge, so the more intensely illuminated parts of the target emit more electrons than the poorly illuminated parts. At any moment, therefore, the target is an accurate electrical map of the scene framed by the lens: areas of high positive charge correspond to brightly lit parts; areas with lower charge correspond to darker parts.

The electron gun, 6, converts the electrical charge into an electronic signal. It focuses a fine beam of electrons on the target, and scans across it, progressing from top to bottom.

As electrons from the gun strike the target, many go to neutralize the areas of high positive charge, but more can pass through areas of low positive charge, and hit the electrode. When the target has been scanned, the varying stream of electrons reaching the signal plate reproduces the areas of light and shade in the scene framed by the lens, but in the order in which they were scanned by the beam.

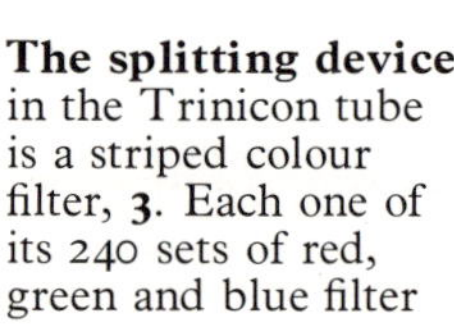

The splitting device in the Trinicon tube is a striped colour filter, **3**. Each one of its 240 sets of red, green and blue filter stripes blocks all colours except its own, which passes through, aligned for the second step in the process.

Each colour stripe aligns the electrons it transmits to strike a specific tooth on the Trinicon tube's comb-shaped electrode, **4**. Each tooth, when struck, transmits to the target an electrical signal corresponding to the colour received.

give a sharper picture and a faster response. They are so compact that four fit into a single camera: one to transmit the luminance (black and white) signal and one for each colour signal, red, blue and green.

In the sixties a single tube fitted into a compact camera made the first portable black and white video camera, and just under a decade later today's small video cameras became available, with all their colour circuits combined in a single tube. In a professional camera this will be a Plumbicon or Leddicon – highly sensitive and very expensive. Semi-professional versions will have a cheaper vidicon, or the Saticon, which is replacing it. This still suffers from lag, but perfection is not for away.

In many 3-tube cameras, light from the lens is split into primary colours by passing through a system of prisms. Each colour is focused into its own tube, and produces a picture and video signal. These are amplified and transformed into a single signal in an encoder.

The changing flow of electrons produces a video signal, which is passed down the cable to the VCR and recorded on the magnetic tape in the video cassette.

The video signal is re-created when the tape is played back through the TV set. The three colour signals are split up and fed into an electron gun in the TV tube, **7**. This fires three electron beams, which pass through a grille, **9**, aligned with coloured phosphor stripes, **10**, which coat the screen. The red phosphors, for example, glow red when struck by electrons from the red gun, the brightness of the colour depending on the intensity of the signal.

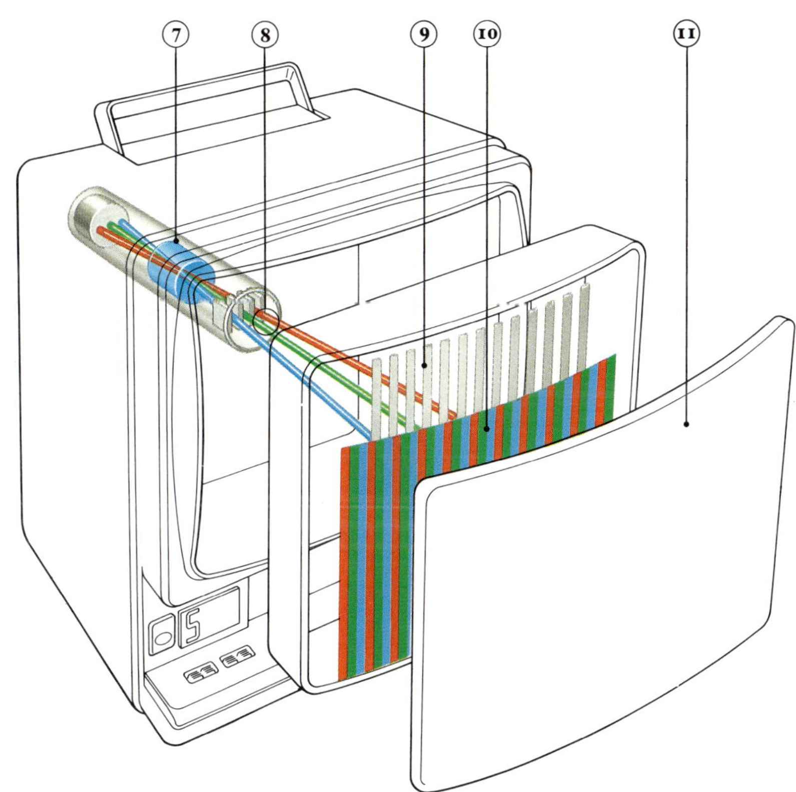

The TV screen, 11, glows brightly in areas where the electron beam striking the phosphors is intense; where the signal level is lower, fewer electrons strike the phosphors, and that part of the screen appears darker. These areas of light and shade correspond to those on the camera target, so the TV screen turns its electrical map into a copy of the scene.

The electron beams, 8, scan the screen at the same synchronized rate as their counterpart in the camera tube, but the number of electrons hitting the screen varies according to the recorded signal. The picture is scanned 25 or 30 times per second, depending on the transmission system.

Caring for your camera

A video camera is much more fragile than a cine camera. Its weakest point is the tube, which is not only the camera's most vulnerable part, but also the most expensive.

Dropping the camera is obviously likely to break or damage the tube. If you use the camera on a tripod, make sure the mounting is secure and all the adjusting screws fully tight after every shot. An unattended camera can flop over on a loose mounting, breaking or damaging the tube or lens.

Apparently trivial bumps, and even prolonged vibration, can disturb the precise physical alignment between the tube and the optical electronic focusing adjustments, resulting in a fuzzy picture. Three-tube cam eras are especially prone to misalignment. Leaving the camera in a face-down position can cause flakes from the internal coating of the tube to fall on the screen and form permanent black spots on the picture.

Video cameras deteriorate in heat or cold; so be careful not to leave a camera for long periods in places where the temperature is likely to fall to near freezing, nor to set it down near a radiator or other source of heat. Leaving it on the seat of a parked car could mean exposing it to overheating, if the sun shines on it through a window.

Damp is a great enemy of all electrical equipment, causing short circuits which can result in serious damage. If you have to shoot in the rain, use an umbrella or plastic bag to cover the camera. Special hooded raincoats are available for location work which cover camera and operator, but leave lens and eyes exposed.

Sudden changes in temperature cause condensation, so keep the camera in its case when not in use. It should be as horizontal as possible, and surrounded by packs of silica gel for insulation. These need replacing from time to time, or they can be recharged by warming in a low oven.

Keep the camera away from any strong magnetic fields produced by power transformers, for example, and beware of salt and sand at the seaside. They can play havoc with circuits and lenses.

Video cameras give their best results in bright sunshine, yet this can be a deadly enemy to the tube, particularly its light-sensitive surface. Even bright indoor lights can deliver enough energy to leave a burn inside the tube, if they shine in through the lens, so never rely on the automatic iris to provide complete protection; replacing tubes is costly. It is advisable not to record brilliant sunsets, or scenes with bright lights.

Whenever you finish a shot, close down the iris and replace the lens cap: it is just as easy to burn the camera tube when the camera is not switched on and the auto iris not working. Some cameras have an extra internal shutter between lens and tube. This should be closed when the camera is switched off.

Never tilt the camera upward at the end of a shot. This is a natural tendency with pistol-grip cameras, to relieve the strain on the wrist, and it can easily result in sunlight or artificial light entering the camera tube.

Ninety per cent of apparent camera faults are not faults, but mistakes. Even if you are an experienced camera operator, if a picture fails to appear in the viewfinder, check the connections and make sure the lens cap is not still in place, and that the power is on. Of the remaining ten per cent of problems, fully four-fifths are cable faults, so learn to carry out a simple continuity check, using a multimeter.

The remaining two per cent of problems are caused by electronic faults, and these usually require specialist help. However, tube burns, all too common, are caused by bright light producing a dark spot on the surface of the tube. Depending on the severity of the burn, you may be able to remove it by pointing the camera at a brightly lit white surface and leaving the system in the record mode for an hour or so. If this has no effect, try leaving the camera operating for a day or two, with the lens cap in place.

Some spots on the picture may be caused by specks of dirt on the lens, or even on the target surface of the camera tube. You can clean these surfaces yourself, but always use special lens-cleaning fluid and implements. Apply no pressure; this could remove the lens coating or disturb delicate adjustments. Keeping the lens cap on except when recording will reduce the chances of dirt, scratches and even finger-prints damaging the clean surface.

Remember that tube performance can deteriorate if the camera is out of use for too long. Use it for at least a couple of hours every six months to check that all is well. If it is not, take the camera to a dealer for advice and attention.

Caring for connectors
Cable faults account for almost 8% of all camera problems. These usually occur in the connections which link the ends of the cables to the equipment. Always connect and disconnect cables by holding the connector, not the cable, and never use force. Many camera connectors are designed in such a way that they cannot be disconnected accidentally. The 10-din camera connector, *right*, should be gripped by the O-ring, which releases it from the socket.

Inspect connectors regularly to make sure they are clean and that the pins are not broken. Camera cables have multi-pin connectors, and each pin carries a different camera function, *far right*; so if you drop or tread on a connector, make sure none of the pins is broken or misaligned.

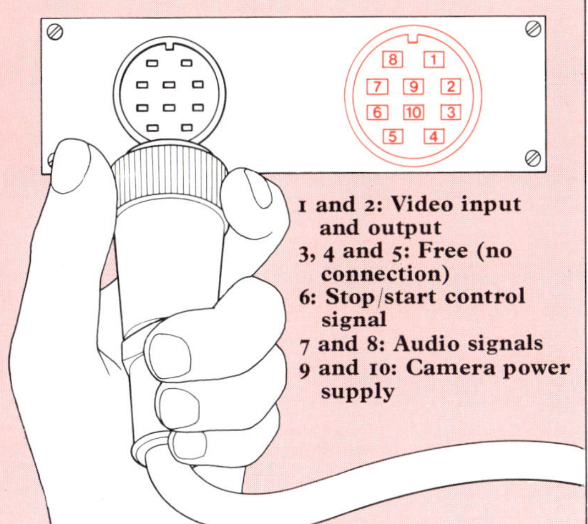

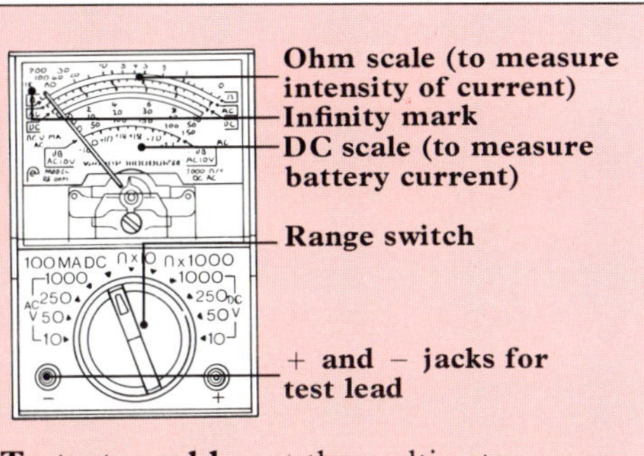

To test a cable, set the multimeter range switch to low resistance scale, and put a test lead probe on each of the cable connector terminals. If the needle deflects, indicating a finite resistance, the cable is intact. If it stays on the infinity mark there is a break in the cable, or one of the pins is not properly soldered to the appropriate wire within the cable.

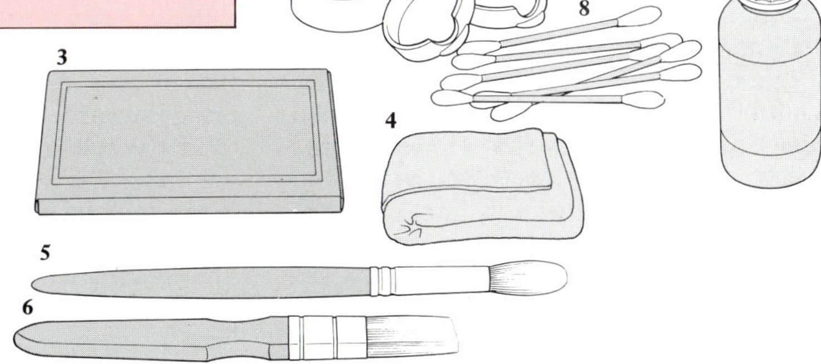

Keep your camera clean. Blow dust off the lens using a compressed-air product, **1**. Reducing contact minimizes the risk of scratching the surface. Use a large can, and hold it upright, to avoid leaving behind a smudge of vapour. A blow brush, **2**, is convenient to carry around, and both dislodges and clears dust. If the dust is too fine or greasy to blow away, use a cleaning tissue, **3**, or a well-washed, soft chamois leather, **4**, with a little lens fluid.

A sable artist's brush, **5**, is a useful dust-brush, and a stiffer brush, **6**, is needed to remove grit and fluff from the bodywork. Fluid cleaners formulated to dissolve clogging dirt in connections, and provide a protective coating for metalwork, **7**, should be applied sparingly with a soft cloth. Cotton buds, **8**, are invaluable for removing odd specks of dirt and grease.

The video director

With the simple addition of a camera to your video system you can provide your own input, and make your own recordings on any theme, any subject, any location which interests you. You can create a record on video tape of baby's first words, or an imperishable reminder of a family Christmas, to a travelogue of a specially exciting vacation, or to a commercial for your own business.

You may not have the expensive equipment or the teams of specialists the professional broadcasters use – but the next sections of this handbook show you how to achieve professional results from the simplest and cheapest equipment. The scope is limited only by your imagination and the originality of your ideas.

The first thing to remember about making recordings is that there is a minimum of complicated rules. This is the beauty of video: because the cost of the materials you use in shooting is so low, you can afford to practise and to experiment on a scale which is becoming far too expensive for cine film enthusiasts. Moreover, no one can say that a particular recording has to be made in a particular way, because that way is right and every other way of making the same recording would be wrong. There is only one hard and fast rule to remember: if it works, it is right. If the video tapes you make keep your audience interested, if the pictures are sharp and colourful and well composed, and if they fit together in a sequence that tells the story you want to tell, you have succeeded.

The following pages explain how to use a video camera, and describe the techniques you need to know to create a video production.

Although basic camera techniques are easy to master, putting them together into a good video production is difficult. However, it is usually simple to see what you have done wrong and to ensure that you don't make the same mistake twice, because video's instant replay facility allows you to play back the results within seconds of recording them; a great advantage. By the time a cine film comes back from the processors it is usually too late to reshoot the scene the way you realize it should have been done. One of the difficulties every prospective video director has to overcome is that, if you are eager, you will become your own sternest critic. Yet, because video recording is so creative, because the finished result is so much your own work, there is a powerful satisfaction in seeing your family and friends entertained by something you have conceived, planned and shot according to your own ideas.

Shooting a video production is entirely different from shooting a cine film. Neither is it like shooting professional-quality still pictures. This is because the sequence matters more than the individual shot, which means you can never take a picture in isolation, but only in relation to the shot that preceded it and those you hope will follow it. You cannot cut video tape and splice it together in a different order afterwards as you can cine film or audio tape; therefore you will have to shoot your pictures in the order they are to appear in your finished production.

As a home video director you have to be a researcher, planning what you are going to shoot, and why. You have to be a scriptwriter, working out the sequence of shots with which to build up the finished recording, as well as what you and your subjects are going to say. You have to be the camera operator, handling a piece of equipment that is often heavier and trickier to operate – to begin with at least – than a stills camera. You have to hold the camera steady, not for the fraction of a second the shutter is open to capture the exposure, but to follow the whole action you want to record, smoothly and professionally. You have to be a sound engineer, and a lighting expert, and an editor – because you need to know when and how to finish a shot, and how to start the next one.

This is where our handbook can help: it can give not rules, but practical advice, all of it derived from meeting, and surmounting, the obstacles everyone finds in trying to make recordings. Each two-page illustrated article takes the subject one step further: you need less expertise, fewer ambitious ideas and less experience to make a first tape of a simple subject such as a child opening Christmas presents than you would if you were intending to tape a friend's wedding as a present for him, or trying to capture for posterity the most exciting moments of your local club's tennis tournament. By starting off with the simplest of objectives and the simplest of equipment, it is easy to become expert in handling the camera and recorder, and then to progress to something a little more ambitious – and even more professional.

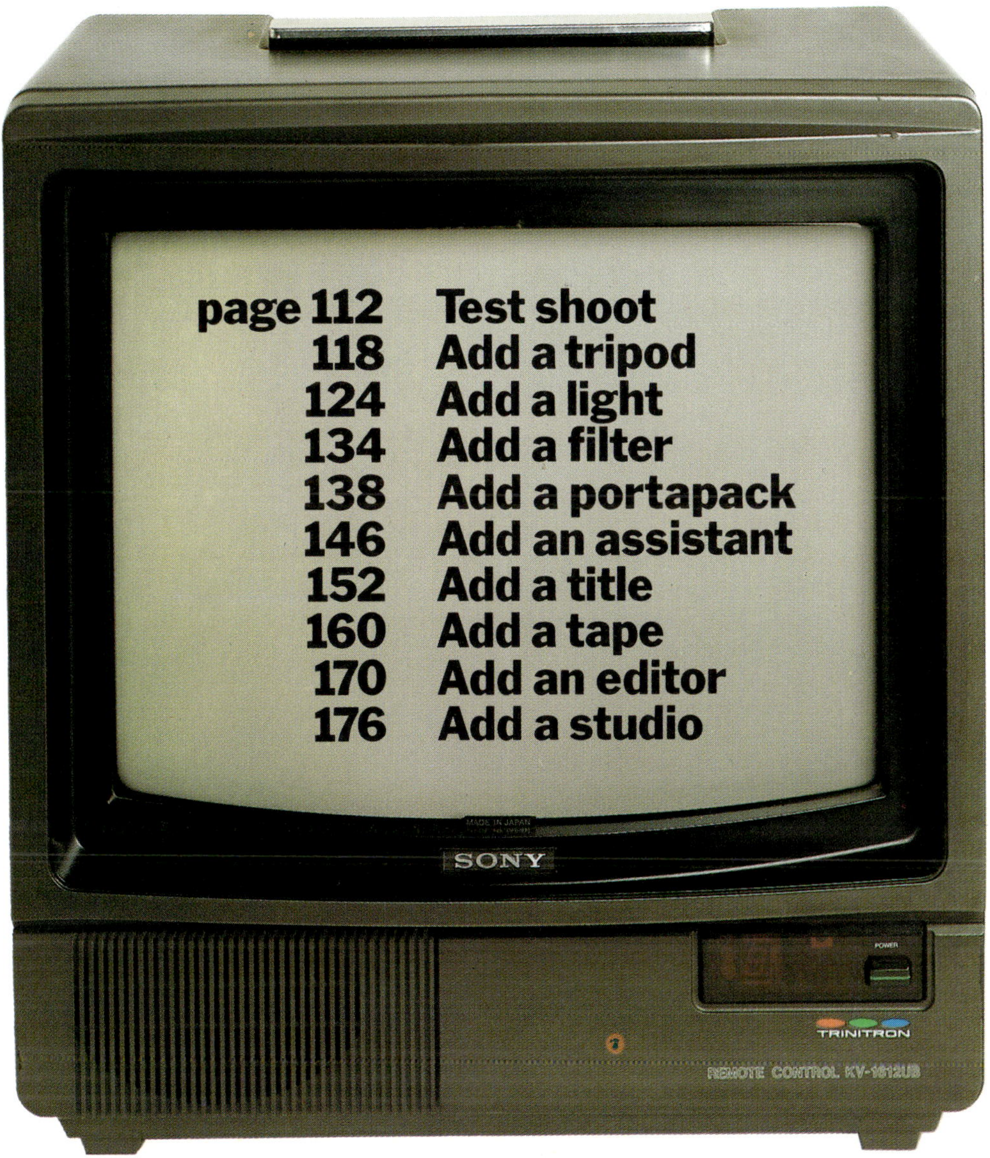

In each of the next sections we guide you through the making of a series of progressively more inventive video tapes. Each one will introduce a different facet of the video-production business – sometimes a different piece of equipment, and sometimes a different technique. However, the emphasis throughout will be on ideas, and on ways of using the new methods and the new assignments to make better productions. By the end of these assignments, while still making use of the cameras, recorders and other equipment now manufactured for the home user, the ideas and methods we feature will be those of the professionals. By the end of this chapter you should feel at home in a professional television studio. Some of the differences, such as the cost and complexity of the equipment, will be obvious; but so, by then, will the similarities. You may be working with video designed for the so-called amateur market, but the results you achieve, and the methods by which you achieve them, will be totally professional.

Handling the camera

Most people have used a camera for taking family snapshots. In some ways, using a video camera follows the same principles. The delight on the face of a child unwrapping a Christmas present; the animation of a party game, the colours and textures of the decorated Christmas tree – subjects like these offer opportunities for captivating still photographs and moving sequences alike.

Yet introducing movement can be a blessing and a challenge. While it allows you to record the whole of an action, rather than forcing you to pick a single moment to tell the story as best it can, good technique imposes its own discipline. When taking still pictures it is normal to tense up at the moment of pressing the shutter release to prevent camera shake; but to do this with a video camera would be fatal. Shooting a single sequence can take a minute or more, you can cope with it only by being as relaxed as possible throughout; by keeping your movements smooth and fluid, and by thinking ahead so that you are not caught unawares in the middle of a difficult shot.

The first step is to get the feel of your camera, learn to handle it, and become accustomed to the weight, the balance, and the position and operation of all the controls. You will get poor results if you grip the camera handle as if it were a loaded revolver, force your eye tightly against the eyepiece of the viewfinder, aim it and then shoot as quickly as possible. The best shots result from following through the action of your subject; so you need to be able to handle the camera without having to stop and remember which button to press.

Once operating the camera becomes second nature, you can concentrate completely on recording the best possible picture sequences. Avoid waving the camera around like a garden hose, trying to capture everything happening around you in a single protracted shot. Try it once, then play it back through the television. You will immediately see how distracting it is to watch a swaying, swerving picture on a stationary screen. Instead, keep the camera as steady as possible for each shot and try to move it only in between shots. At first, use the zoom only to frame each shot properly, so that you are close enough to capture the interest, but not so close that you cannot show all the parts of the picture needed for good composition.

Setting-up check list:
1 Connect camera to power adaptor
2 Connect power adaptor to AC supply; the Video Out and Audio Out plugs to the Video In and Audio In sockets on the VCR (check operating manual)
3 Connect the VCR to the TV set using a coaxial lead
4 Put the lens cap on to the lens to protect the camera tube
5 Set the colour temperature control to Indoors/Tungsten lighting
6 Switch on camera adaptor, VCR and TV
7 Point the camera at a well-lit subject and remove the lens cap
8 Open iris (check operating manual)
9 Check image on TV screen and in camera viewfinder
10 Zoom in on the subject, adjust focus and zoom out to frame a shot
11 Check the colour on the TV screen and adjust Tint/White Balance/Fine Tuning (check operating manual)

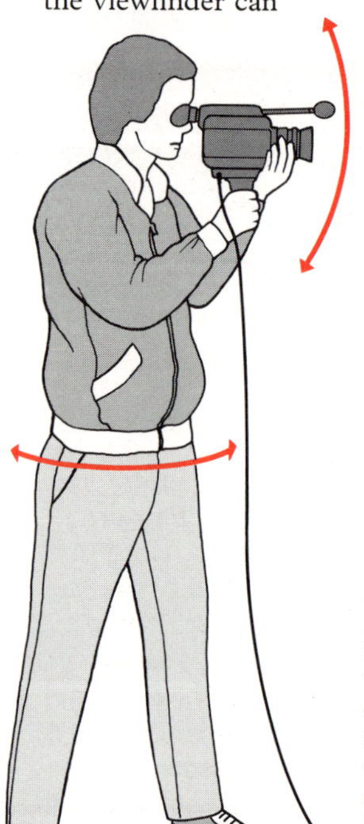

The pistol-grip camera is supported by the eyepiece resting against the forehead and the right hand holding the hand grip. Try and keep both eyes open, so that the eye which is not looking into the viewfinder can keep a check on what is happening outside the camera lens's narrow field of view.

Try to hold the camera as steady as possible, but do not lock yourself into a rigid position. You must try to relax as much as you can without shaking the camera. Keep your arms bent and wrists flexible. If your legs and feet are kept slightly apart you will be properly balanced, forming a human tripod. When you need to move to keep your subject in shot, swivel from the hips and bend from the waist. If your knees are slightly bent the movements will be smooth.

When you are ready to play back the recording, replace the lens cap, close the iris, and switch off the camera. Place it carefully either in its box or somewhere where no-one can trip over its connecting lead.

Framing and composing
Every picture you shoot needs careful framing. You must include all of the subject you need, to avoid missing important details, but beware of diminishing the subject by bringing in too much of the surroundings. Do not always put the subject right in the middle of the screen so the picture is split into equal halves and quarters; and do not tilt the horizon, or the horizontals or verticals of a room, unless for deliberate effect. Avoid tangents, or parallax, errors: a shot of grandma in a chair with a vase of flowers on a table behind her can look as if the flowers are growing out of her head; move sideways, bringing the flowers to one side of her. If there are several subjects in a shot: a baby, some toys and the cat sleeping on the rug, find a shooting position that brings them into an interesting group.

Study the framing opportunities (the squared-off shots) the camera operator sees in the family Christmas scene, *below*.

Making a test shot:
1. Load the cassette into the VCR
2. Check tape is at the correct starting point
3. Pick up the camera, hold it correctly and position the viewfinder
4. Set VCR to 'Record' (check operating manual)
5. Press camera operating switch and shoot test subject for 20 sec; press camera operating switch to stop
6. Close the iris, replace the lens cap, switch camera off and put it down carefully where there is no danger of anyone tripping over its connecting lead
7. Press the rewind button on the VCR to wind the tape back to its starting point
8. Press the play button on the VCR and check the replay of the shot on the TV
9. Repeat the process with a variety of different subjects of different colours to see how realistic they look on the TV screen.

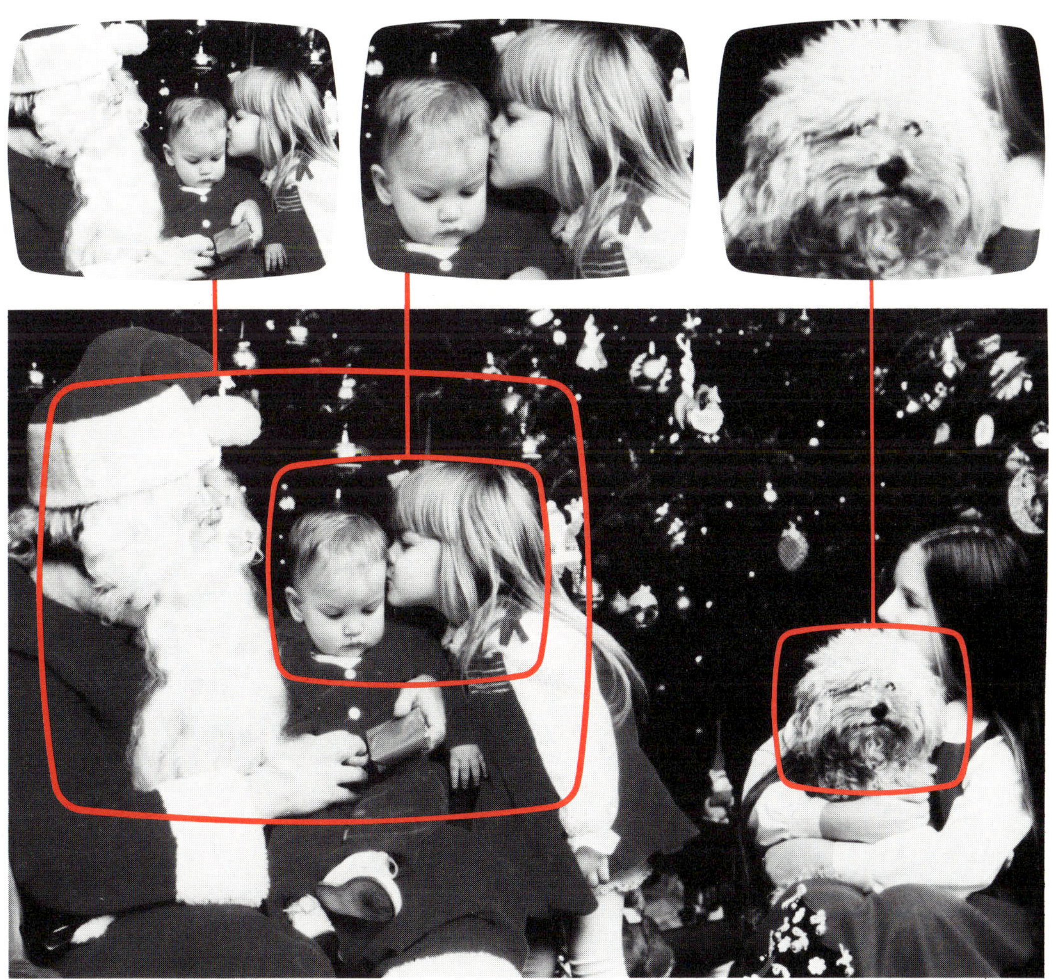

Camera angles

Keep your first recording simple; rely on scenes, situations and people you know. Shooting a family occasion may seem a boringly obvious subject, but it has many real advantages. Your family will cooperate while you try to get things right, they will be the most appreciative audience you will ever have, and your video tape will be a record of people and events which might otherwise be forgotten, for nothing can bring the past alive as effectively as moving pictures.

Start as you mean to continue, with a thoroughly professional approach. Remember that those who watch your finished video tapes may not know much about recording techniques and problems, but they are used to a diet of highly professional television every evening. So begin with a proper story line in your mind, if not on paper. What are you going to show in each sequence? How are you going to begin? What will your opening shot be?

Try and establish a length for your recording, and a rhythm. Usually it is best to begin fairly slowly, to attract your audience's attention and involve them in what they are seeing on the screen. You can build up pace through the middle of the recording, slowing down again before the end to prevent a finish which might be too abrupt. Remember that there is more to planning a recording than deciding the basic shape: you need a few seconds' pause every so often.

Your sequences will be made up of individual shots. Try to make each shot as interesting as possible: think of unusual viewpoints, or different camera angles, to enliven an essential shot which might otherwise lack visual interest. Make sure, however, that every shot you use is justified in terms of the story you are trying to tell. Your audience will see only the pictures you choose to show them in the order you record them, so a sudden change of subject, or a jump in time or place, may be distracting or confusing unless your sequence explains it properly. Try to vary the types of shot you use to keep your sequences of pictures as interesting as possible.

Finally, remember that the sounds picked up by the microphone on your camera will be the loudest noises, and these may not relate to the pictures seen through the lens. Always try to ensure that the sounds your audience hears are in character with what they see.

Indoor lighting
Generally speaking, the higher the light level, the better the pictures. Video cameras give their best results out of doors on sunny days; so if your first indoor test shots seem gloomy and flat don't be disappointed – remember how fierce the lighting has to be in a TV studio.

Try shooting your subjects by a window, where they can be illuminated by direct daylight. Adjust the colour temperature for daylight if you do this, and don't point the camera at the window or your subject will appear as a silhouette. Alternatively, you can use what daylight there is and back it up with artificial room lighting – though this can cause colour adjustment problems. Do you set for room light or daylight? Experiment, using your colour TV screen.

Try bringing all your lighting resources to bear on the subject: table lamps, standard lamps and, especially, spotlights or desk lamps with hinged arms – but never point the camera at any light source.

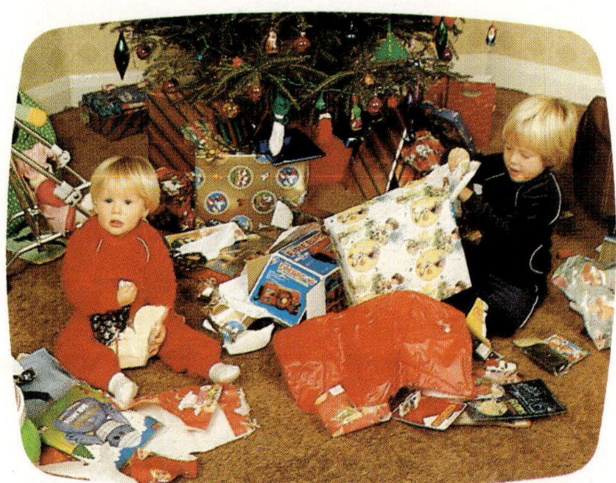

When taking a shot while standing, brace yourself, and the camera, against walls or doorposts, tables or the backs of chairs. This will help keep the camera steady, especially when zooming into close-ups or panning across the scene.

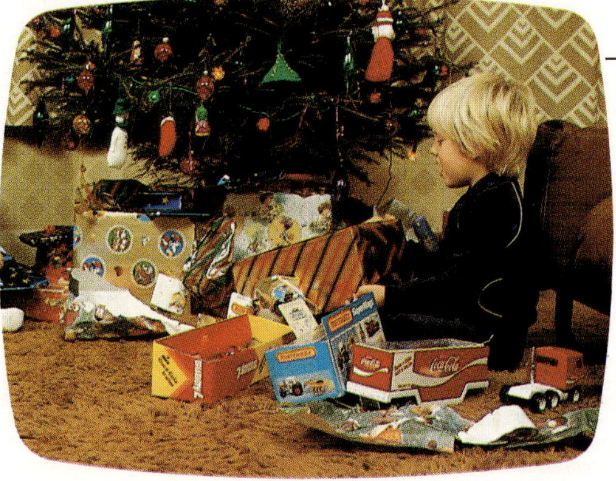

Stand on a chair for an interesting high-angle shot, or sit comfortably on the top step of a step-ladder. Make sure the ladder is really secure before you mount it, and brace the camera firmly on your shoulder, or steady it on your knees.

Starting and ending
Always begin a shooting sequence with a picture that will introduce the subject you are about to record. A good opening shot for this family Christmas tape might be a close-up of the wrapped-up presents at the foot of the Christmas tree, or the star at the top coming into focus. The closing shot should round up all you have been showing in your sequence. For example, the Christmas tree star might go out of focus and then fade gradually to black.

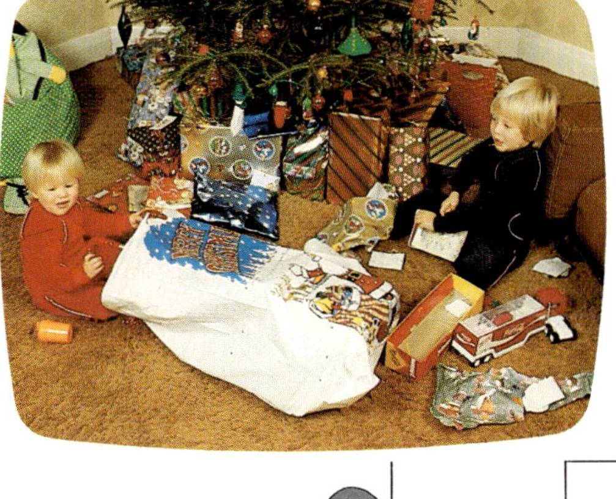

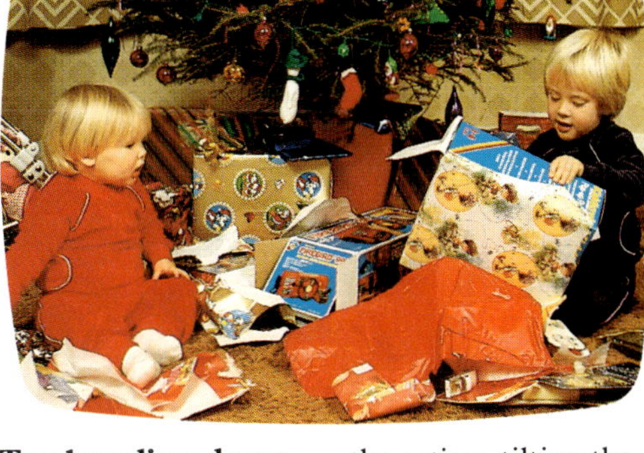

Try kneeling down, with the camera in your lap, to get a low viewpoint. Look up at the action, tilting the viewfinder upwards so that you can see the picture. For a slightly higher viewpoint, shoulder-mount the camera.

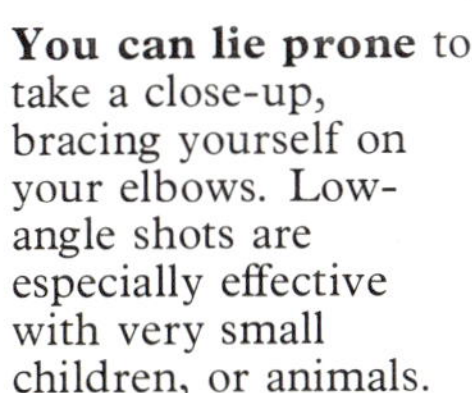

You can lie prone to take a close-up, bracing yourself on your elbows. Low-angle shots are especially effective with very small children, or animals.

You need not take every shot from a standing position. A low viewpoint can portray a child's-eye-view of the action; from a high viewpoint you can take in most of the room, or pick out close-ups, clear of people's heads.

Developing technique

Always aim to communicate as directly as possible with your audience. That means building up a sequence of shots in such a way that the viewer is never distracted by the mechanics of the system, nor by mistakes in shooting: wobbles of the camera, for example, or subjects which disappear out of one side of the picture frame, or wander in and out of focus.

The screen on which your recording will be watched is a fraction of the size of the screen you would use for Super 8 home movies. In addition, video produces pictures by electronic means, so the fine detail is not as sharp as in cine film. For these reasons, the best video pictures are usually close-up shots where the action fills a large part of the screen. Yet your recording should be more than a series of tight close-ups. After a while the audience will lose track of where the different close-ups fit into the overall view. Wider shots will remind them of the context. If you shift to a new location, a wide shot will tell your audience where they are now and what they are looking at, so they will then understand a set of close-up shots. You could, however, start on an intriguing close-up, to add mystery or drama.

Vary the subject matter of the picture. Cut from an active shot of someone doing something to a passive shot of someone watching them. Although you will be looking for candid material (that is, people forgetting, even for a moment, that they are on camera), include the occasional deliberately posed shot to add to the variety and the interest. Use such shots sparingly: moving snapshots are a waste of video's possibilities.

Widen your range of shots by including simple camera movements. If someone moves from one side of the room to the other, follow him or her as smoothly as you can. Use your other eye to check what is ahead of the camera, and frame the shot to allow space in front of the subject, so that the camera 'leads' the person through the movement.

Plan each sequence carefully. Stop the tape when each shot is finished, and look around for the unexpected: if you can fit an extra shot into your sequence, shoot it, but remember: however interesting the subject, it must work as part of the whole. Finally, always remember to shoot as though you are telling a story: keep a definite beginning and end in mind.

Panning

Swinging the camera horizontally across a scene, or panning, is useful where you cannot show a complete scene in a single shot – either because the angle you have to cover is too wide, or because you cannot get far enough away with the camera. If, for example, you wanted to show everything happening in a room you would either have to show it in a series of single shots, or pan right across the room in one smooth movement. The movement must be as smooth as possible, starting and finishing slowly and without jerking the camera. Do not overshoot your finishing point, and never move back in the opposite direction as this is very distracting for the audience. Unless you are deliberately looking for a fast blur effect, pan as slowly as you can. Try a few practice pans and you will find that the one that feels too slow as you record it will be the one that looks right on the screen. In time you will learn how to judge your speed of movement exactly.

1

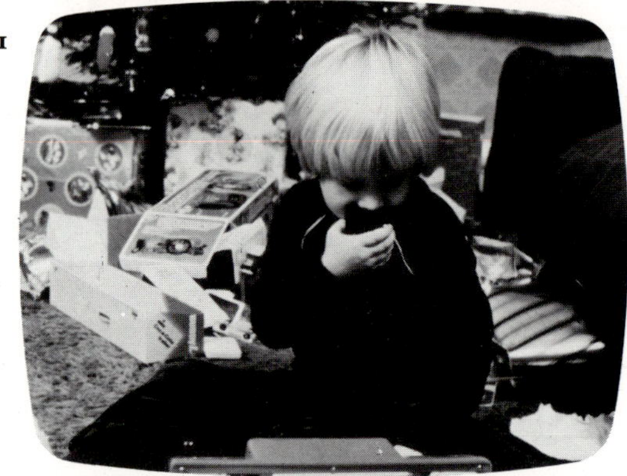

Use camera movements to maintain interest. Instead of cutting from a close-up of one subject – such as the child, *above*, to another, move from one to the other by zooming back and linking them in a wider shot, *below*.

2

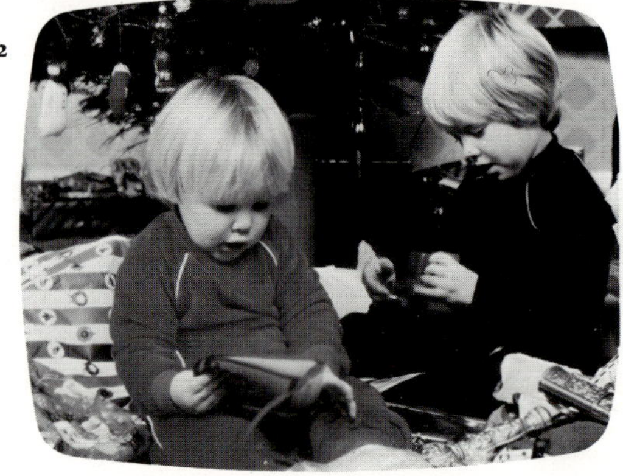

Defining shots
Extreme close-up (ECU)/Big close-up(BCU): eyes, nose and mouth from the middle of the forehead to just above the chin
Close-up (CU): the entire head down to just below neck level, *below*
Medium close-up (MCU)/Chest shot/Bust shot: head down to just above the elbows
Mid-shot/Medium shot (MS): head to waist
Medium long shot (MLS): head to just below knee level
Long shot (LS): subject in full length

3
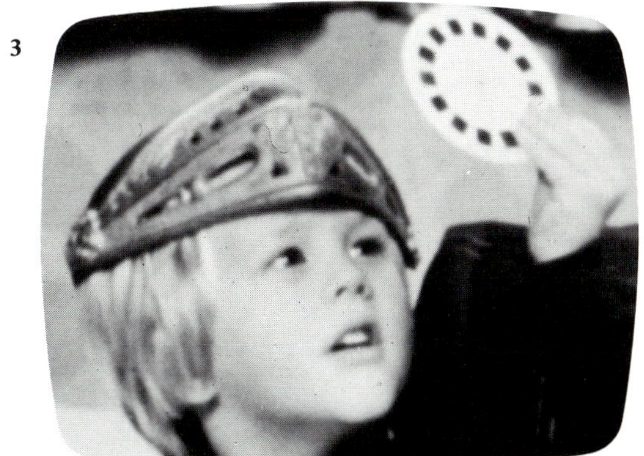

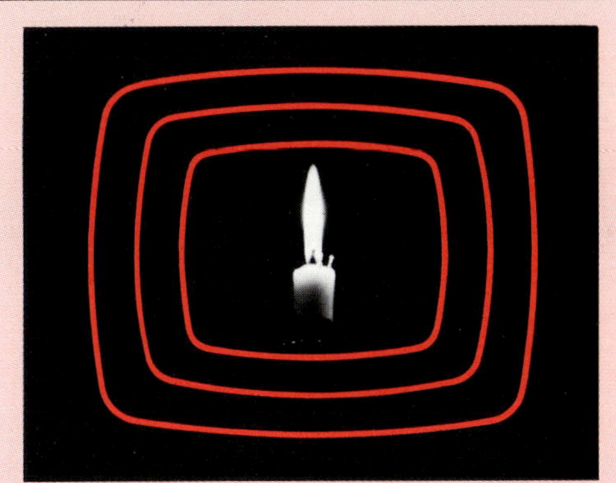

The zoom lens is so constructed that if a subject is correctly focused in close-up, it will remain in focus as the lens is moved out to the wide-angle position. This is a useful facility: it allows you to begin with a wide-angle shot showing the entire subject, such as a Christmas tree, then zoom in to show a detail – perhaps the star at the top. Alternatively, you can begin with the detail, *above*, and pull out to show it in context.

Subjects such as the Christmas tree, *right*, need a vertical tilt rather than a horizontal pan. This could start at the star on the top of the tree, move slowly down to show the decorations, and end at the parcels at the foot. Hold the start of the shot for a few seconds, tilt the camera slowly and smoothly and, taking care not to overshoot the finishing point, hold the shot there briefly. You could end the sequence by tilting upwards from the foot of the tree.

If an activity you are recording, such as a child unwrapping a toy, goes on for too long, what do you do? Stopping the camera momentarily causes a jump in the action so instead, bridge the gap with a cutaway: focus on another subject, then pick up your shot again.

Camera stability

The most obvious difference between home video recordings and those shown on television is that professional video pictures are steadier. A professional camera operator can hold a camera on one shoulder and still record a far smoother shot than the beginner, but for really smooth sequences the camera is mounted on a tripod. The tripod screws into the base of the camera body – pistol grips usually detach or fold away to allow this. Some have a screw for the tripod in the bottom of the pistol grip but they are less stable since they may move during shooting.

Once the camera is fixed so that it cannot move while the shot is in progress, your work will look a great deal better. However, there are disadvantages to using tripods: a carefully composed shot can be ruined if the subject moves out of frame, yet moving the camera from shot to shot takes time, so you could miss some of the action. Also, in a crowded room someone may trip over one of the tripod's legs and ruin the shot, or damage the camera.

These disadvantages can be minimized by choosing the right tripod for the job. Never economize by using a tripod designed for a still camera: a heavy video camera may be unstable on a lightweight tripod. Moreover, the head is not designed so that you can move the camera smoothly while recording a shot. Buy a tripod which has been designed for video use, with a head which allows you to pan the camera smoothly, to tilt it up and down, and to follow the movements of a subject in close-up. A tripod which suits your camera and allows you to do all these things will enable you to produce really steady zooms, to change focus to shift the viewer's attention from a close-up object to one further away, and even to fix the camera in position and experiment with trick frame effects. For instance, you could stop the camera and restart it when the subject has moved, or gone out of shot altogether. If the camera is absolutely steady on its tripod, the background will remain stationary while you create interest with a moving subject.

The new-found steadiness of your tripod shots will make any hand-held shots you include in a sequence seem even shakier by contrast, so if you need to record hand-held shots, wear a body brace. This is a metal prop that can be fixed to the bottom of a shoulder-mounted camera and rests on the abdomen.

A tripod head, 1, may be a friction type (comprising a fixed lower plate and a rotating upper plate), or a more sophisticated fluid head (so called because the upper plate rotates on a fluid base). A side screw, **2,** releases the head for panning; a handle at the back, **3,** is twisted to tilt the head up and down.

The height of a tripod may be adjustable from 22 in (56 cm) or less to over 70 in (1.78 m). There may be a pedestal, a centre shaft which is cranked up and down, **4,** and the legs may be extensible, **5.**

The feet, 6, may be converted from a non-slip rubber dome to a spike.

Some tripods can be locked on to a set of wheels, or dollies, **7,** which may be collapsible and fitted with brakes, direction locks and guards, **8,** to lift stray cables. A tripod head or camera may be fixed to the centre joint, **9,** or this can carry a portapack.

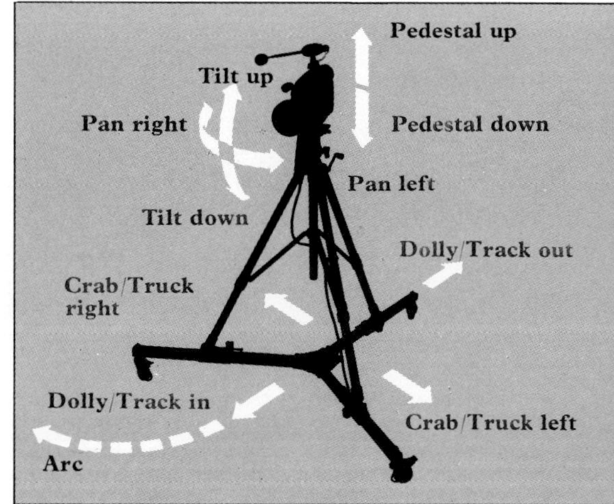

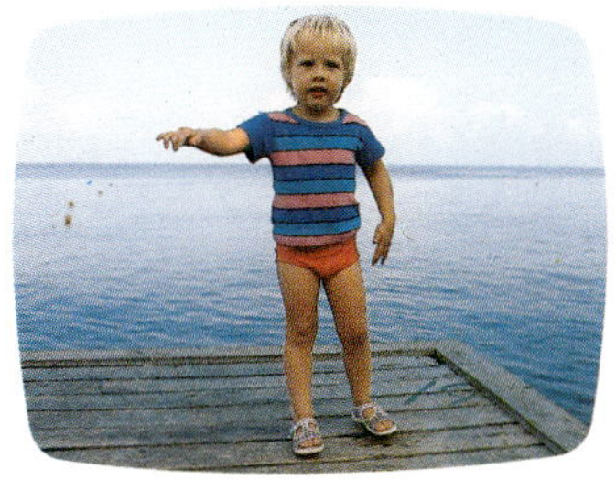
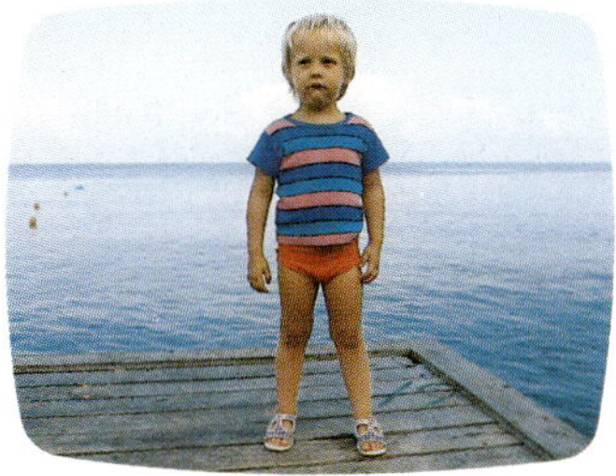
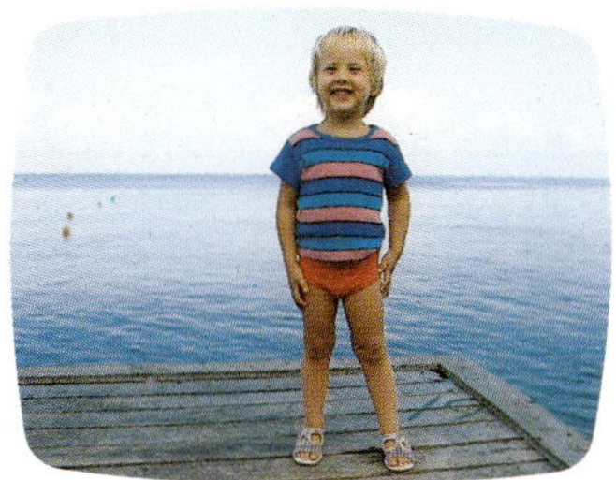
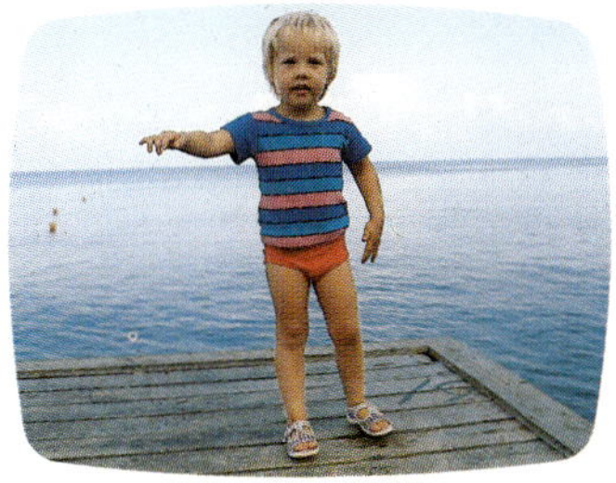
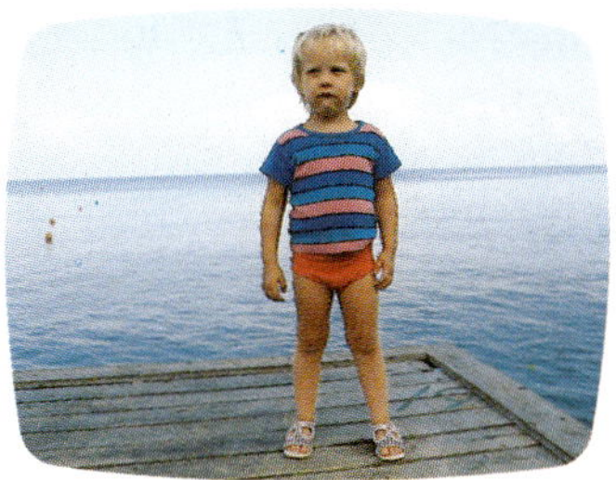
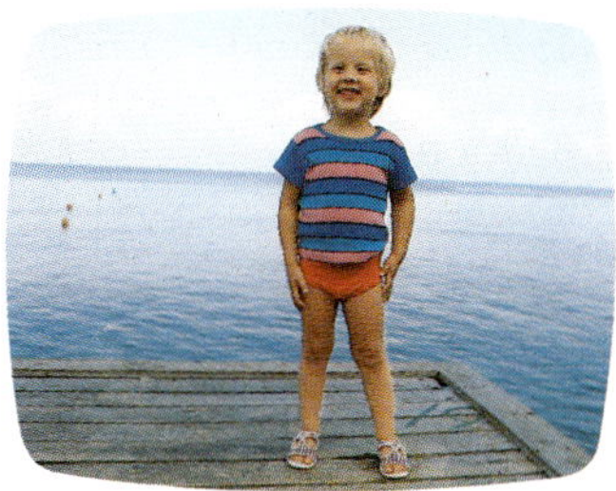

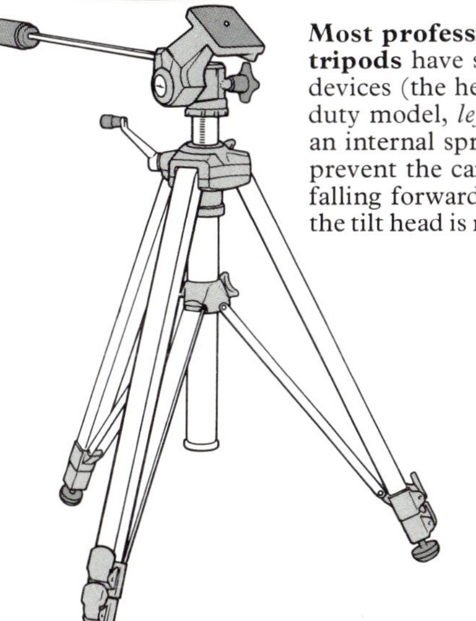

Most professional tripods have safety devices (the heavy-duty model, *left*, has an internal spring) to prevent the camera falling forward when the tilt head is released.

Most good tripods have a spirit level on the platform so that it can be set level on uneven ground. If you then set the head level (some have a second spirit level for this on or near the head) you can keep the horizontals level. Compare the horizon in the hand-held sequence, *above*, with the tripod-mounted version, *top*.

Tripod legs are individually adjustable for shots on uneven ground. Many have legs that splay out, *right*, for low-level shots.

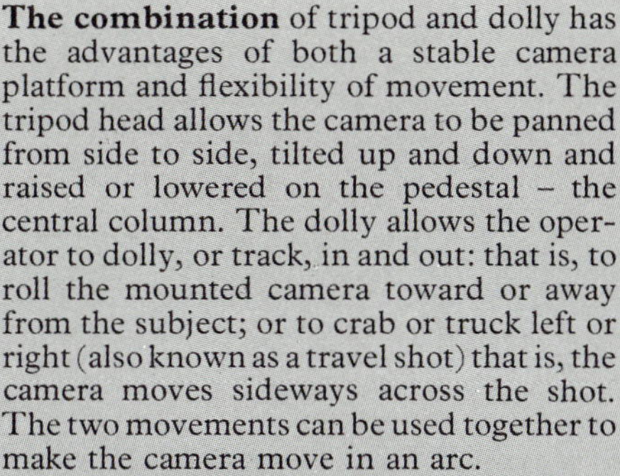

The combination of tripod and dolly has the advantages of both a stable camera platform and flexibility of movement. The tripod head allows the camera to be panned from side to side, tilted up and down and raised or lowered on the pedestal – the central column. The dolly allows the operator to dolly, or track, in and out: that is, to roll the mounted camera toward or away from the subject; or to crab or truck left or right (also known as a travel shot) that is, the camera moves sideways across the shot. The two movements can be used together to make the camera move in an arc.

For steady movement, the surface across which the dolly is moving must be completely smooth.

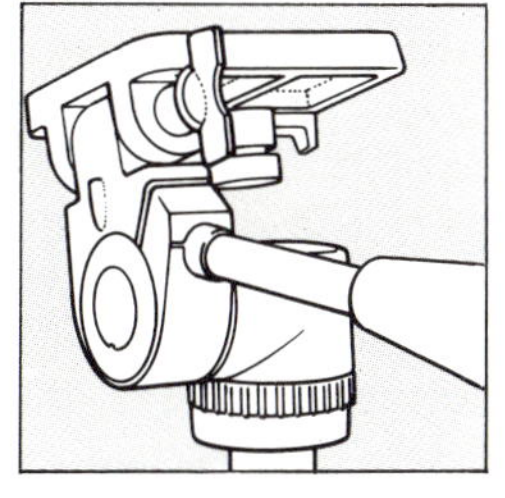

Cheap friction heads with no cushioning between the plates, tend to be jerky. Moving from pan to tilt entails loosening both controls while supporting the camera.

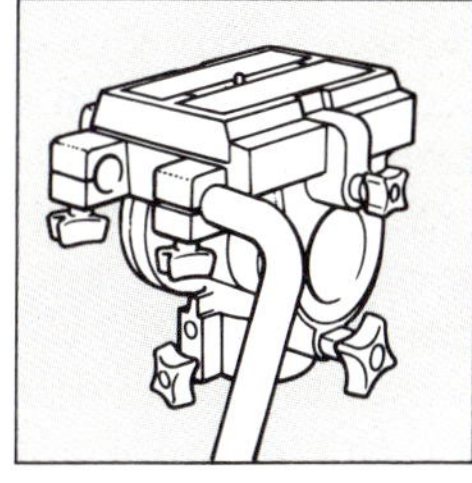

Fluid heads give a smooth action. When panning, the start and stop are barely perceptible. Moreover, you can release the controls to pan and tilt at the same time.

Camera mobility

All tripods designed for use with movie or video cameras have certain features in common. They all have three extensible legs, adjustable so that the camera platform is flat even when the tripod is standing on uneven ground. They have feet ending in rubber pads or screw-out tips to help grip on smooth surfaces. They usually have a pedestal to enable the camera to be raised or lowered. A pan-and-tilt head allows the camera to be swivelled from side to side or tilted up and down; locked in one position or freed in either, or both, planes, and the adjustment is variable to allow variation on the amount of frictional drag on the camera head.

It seems logical that in order to achieve the smoothest camera movement, there should be no friction at all on the camera head. This is not so. If you free the pan adjustment the resulting movement is anything but smooth. With nothing to push against, the slightest tremor in the hand is immediately transferred to the camera and the shot. Moreover, it is difficult to end the pan exactly where you want it to end. If you tighten the friction adjustment so that the camera is difficult to move, this smooths out the movement.

However, frictional drag on the head causes difficulties at the beginning and end of a pan. Sliding friction is less than static friction, so you find yourself increasing pressure on the camera until it suddenly starts to move and the result is a jerky beginning to an otherwise smooth pan. Similarly, toward the end of a pan you use less and less force until the head suddenly sticks: another jerk. A fluid-head tripod solves both problems: it uses either a hydraulic head or a nylon semi-fluid head to produce a smooth transition from beginning to end of a movement.

The camera can be moved bodily in relation to its subject. It can either dolly in or track from a wide shot to a close-up without zooming, or it can dolly out or track away from the subject, or even crab or truck sideways across it. For all these movements the camera needs to be supported on wheels on a smooth surface. If you try to move a dolly across a rough surface, the surface defects will be obvious on the screen. Professionals often ensure smooth movement by laying lengths of miniature railway track for the camera to run on. You can achieve the same effect by having someone push you along in a wheelchair or on a suitable cart.

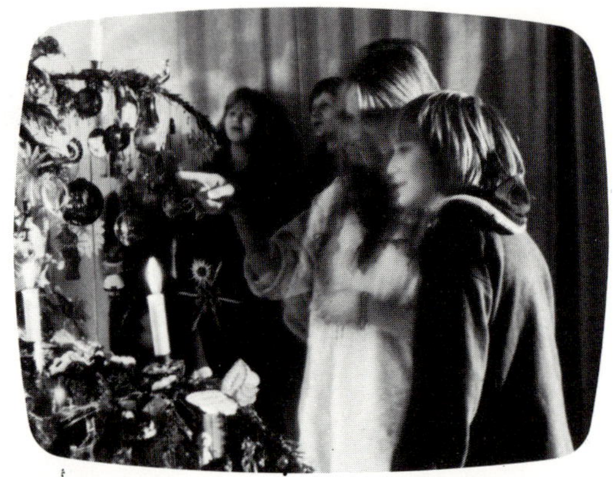

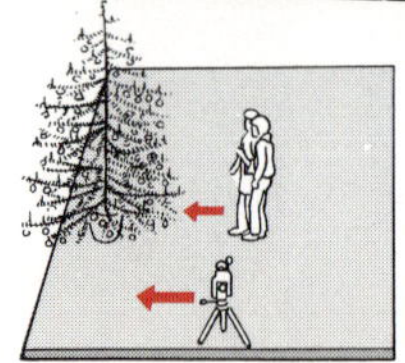

If your subject has to move through a shot, follow the movement by locking the dolly wheels in the transverse position and trucking parallel.

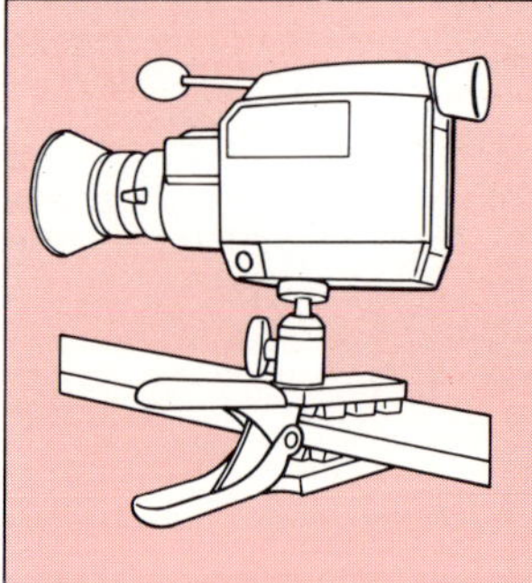

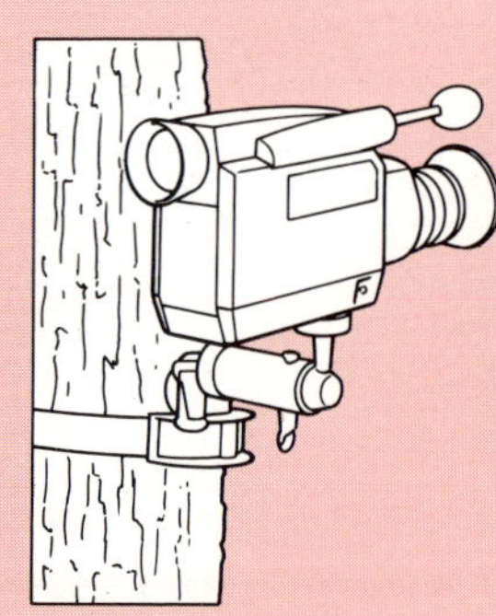

For inaccessible shots, a tripod head fixed to a spring clip, *above left*, or even a webbing belt tightened between two supports, *above right*, will anchor the camera.

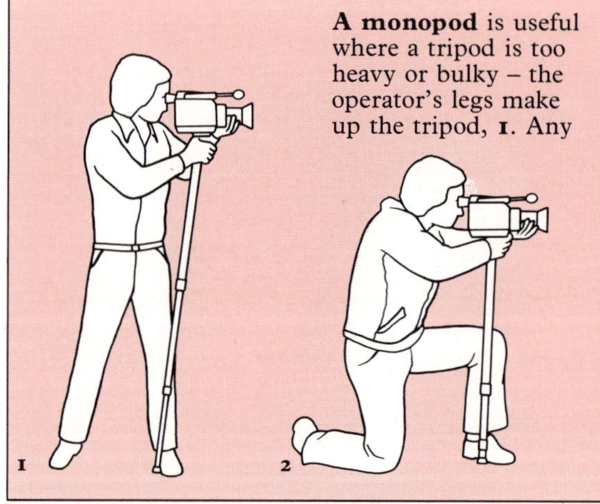

A monopod is useful where a tripod is too heavy or bulky – the operator's legs make up the tripod, **1**. Any

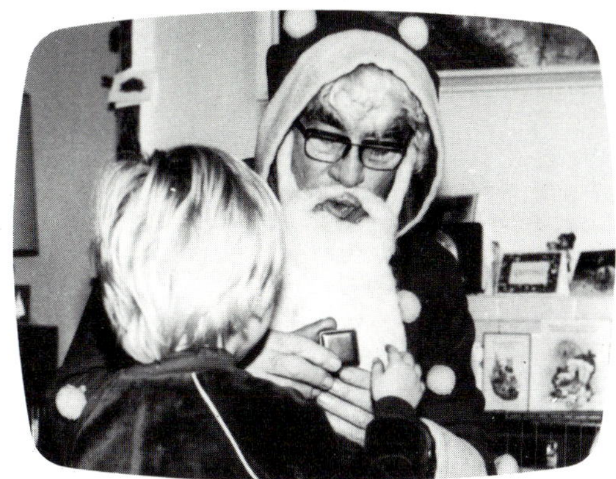

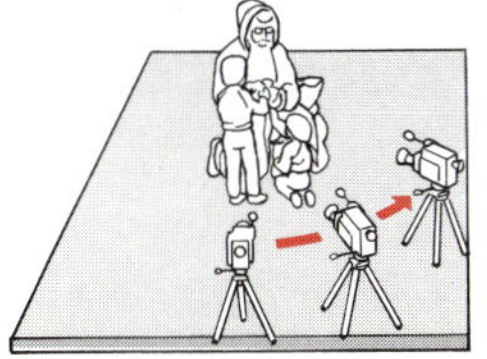

Arcing enables you to alter perspective without changing focus. You use the crabbing movement to describe an arc around the subject.

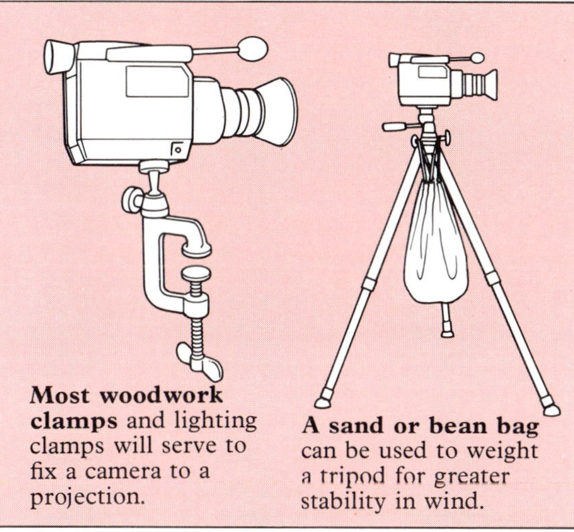

Most woodwork clamps and lighting clamps will serve to fix a camera to a projection.

A sand or bean bag can be used to weight a tripod for greater stability in wind.

comfortable position may be adopted, but keep the camera steady by bracing the monopod against a knee, **2**, or a foot, **3**. A strap around the neck gives extra support and rigidity if you slip your right arm through it.

Monopods are lighter, more compact and less expensive than tripods. They may extend to 70 in (1.78 m) and fold to as little as 25 in (0.63 cm). Telescope the monopod, if shooting in a vehicle, and brace it against the upholstery.

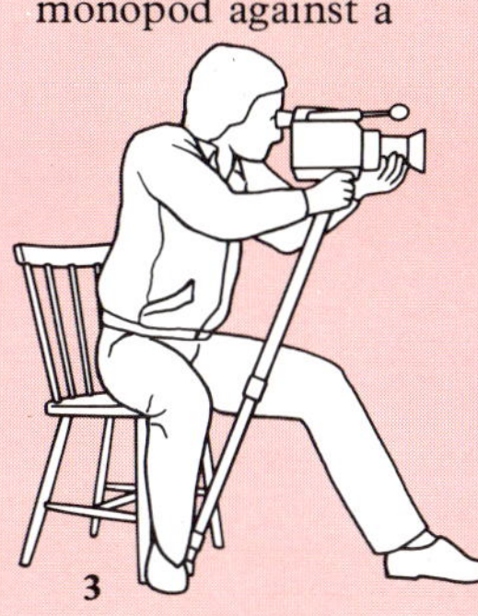

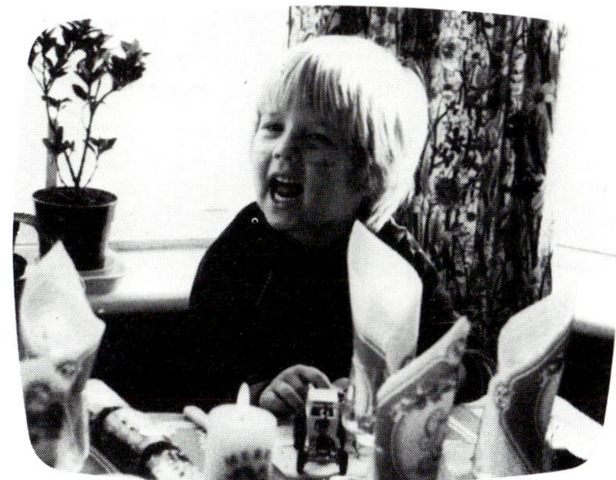

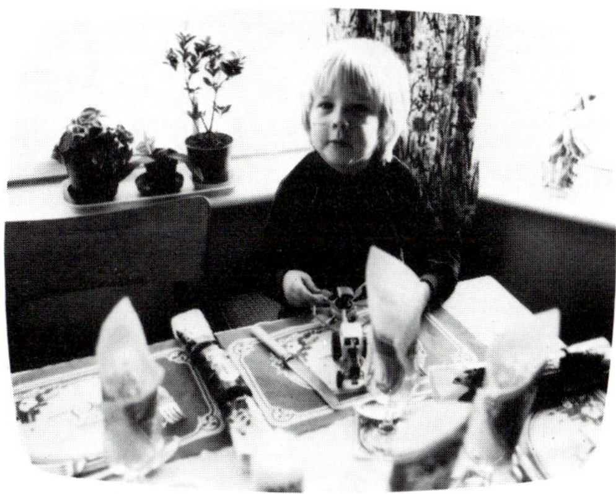

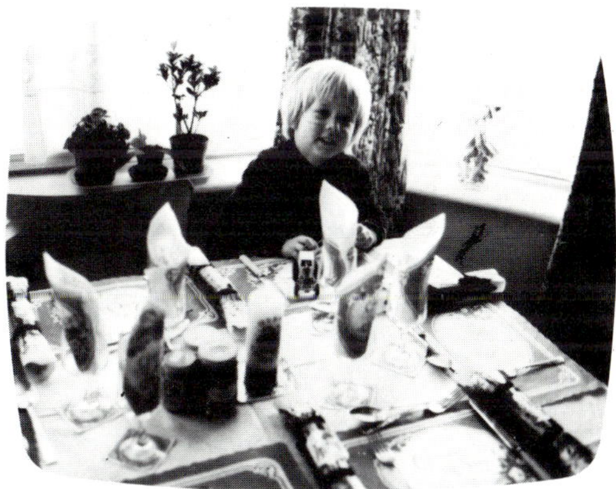

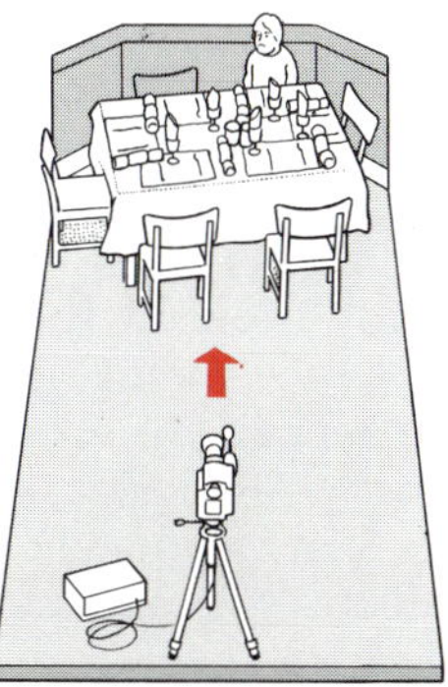

Move from a long shot, *bottom*, to a close-up, *top*, by dollying or tracking: moving the camera toward the subject without zooming, *left*. This gives you more range than the zoom, but you have to change the focus as you move in or out, especially when shooting indoors with a short depth of field.

Linking shots

Once you have mastered the basic camera shots, you may find you need something more concrete than a mental picture of the shape a recording is to take. You need a plan on which shots can be worked out in advance and consulted when shooting – in short, a storyboard.

The problem with planning a video production is that you are working in two media at once: words and pictures. You can write a script which reads well but which is hard to translate into pictures; similarly, the most perfect picture sequence may be difficult to link to any kind of commentary because of sudden changes in the development of the visual sequence.

The storyboard is a device for bringing words and pictures together before shooting begins. In a simple story, such as a record of a family Christmas, a detailed script would be unnecessary. Only a story outline would be needed, as a basis for the order of shots.

To work out such a plan for any video production, ask yourself a set of standard questions. What is the purpose of the recording? (A record of our family celebrating this particular Christmas – this is the *objective*.) At whom is it aimed? (At members of the family, and close friends – this is the *audience*.) What topics will it cover? (The family opening presents; a tour of the house showing different members of the family playing with their presents, preparing a meal; eating; playing games – this is the *theme*.)

The storyboard takes this planning a stage further. The theme of the production is written down one side of each page, and opposite each section of the theme is a series of sketches, one for each shot. These are to remind you how each shot is to be framed, how it will fit into the sequence and what it should show. You may have to guess at certain shots: you may plan a cutaway shot of the cat playing, but it will be possible only if the cat cooperates. The value of the storyboard lies in helping you to decide what kind of shot to look out for.

As you become more ambitious, the storyboard may become more detailed. In a tightly planned production, the timing and length of each shot can be added. The script for a voice-over can be written alongside, so that words and pictures can be matched exactly. The storyboard allows you to try out different approaches before you begin.

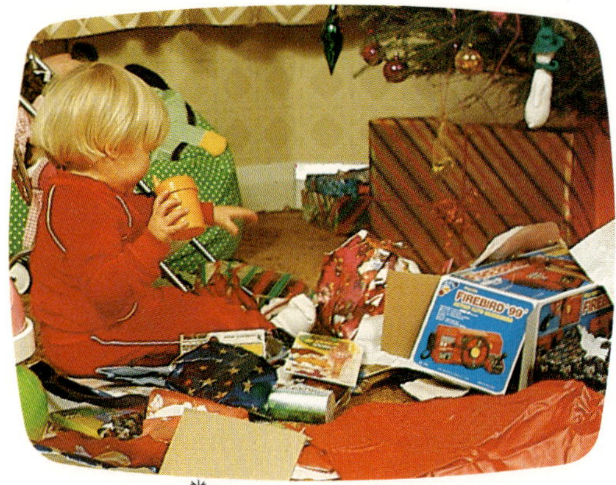

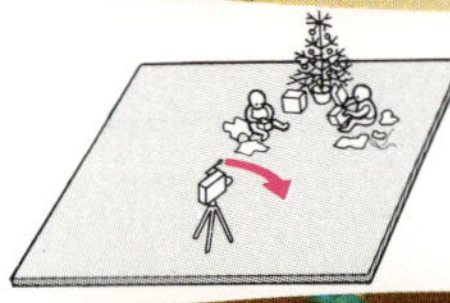

1. Establishing shot: LS (to show whole scene. LS cramped in a small room) therefore:

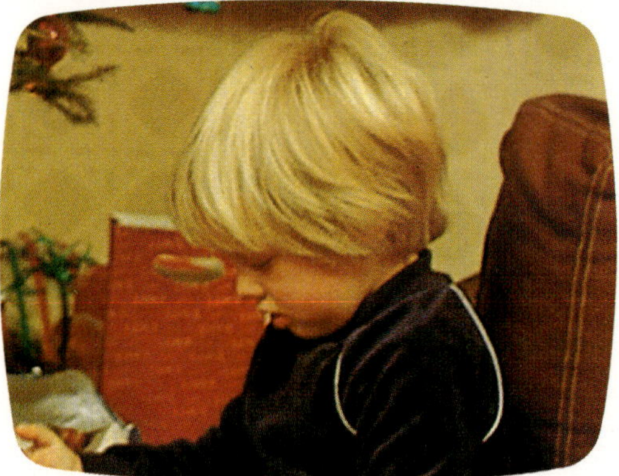

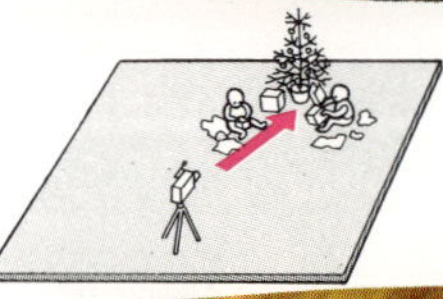

4. MCU (no closer if he is likely to move; stop camera by pressing 'Pause' button);

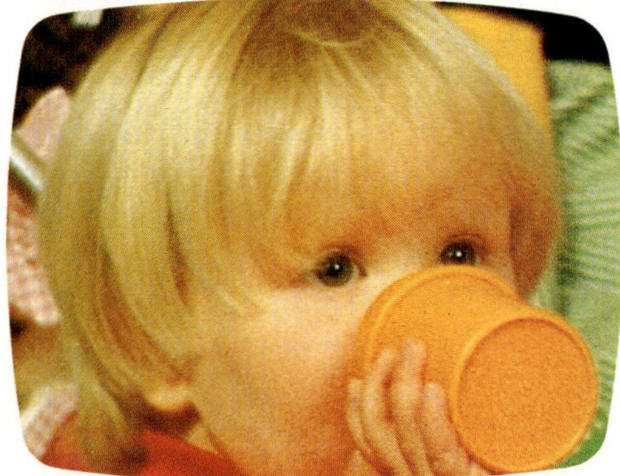

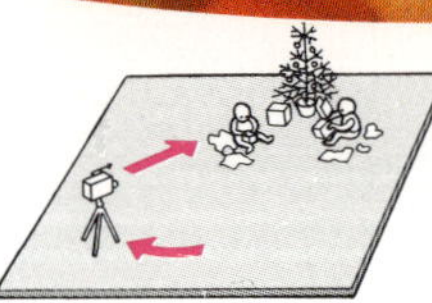

7. ECU first child (don't linger too long; be ready to change focus, move or stop);

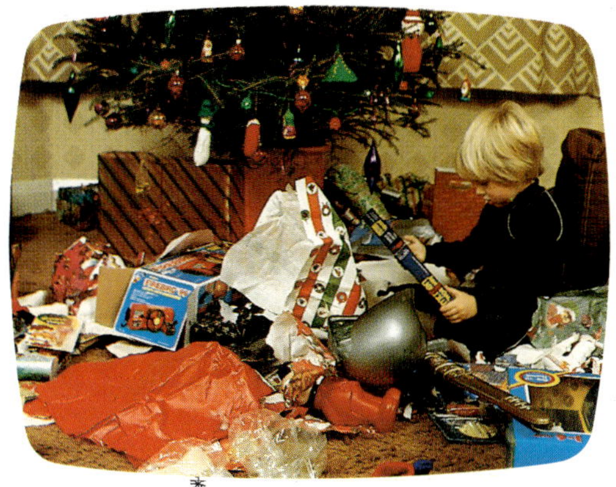

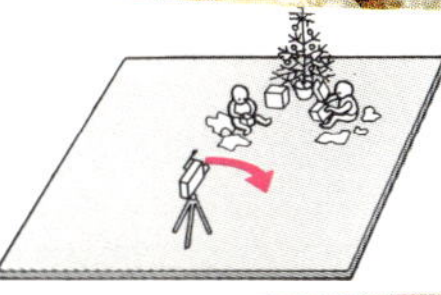

2. PAN to reveal second child (finish panning when in centre of frame);

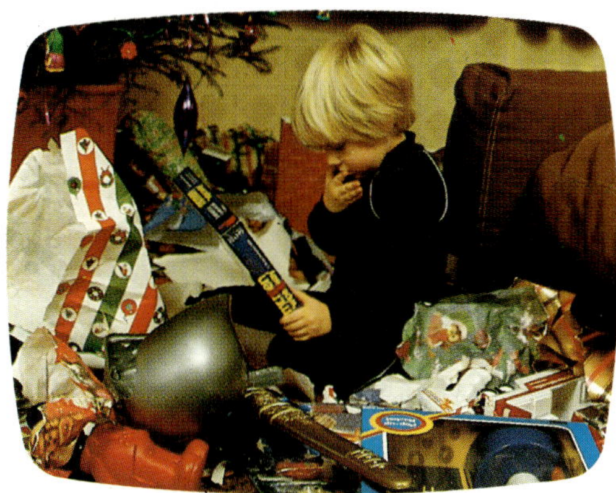

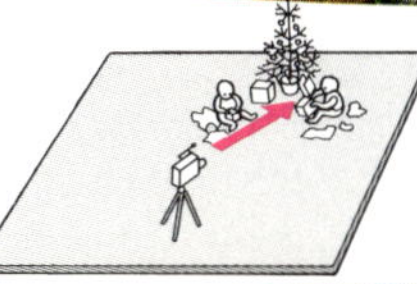

3. ZOOM in on second child (decide now where you want to finish zoom);

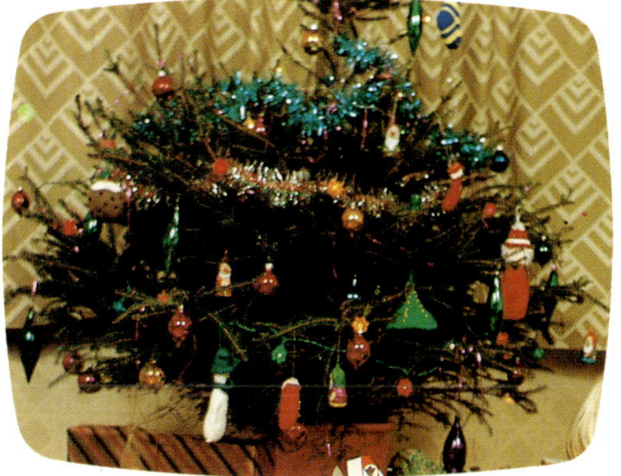

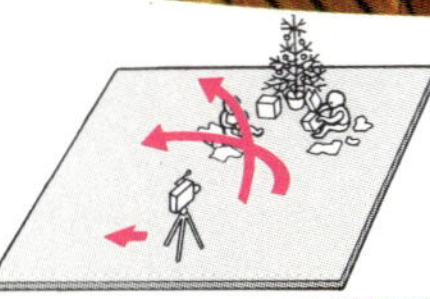

5. TILT up from cutaway shot of tree; (decide direction of next move now);

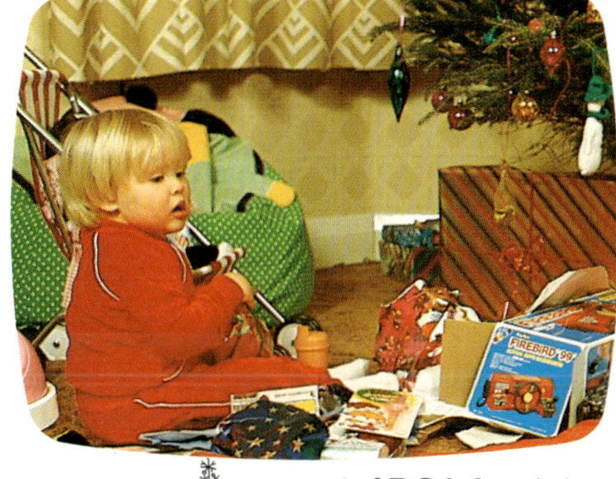

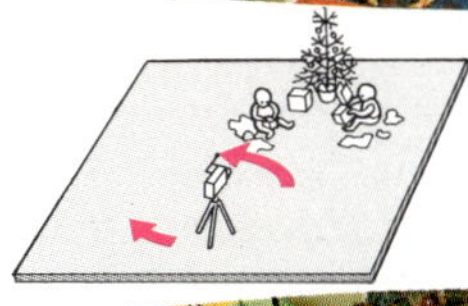

6. ARC left to bring in first child (stop camera; move tripod in close; crank pedestal up slightly);

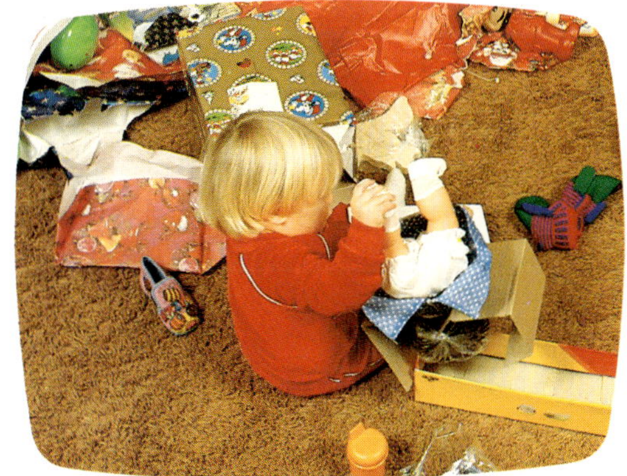

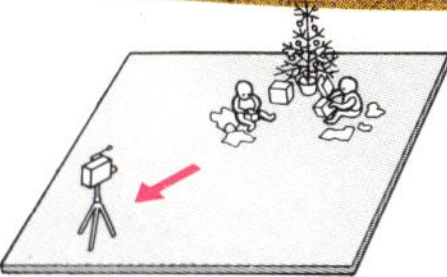

8. Crank pedestal up. DOLLY OUT to reveal first child among toys and wrapping paper;

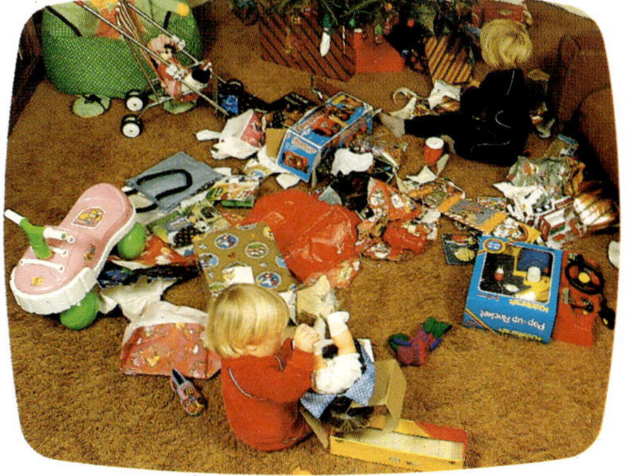

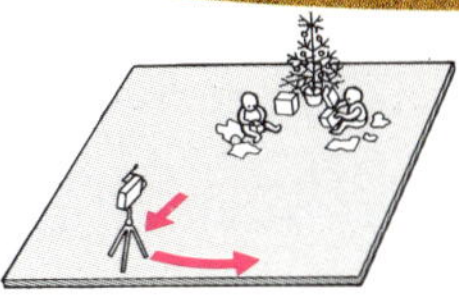

9. TILT up to reveal second child; CRAB right to centre of room; ZOOM out (fade out and stop).

Studio lighting for lifelike pictures

Light is the most important requirement for good video pictures. Your own tapes will be at their sharpest, brightest, most colourful and most lifelike on a sunny day; and on a cloudy day, which seems gloomy by comparison, you will still produce better pictures than you can shoot indoors with domestic lighting. This is because the human eye is much more versatile than a video camera. Our optical system reacts so well to differences in light levels that we hardly notice how much darker it is in a relatively well-lit room than outdoors. A video camera does.

To record an indoor scene successfully, therefore, you need extra lighting. The more lights you can afford the better, but lighting needs care if you are going to achieve the best results with your equipment. The sun is easy to work with because it radiates even light of a constant colour temperature over the whole scene. The different types of artificial lighting radiate light of different colour temperatures, which can play havoc with the colour balance of your picture. Two lights of different colour temperatures will tend to give it a colour cast that can be eliminated only by careful adjustment of the fine-tuning control on the camera. Moreover, every extra light you point at your subject tends to cast deep shadows elsewhere. The only reliable way to produce good indoor pictures is to use several lights of identical colour temperature in a planned scheme.

There is more to lighting than producing bright pictures. Like film, video is a two-dimensional medium and careless, over-bright lighting can make every figure look flat. Good lighting also has to help create the illusion of depth by emphasizing the shape of the subject and the distance between it and the background.

Another problem is that, however carefully you work out your lighting with a subject sitting still, and however closely you stick to the rules, you may find difficulties when your subject moves. That is why every lighting set-up has to be a compromise.

The television screen is the acid test of whether the lighting is doing its job or not. Are the pictures bright enough? Are the colours lifelike? Run through the sequence of actions with your subject, if you are shooting a rehearsed sequence, and experiment with the lights so that you can get the effects you want throughout the scene.

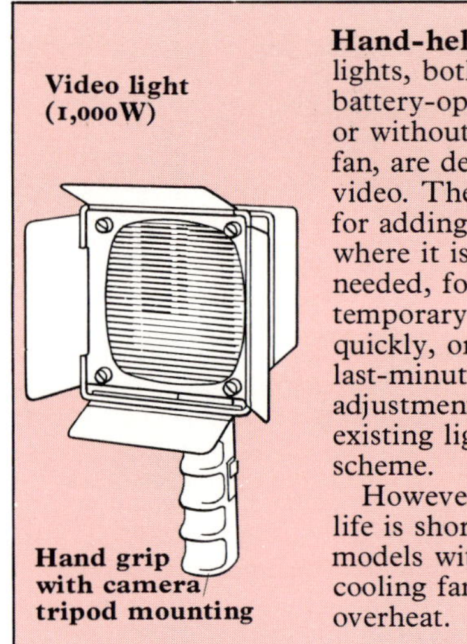

Hand-held quartz lights, both AC and battery-operated, with or without a cooling fan, are designed for video. They are useful for adding extra light where it is especially needed, for setting up temporary effects quickly, or making last-minute adjustments to an existing lighting scheme.

However, battery life is short, and models without a cooling fan tend to overheat.

Photofloods radiate a soft, even light which can give your pictures a subtle thirties film quality. If you mix them with other lights, beware of colour temperature problems.

Quartz film lights are ideal for video. They can be adjusted to give a bright spot of light, or a soft pool, so they are useful in most situations. The four

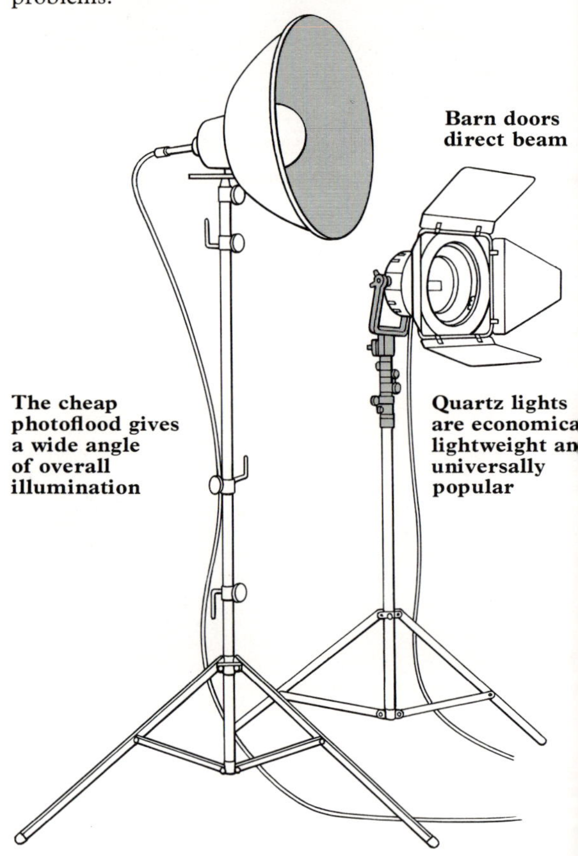

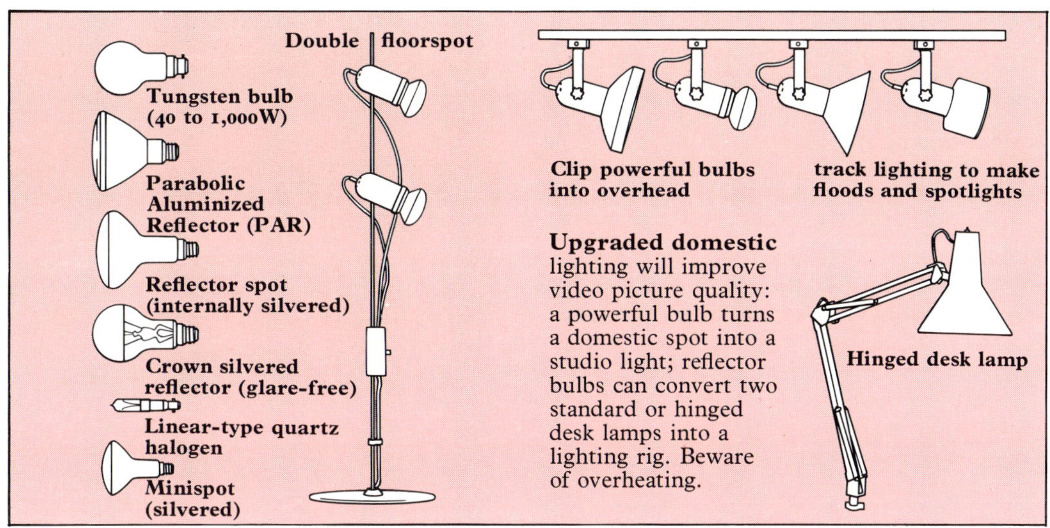

Upgraded domestic lighting will improve video picture quality: a powerful bulb turns a domestic spot into a studio light; reflector bulbs can convert two standard or hinged desk lamps into a lighting rig. Beware of overheating.

barn doors, usually attached, limit the area of illumination. A set of three quartz lights will achieve most effects, with no colour temperature problems.

A fresnel lens can be mounted in front of a light to focus the beam, or to soften the edges of a spot if it proves to be too harsh for the effect you are trying to create.

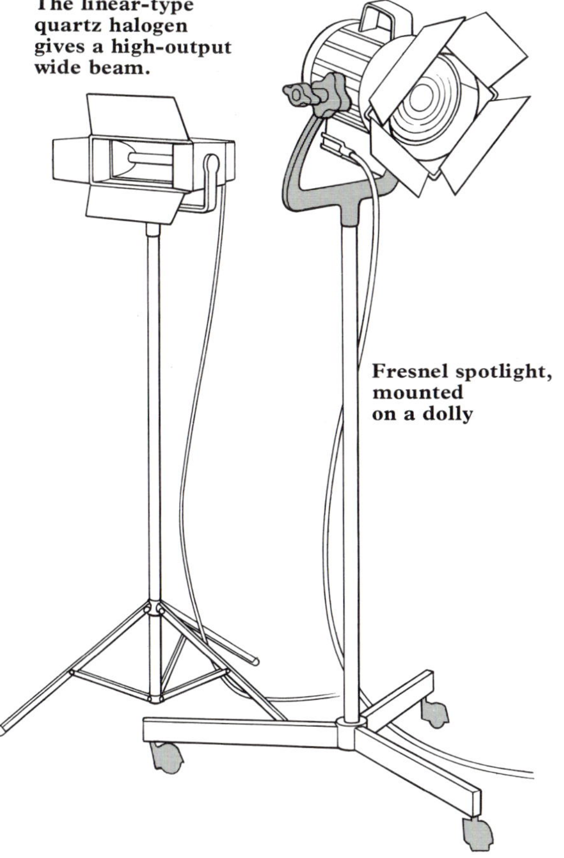

The linear-type quartz halogen gives a high-output wide beam.

Fresnel spotlight, mounted on a dolly

Colour temperature check list

Natural light	Colour temperature °K	Artificial light
Heavily overcast sky	10,000	
	9,000	
Haze	8,000	
Lightly overcast sky	7,000	
		'Daylight' fluorescent lights
Sunlight/ blue sky	6,000	
North daylight	5,000	
		'White' fluorescent lights
	4,000	'Cool white' and 'Warm white' fluorescent lights Photographic incandescent bulbs Quartz halogen lamps
	3,000	
		Photofloods Domestic incandescent bulbs
Sunrise/ sunset	2,000	
	1,000	Candle-light

Deploying lighting

The professional way of arranging lights to illuminate a subject brightly while creating an illusion of depth, is called the three-point method. The first light is the key light. It takes over the role of the sun in providing most of the illumination, so it needs to be in front of the subject, and slightly to one side. It should also be set fairly high, though not so high that it makes the eyes vanish into pools of shadow. Usually an angle of 15 to 45 degrees above the subject will be right. This should give sufficient brightness to show up your subject clearly enough.

Sunlight is reflected and diffused in many ways, so the shadows in areas out of its direct range are also quite well lit, but the key light will cast deep shadows on the other side of your subject. As a result, the contrast between over-bright highlights and black shadows may be harsh, and you will want to soften it with a less intense beam from a second light. This is called the fill light. It must be bright enough to eliminate the deepest shadows so that in the picture they look like shadows, not deep black patches.

The job of the third light, the back light, is to make the subject stand out from the background. Again, it needs to be a softer light than the key light.

Your three-point lighting rig need not be expensive. If the room you are using as a studio is fairly well lit you may be able to manage with domestic lighting, or just one professional studio light. A reflector bulb screwed into a hinged-arm desk light makes a serviceable key light; for the fill light you can use overhead lighting – perhaps track lighting fitted with photoflood bulbs. Table lamps, standard lamps and desk lamps can all double as back lights.

If you decide to invest in a studio light you will find quartz film lights the most versatile. They can be adjusted to give a bright spot of light, or a softer pool, and usually have a set of barn doors which can be adjusted to rearrange the area of illumination. A quartz light fitted with a fresnel lens will throw a perfectly even pattern of illumination. Photofloods are good fill lights, but will be incompatible with the colour temperature of quartz or tungsten key and back lights.

Keep the camera switched to the 'Standby' or 'Pause' mode, or put the lens cap on, while moving lights around. A starry reflection from a light could damage your camera.

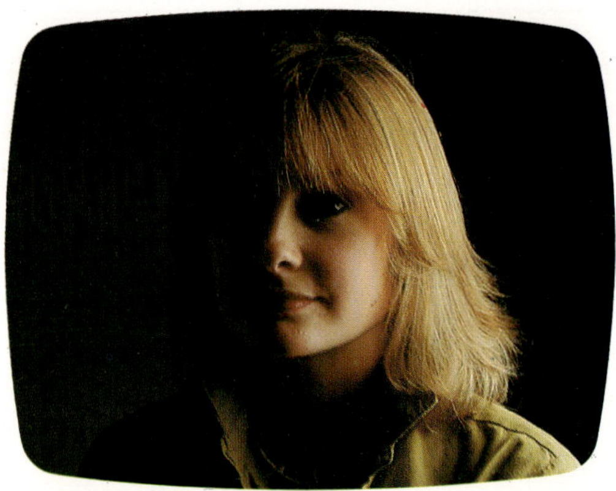

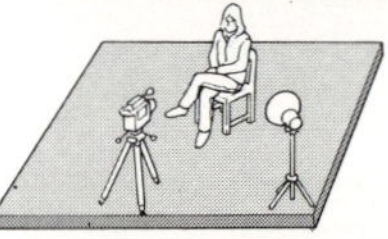

A key light in front of a subject flattens its shape and form. Placed to one side, it casts hard shadows.

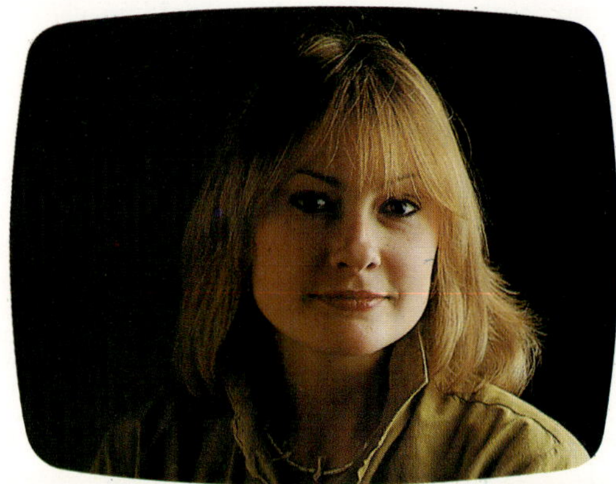

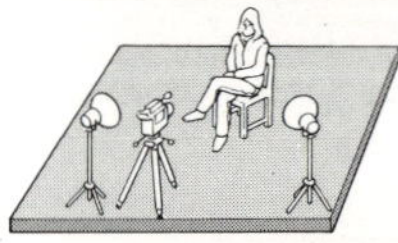

A fill light opposite the key light restores the shape of the subject and softens the shadows.

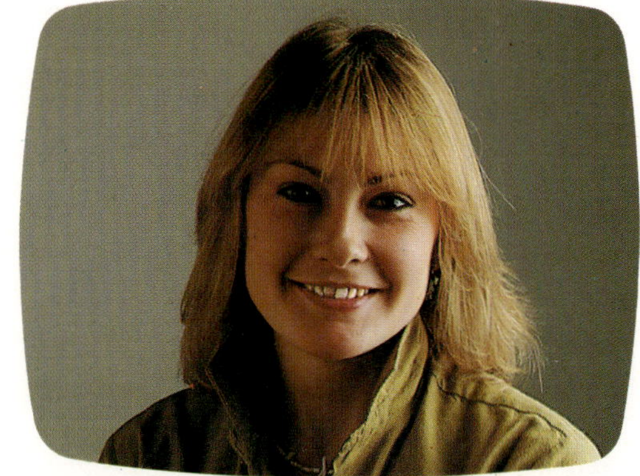

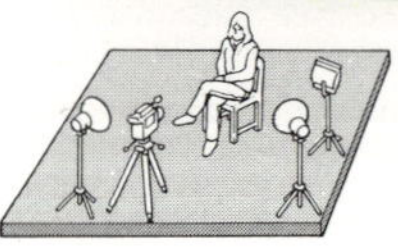

A back light on the same side as the key light, but above the subject, separates it from the background.

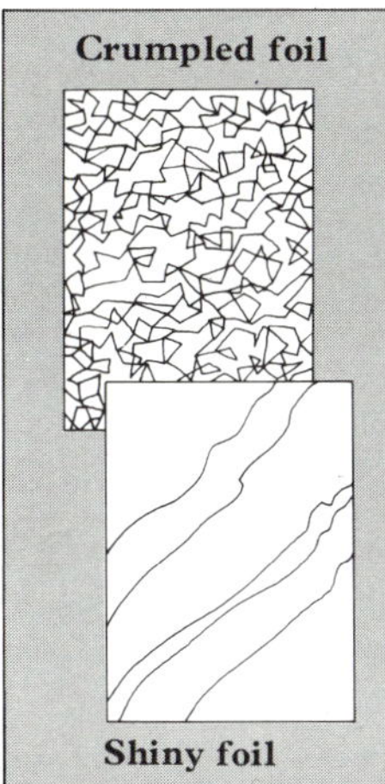

The effect of any video light can be enhanced by using a reflector. You can make one very easily by gluing kitchen foil to a board. One side is made smooth and shiny for direct reflection and the other rough (by crumpling the foil) to give a more diffused reflection. Either side can be used to reflect light back into areas of shadow.

Eliminate unwanted reflections by tilting mirrors and coating reflective surfaces with matt spray, water mist or black tape, *below*. Lights may be diffused through a silk or paper screen, *left*.

Silk screen

Tracing-paper screen

Matt spray

Water spray

Black tape

Daylight streaming in through a window may serve as a boost to domestic lighting, but since the two have different colour temperatures, check the scene carefully on the TV screen before you begin recording, and eliminate any colour problems.

Alternatively, you can boost the daylight by beaming two photofloods through a window, *left*. Here, the light from the photofloods outside is being diffused and softened by a tracing-paper screen fastened over the window. The beam from the photoflood in the room is directed at the ceiling, and so diffused and reflected.

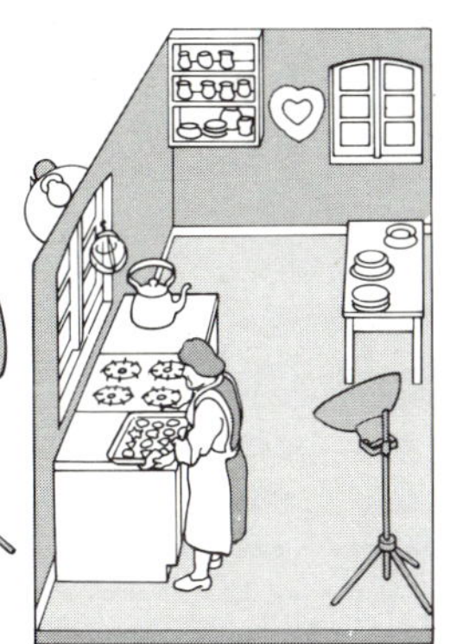

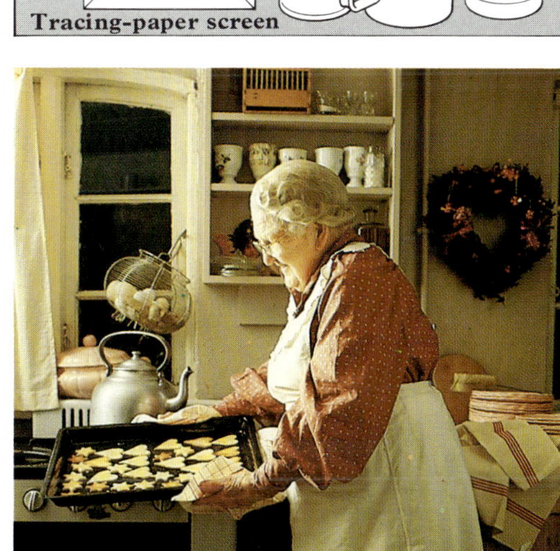

Large reflective surfaces diffuse light by bouncing it. The two photofloods in the scene, *below*, are directed not on the subjects but down to the white napkins and glassware. These reflect the light upward, indirectly illuminating the child and her grandmother, *right*. A third photoflood reflector bulb, hidden by the decorations around the overhead light, directs a beam downward, flooding the table area, which reflects diffuse light all around the room.

The candles' warm glow is mirrored by reflective surfaces all around the room.

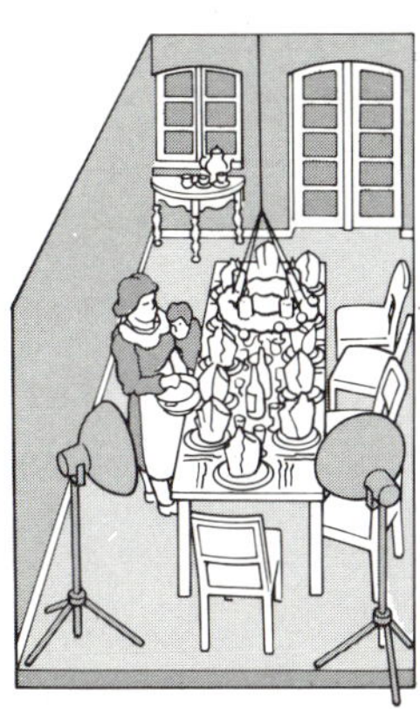

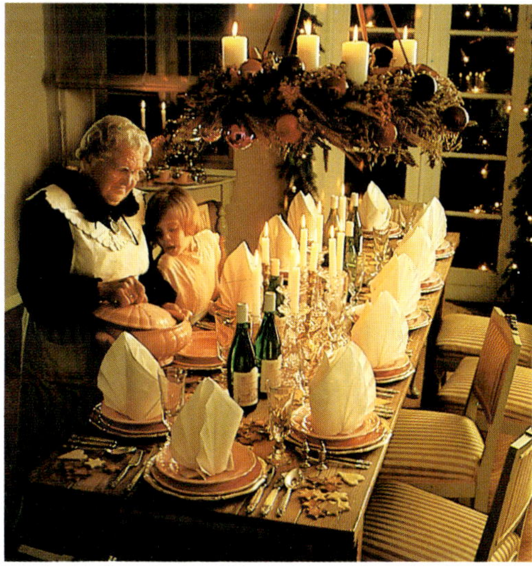

Breaking the rules

The conventions of good lighting are broken time and time again on television with perfectly acceptable results. Remember that the rules are there for guidance, and that if you break them, you must have a particular effect in mind.

Putting the back light on the same side of the subject as the key light is the usual way of arranging the secondary lights, but putting it on the opposite side of the subject heightens the dramatic effect. Using two back lights helps to bring extra sparkle to the hair. When you place the fill light, take care not to put it too far to one side of the camera, where it may cast unwanted shadows which can look confusing against shadows from the key light. Instead, try moving it closer to the camera.

The brighter the subject, the less the camera will see the background, particularly if it is dark. To bring out background detail, lighting shone directly on the background will light it up neutrally, but if you want more detail, try shining a light sideways across the background.

Use natural light to aid your indoor shooting, but never use an outside window as a back light unless you want the subject to appear as a dramatic silhouette against the brilliant outside light. While indoor lighting is consistent in quality, intensity and colour temperature, outdoor light tends to vary. Professional film units often insure against this by shooting at night and putting an additional light outside the window to give an illusion of daylight.

In the same way you can create an illusion of night by boosting domestic room lighting with more powerful illumination. If a subject were to sit reading by the light of a small standard lamp, the light level of such a lamp would be too low to produce a good video picture. If you step up the illumination with a hidden light thrown down on to the subject from roughly the same direction it will reinforce the shadows cast by the lamp, and look just as realistic. You can eliminate the harshest shadows with hidden fill and back lighting, leaving the reader seated in a pool of warm light enclosed by a soft darkness.

A carefully planned lighting scheme may lose its entire effect once the subject moves, so think ahead. Light the whole area in which the movement will take place; follow the subject with a hand-held light, or track behind with a light mounted on a dolly.

Lighting practice check list

1. DO NOT use a powerful light bulb in a household light fitting for more than 10 minutes – it will overheat.
2. DO NOT overload power supplies – lights consume a great deal of power.
3. DO NOT shine bright lights directly into your subjects' eyes.
4. DO NOT move lights without switching them off – they may shine into the lens.
5. DO let your lights cool before you move them – they will last longer.
6. DO NOT mount a light on the camera – it might reflect off a shiny surface into the lens.
7. DO NOT put lights behind the camera operator. They cast shadows on the subject.

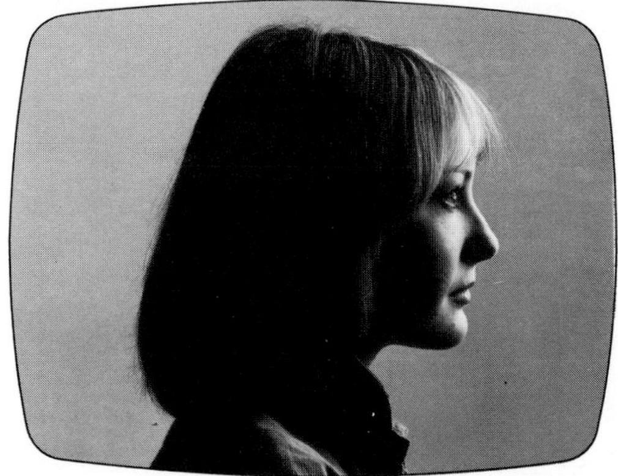

Lighting sets the atmosphere of a scene and tells the viewers much about your subjects' personalities. A cameo shot, *above*, dramatizes a face. This subject is lit by a single key light which, positioned just to her left, highlights her profile.

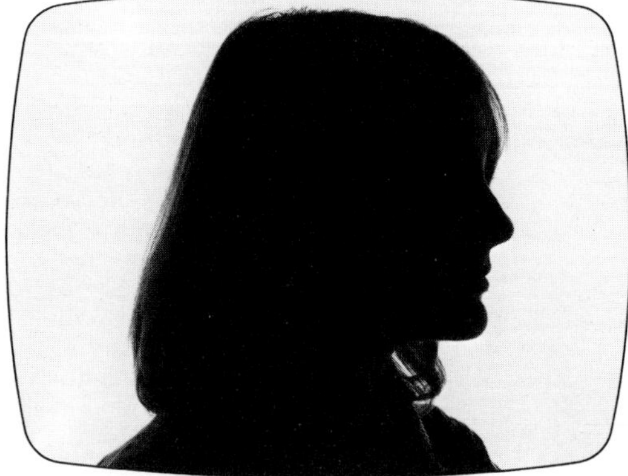

A silhouette is a powerful means of eliciting the desired emotion and it can reveal character before the subject walks or speaks. The background light is diffused through a screen to create even illumination behind the silhouette.

Snoots and reflectors
A reflector is a metal shield that fits on a light, and directs the beam; there are many shapes and sizes, **1**. A snoot, **2**, is a metal cone used to narrow a spotlight beam; a parabolic reflector, **3**, throws a near-parallel beam for a spotlight effect. A diffusion reflector, **4**, spreads the beam more widely; a metal cap in the centre reflects and diffuses the beam. A spotlight, **5**, has a reflector and fresnel lens to concentrate the beam on a small area.

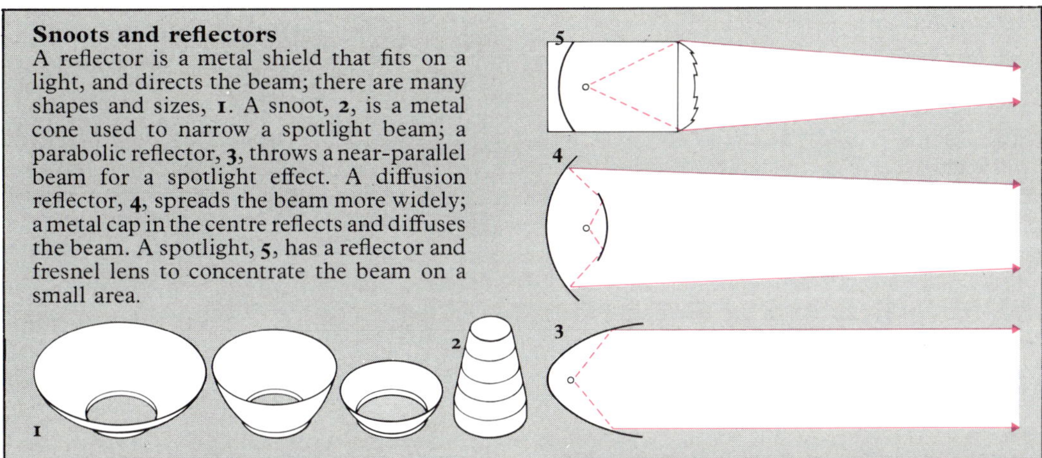

Colour temperature incompatibility may be exploited to create atmosphere, *right*. In this scene the pale daylight is augmented by candles set in trees, in the windows and on the ground.

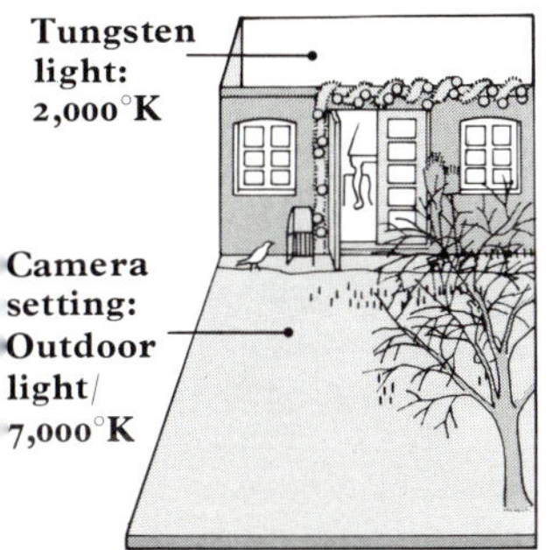

Experiment with the colour controls to exploit the visual potential of such a shot. Setting the white balance to outdoor lighting will emphasize the glow of the candles; turning it towards indoor lighting tones down the orange and makes the daylight look bluer. Check the results on the monitor until you have the balance you want, but be ready to readjust the colour balance when you change frame.

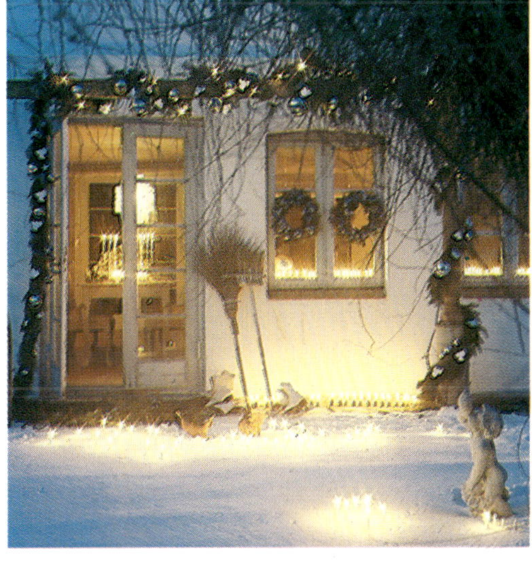

Dramatic scenes can be set up in unusually low lighting conditions, *right*. In this candle-lit scene, the light level seems too low for a video camera, yet to boost it might destroy the atmosphere.

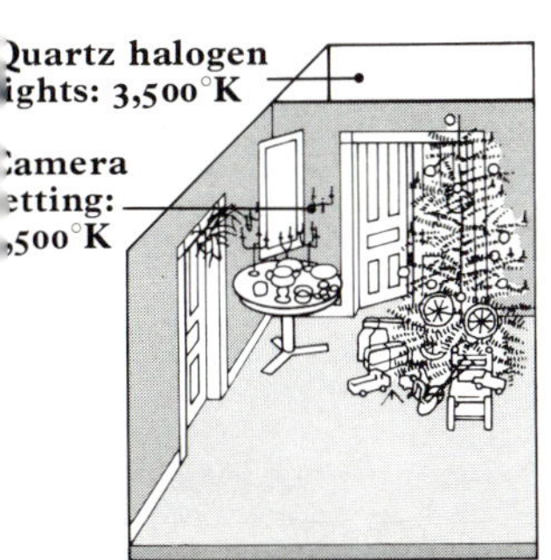

Unconventional but subtle lighting is used to raise the overall light level without spoiling the effect. Two lights set up in an adjoining room are aimed at the white door; using the door as a reflector should increase the light level while diffusing the beams. However, the monitor may still display no more than a white streak on a dark screen and you may have to increase the background illumination subtly by switching on a lamp just out of shot.

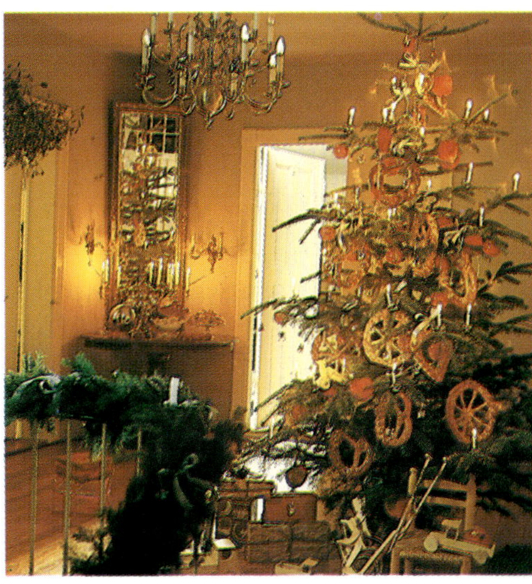

Illuminating movement

Lighting has to be set up to suit the scene you plan to record, and this has a hidden advantage: it obliges you to plan ahead. Imagine, for example, that you want to record a child playing a piano piece. Simply setting up the lights so that you can record the piece as a single shot would give you a record of the player's ability. It would, however, be very boring to watch.

To make it more interesting, you would look for different ways of viewing your subject. An opening shot could bring in piano, player and surroundings, then zoom in to a close-up of the player's hands on the keyboard. Next you could change to an over-the-shoulder shot, showing the music. You might rest on a close-up of the player's face, or the hands from a different angle – perhaps a low shot along the keyboard.

The first of these shots may be close enough to the next for the same lighting arrangement to work for both, but it is more likely that some changes will be needed: a suddenly obtrusive shadow, a key light or back light in the wrong place, which means having to pause while you adjust the lighting.

Since the subject is someone playing a piece of music, this poses problems. Putting the camera into the pause mode means cutting the sound, and missing phrases of the music while you set up the lighting for the next shot. The only way you can avoid breaks in the sound-track is to persuade the player to stop, then start again when you are ready. This demands slick timing from both player and camera operator if the finished result is going to sound as if it were being played through at one sitting.

Yet you must not fall into the trap of trying to get away with two almost identical shots simply to avoid moving the lights. If a shot of the player's hands follows another taken with similar framing, when you cut from one to the other in the finished sequence there will be a jump cut, that is, the picture will seem to jump from the first position to the second, with distracting results. It is far better to wait until you have said all you can with the first shot, then cut to something different, even if it means moving the lights altogether.

Check the lighting before every shot. Make any necessary adjustments with the camera's lens cap in place, before checking on the monitor that all looks right when the lens cap is removed.

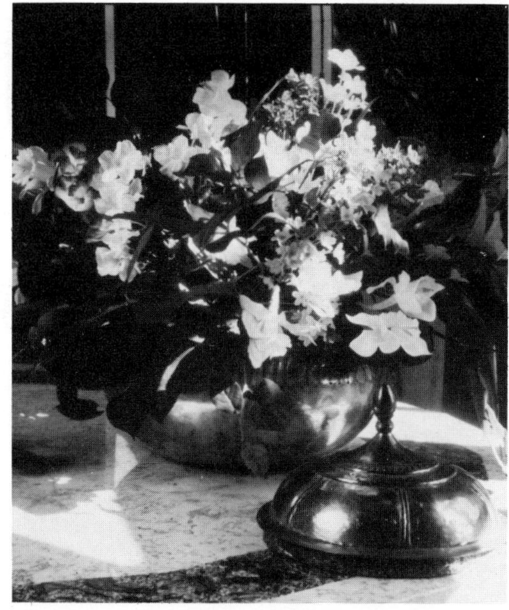

Daylight is the best light for video. Use it to bring out subtleties of colour but avoid deep contrasts of light and shadow. Video cameras see more contrast than there is; they react to the light areas, and shadows turn black.

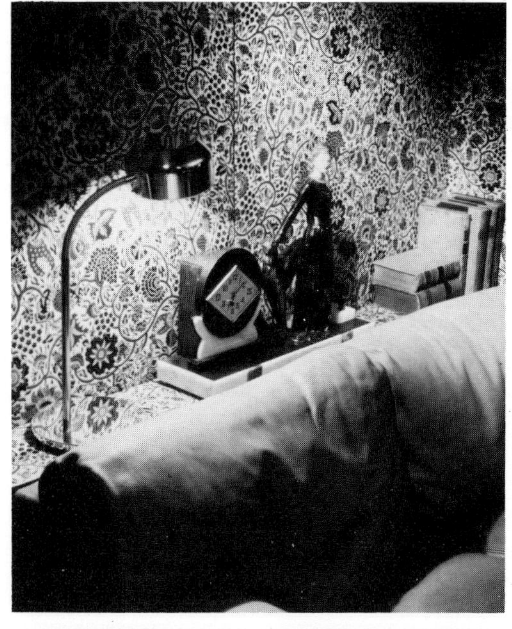

Domestic lighting is less intense than daylight and it is concentrated on a small area. To shoot the scene, *above*, you might need a fill light to lift the overall backgroud level. Add a domestic light, such as an overhead room light.

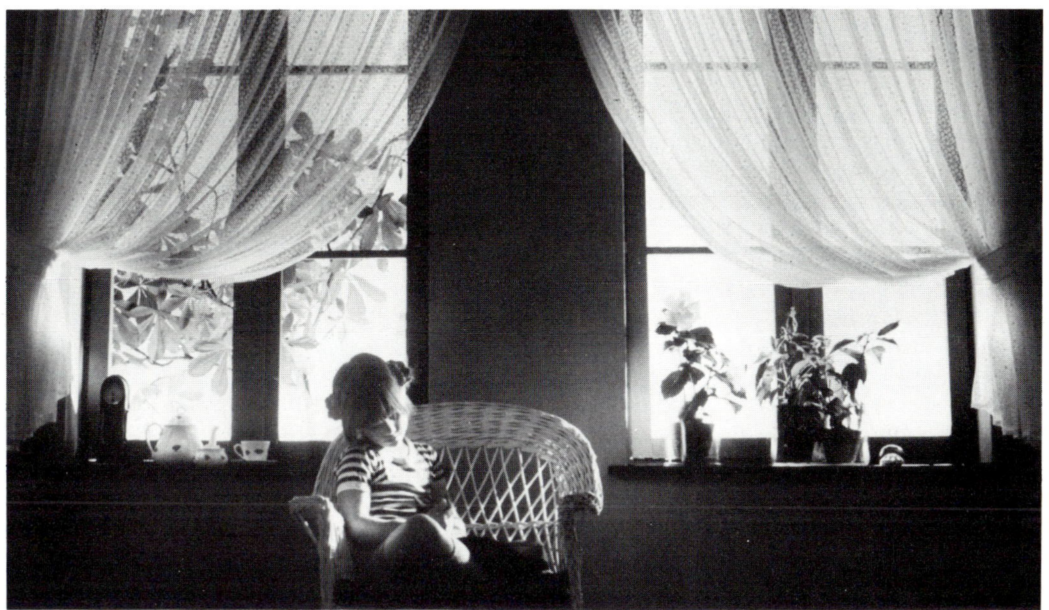

Shooting a subject against a brightly lit background is a lighting nightmare: the lens aperture closes, reacting to the light, so the subject appears in silhouette.

This can be effective if you plan for it in framing and composition. Alternatively, diffusing the light through a pierced screen, the leaves of a tree or a vase of flowers gives interesting results.

If you want to record a subject in detail, powerful lights are needed to compete with the daylight. If the colour cast looks unnatural, reset the camera for the colour temperature of the lights. The best solution may be to shoot as the outside light fades.

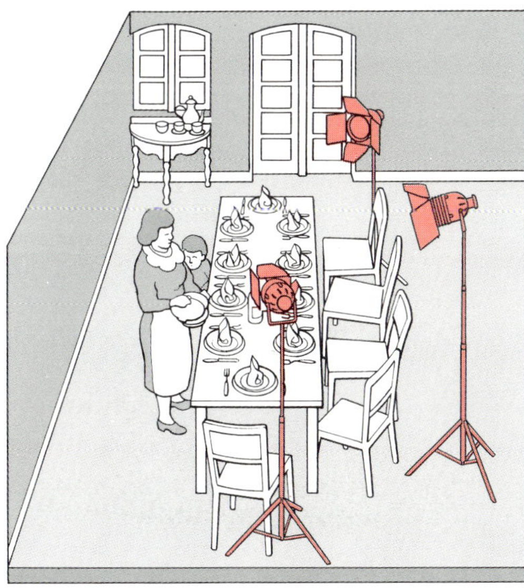

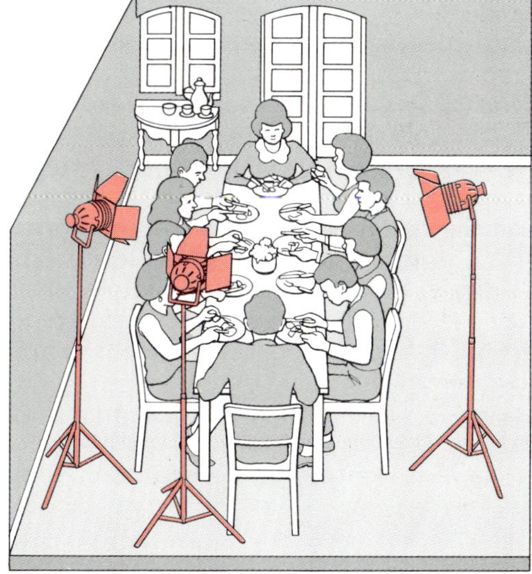

When you light a scene to suit the camera's requirements you commit yourself to changing or moving the lights whenever the subject moves. Here, *above left*, the same two lights will be repositioned so that they stay in the same place relative to the camera and the subject. When the scene changes to a group, *above right*, standard 3-point lighting may not be enough. Add overhead lighting, or reset the studio lights to give a flood and not a spot.

Faces made for lighting

The first thing to remember about make-up for video is that you don't *have* to use any at all. Television make-up artists aim for a much more natural look than in the theatre. There the performers have to look right under much poorer lighting, for an audience which stretches right back to the rear rows of the circle, rather than the other side of the living-room. But make-up is still useful. At the very least it can mask the shine of perspiration created by the heat of intense lighting. At a more ambitious level it can be used to adjust skin tones.

Generally speaking, these considerations are much more important in studio conditions than they are when you are making your own recordings; but make-up can still be used for routine improvements to performers' facial features, and in the much more ambitious area of character make-up clever ideas can change an actor's age or background beyond recognition.

Anyone who appears on television regularly soon learns to apply their own make-up. The first decision is how much make-up needs to be used – skin tones and blemishes can be taken care of with several thin applications of foundation rather than a single thick coating which can produce a mask-like effect. Make-up can be used to tone down over-prominent ears or noses by making them blend more effectively with the colours of the surrounding skin, and faces which become flushed from the heat of the lights can be lightened.

With more skill – and more practice – you can use make-up to correct flaws. Eye sockets which are very deep, causing pools of shadow, can be made to appear less hollow through skilful use of make-up. Eyes which are too prominent or too small, too close together or too wide apart can be 'corrected' by cosmetics, as can wrinkles, or an untidy hair or eyebrow line can be camouflaged. Shiny bald spots, thinning hair and dark five o'clock shadow can all be made much less obtrusive. However, be careful of carrying the improvements too far. Most television directors agree that too much make-up is distracting and far worse than having no make-up at all. The acid test is, once again, your trusty television set. If the changes you make to your face, or those of your performers, look natural and untouched on the screen, you have achieved your purpose.

Begin camera make-up by cleansing the skin with a cream or lotion, then wiping it off with paper tissues. An astringent closes the pores and cuts down perspiration.

For a foundation, use a dry matt cake of compressed powder, applied with a damp sponge, or cream or liquid greasepaint.

Dot the foundation on to nose, forehead, cheeks and chin; then blend it into a thin, even coating over the whole face.

Model the face by shading it in darker foundation colour, and highlighting it in lighter tones. Highlights broaden the features; shading narrows them.

Ideally, eyes are one eye-width apart at the inner corners. Eyes look wider apart if colour is most emphatic at the outer corners. Emphasize far-apart eyes at the inner corners.

Conceal lines and wrinkles by brushing a lighter tone of foundation into them, and a slightly darker shade over them.

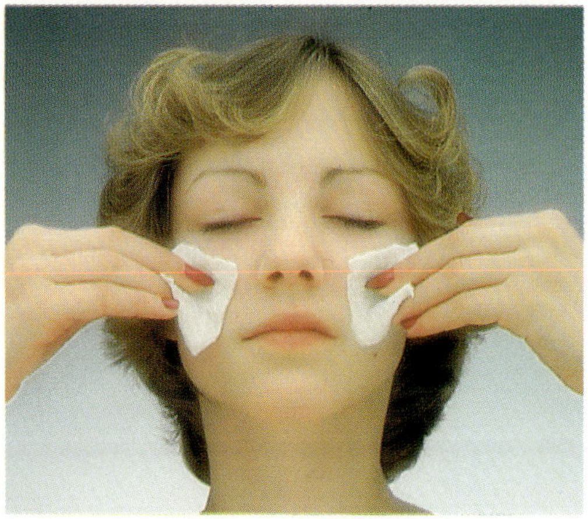

Step 1: cleanse the skin with cold cream or a special cleanser, but wipe it off carefully and close the pores afterwards, or a shine may break out under the lights

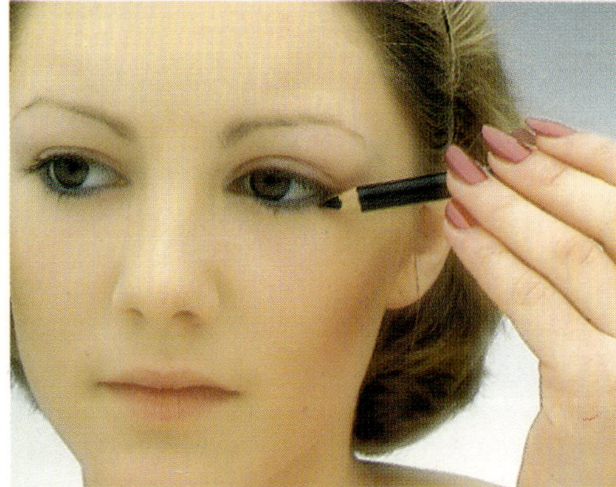

Step 4: modelling the eyes is necessary for the camera. A light eye colour makes deep-set eyes look less sunken; dark and bright colours are used for emphasis.

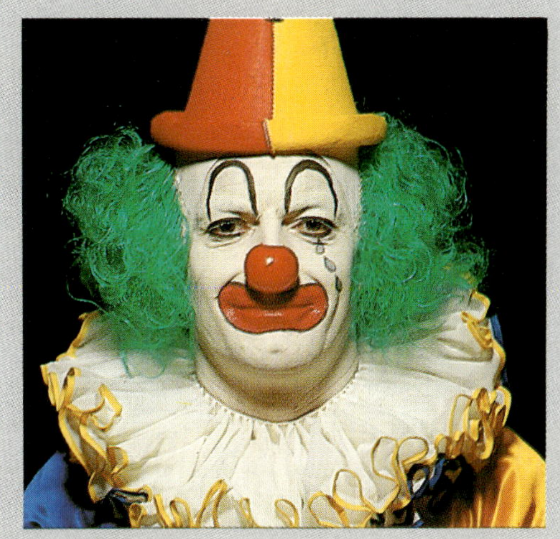

Character make-up
Crêpe hair and greasepaint, bought from a theatrical shop, were the materials used to create this clown. Fix false hair to the skin in tufts with spirit gum.

To give an illusion of age, hair can be greyed with powder or spray tints, or the head balded with a 'bald cap' and false hair added in patches. Shading the facial contours can add years: create wrinkles by rubbing a dark foundation into the facial creases, and shade the hollows of the features, temples and cheeks, under the eyes and at the sides of the nose.

Collodion liquid dries when painted on to the skin, causing it to pucker into 'scars' or, with added imitation blood, 'wounds'. To make *layers* of wrinkles, paint the skin with Sealor and press on to it cotton wool moulded into wrinkles. Seal the 'wrinkles' with Sealor and make up with greasepaint.

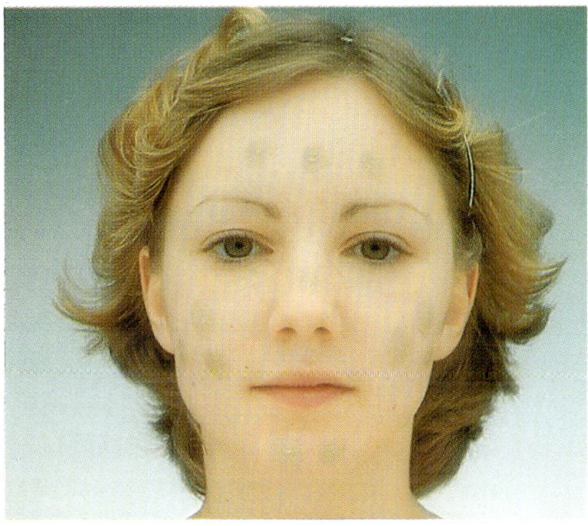

Step 2: apply foundation sparingly in an even layer from hairline to chin, covering surface blemishes. Apply greasepaint with a brush or the fingers.

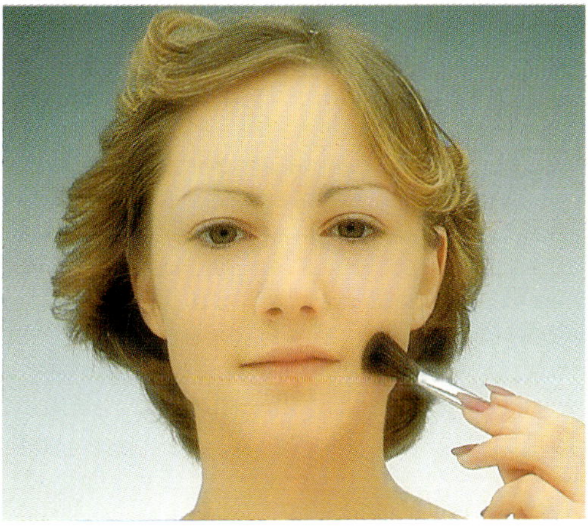

Step 3: shading the face shapes it. A broad jawline will look narrower if the cheekbones are highlighted and a darker shading used down the sides of the face.

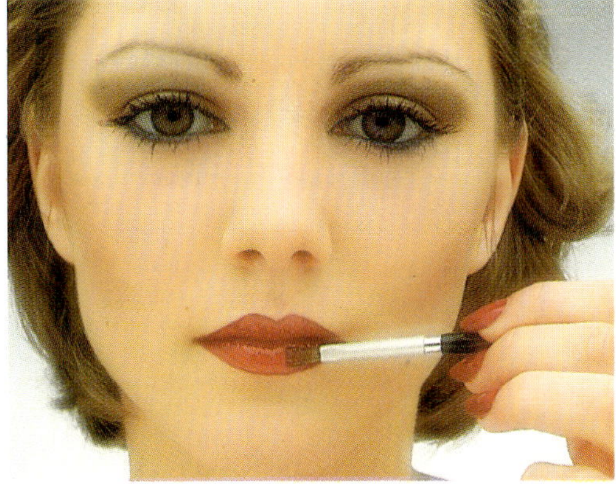

Step 5: contouring the lips gives the face a focal point and balances the impact of the eyes. Avoid bright reds and oranges, they tend to flare on a TV screen.

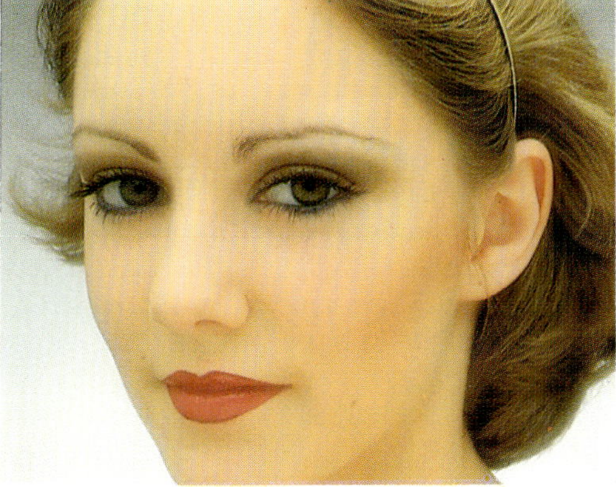

Step 6: camera-ready make-up should be your normal make-up, but think about the shades you use. A colour looks different under daylight and electric light.

Controlling light

All cameras have to cope with light of differing colour temperatures. Most video cameras can compensate for variations in colour temperature by using their white-balance controls. But there are times when a filter, which changes the characteristics of the light reaching the lens, will give better results. Some video cameras have built-in filters to give simple colour correction – from daylight to indoors, for example – but by adding external filters more ambitious shots are made possible.

Perhaps the simplest filters of all are neutral-density filters. They take out an equal proportion of all wavelengths of light, so that the overall colour balance is unchanged. They are used where the ambient light is too intense, causing the highlights of your picture to flare or the shaded areas to lose their detail. In bright conditions this kind of filter allows you to use a wider aperture (lens opening), to keep your subject in sharp focus against a blurred background.

Colour correction filters do a similar job to the camera's white-balance controls or built-in filters, but there are other useful effects. A magenta filter takes out the excess green apparent when copying old slides on to video tape. A blue filter helps to correct the reddish cast you find when shooting under domestic lighting. Amber filters take out the blue cast of cloudy days and give a warmer tint to skin tones for close-ups and portraits. Ultra-violet (UV) filters take out the scattered ultra-violet haze you find when shooting on the beach or in the mountains. Polarizing filters reduce the amount of reflected light, as opposed to direct daylight, allowing you to shoot through glass without unsightly reflections. Also, when shooting on a sunny day, the sky in the area at right angles to the sun becomes a much richer blue, although the colour temperature of the rest of the picture is unchanged.

Finally, there are special-effects filters which can be screwed on to the lens. These include star filters and cross filters which turn pin-points of light into (as their names imply) stars and crosses. Some screw-on special-effects filters are actually lenses, like the prismatic multi-image lens which surrounds your subject with a number of ghost images in the same picture. Sometimes there is a control for rotating the lens so that the ghost images rotate around a main subject.

Colour correction filters

Filter (apparent colour)	**Effect**	**Application**
Neutral density (grey)	Reduces the amount of light reaching the camera lens	Useful where ambient light is too strong or depth of field shallow
Skylight (pink)	Suppresses ultra-violet light	Eliminates bluish shadows cast by a blue sky
Ultra-violet (clear)	Absorbs ultra-violet light	Eliminates blue haze outdoors for clear land- and seascapes
Polarizing (grey)	Absorbs reflected light	Eliminates reflections when shooting through water or windows. Improves clarity and contrast of colour work
Fluorescent (violet)	Eliminates greenish cast	Corrects green cast caused by fluorescent lights. Different filters are available to correct for different fluorescent lights.
Blue (blue)	Absorbs reds, increases colour temperature	Corrects colour balance for shooting under artificial light or in bright conditions if a subject is in shade
Amber (amber)	Absorbs blues, lowers colour temperature	Eliminates any blueness in lighting; useful for portraiture in cloudy weather

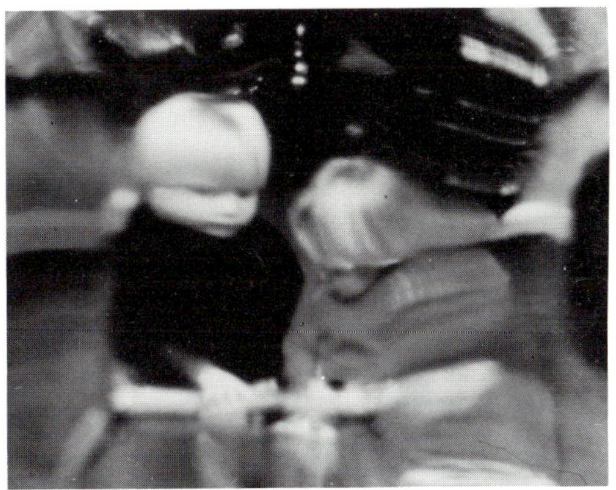

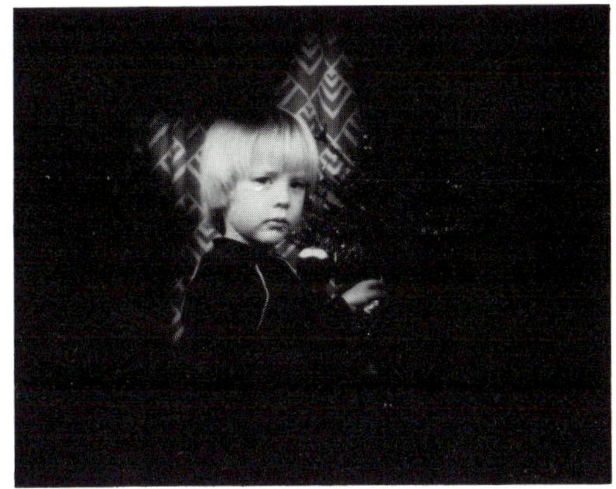

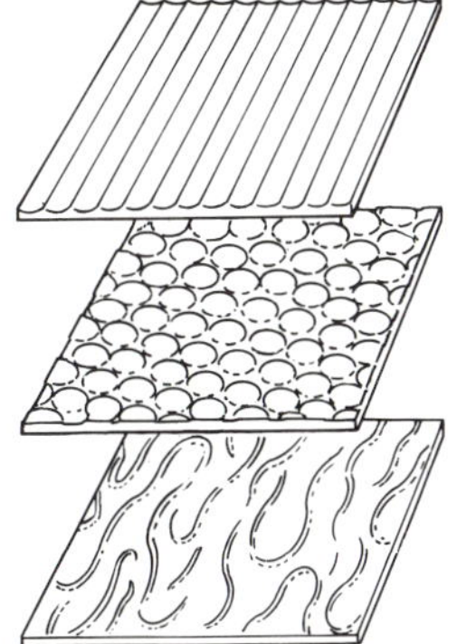

Some filters use the irregularities in the glass to spread light. Soft-focus filters are made from glass in which tiny dimples are impressed in a random pattern. The degree of softening is determined by the size and number of the dimples, while the centre may be clear. A piece of glass held in front of the lens can have the same effect.

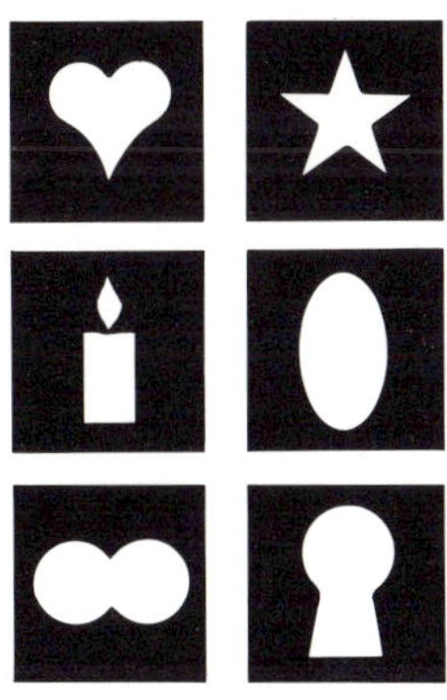

Shots are normally framed in the familier television screen shape. If a sheet of card or metal with a central cut-out (a frame) is fixed in front of the lens, the shot is framed inside the cut-out. For instance, a key-hole-shaped frame gives the impression of looking through a key-hole. Popular shapes are star, circle, heart and binocular.

Macro photography

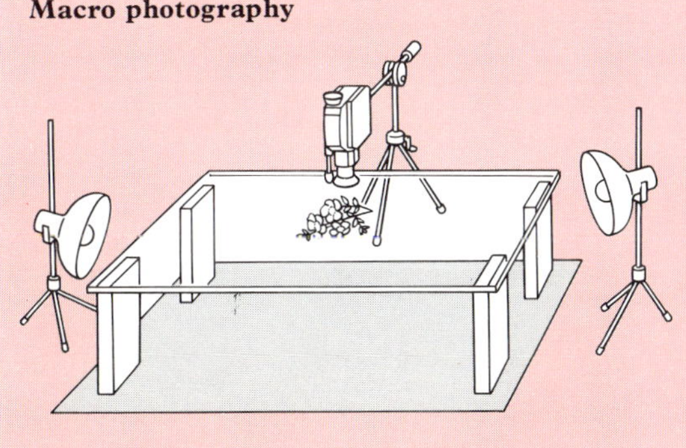

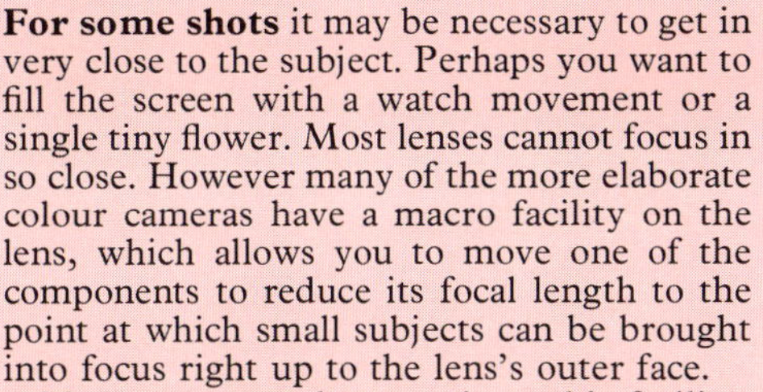

For some shots it may be necessary to get in very close to the subject. Perhaps you want to fill the screen with a watch movement or a single tiny flower. Most lenses cannot focus in so close. However many of the more elaborate colour cameras have a macro facility on the lens, which allows you to move one of the components to reduce its focal length to the point at which small subjects can be brought into focus right up to the lens's outer face.

If your camera does not have this facility, but the lens is threaded to accept standard screw-in filters, you can use a set of the close-up lenses supplied for film cameras. These are slightly more complicated to use than the simple macro position on the lens, but they do have one important advantage: they are fitted on to the front of the camera lens in its normal configuration; so you can still zoom in the normal way, even in super close-up, which is something you cannot possibly do when using a macro lens.

Special effects

By deliberately using the wrong filters, or tampering with the camera's colour balance, it is possible to create special effects. For example, you can set the white balance controls for a bluer light than the one in which you are actually recording. This will give the picture a redder cast which, in moderation, will give your shots a warmer quality, ideal for sequences of people relaxing before a glowing fire. If, on the other hand, you want to record a shot with a colder light, turn the colour balance controls to a lower temperature setting (to an indoor setting, if shooting outdoors or to a tungsten light setting, if shooting under fluorescent lighting). This will give a blue cast to the picture you are recording and can be useful if you want to create 'moonlight' effects under artificial lighting.

Yet, as always when bending the rules, do so only when you have a particular effect in mind. Change the adjustments (or the filters) step by step, and at every stage compare the results by checking the picture on the television screen until you have exactly the quality you want.

Always remember – the more you change things using electronics or optical filters, the more light you need to begin with.

Television studios have complex (and very expensive) special effects generators (SEGs) to produce split-screen pictures, wipes, mixes and other more abstract effects. With a little ingenuity and some extra preparation, it is possible for the home programmer to build in some of these special tricks, using cheaper and simpler methods.

Beautiful colour patterns are created by the feedback caused if you connect the camera to the television, switch on and point the camera at the screen. A glass plate, coated with vaseline around its edges and held in front of the camera lens, produces soft-focus shots with a nostalgic quality. Frosted glass in different patterns and textures creates 'Impressionist' images of subjects.

Many tricks can be done with mirrors: by arranging a large mirror at an angle of 90 degrees between two subjects you can produce a split-screen image of them talking at opposite ends of a telephone wire, for example. Distorting mirrors, or sheets of polished metal, can produce distorted images to simulate madness or drunkenness, or they can be used when shooting dream sequences.

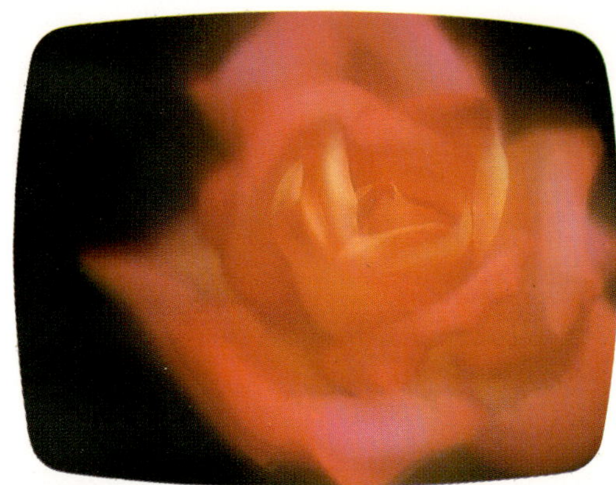

Soft-focus filters are made from glass covered with tiny dimples; the degree of softening depends on their size and number. Soft focus can also be obtained by rubbing vaseline on a plain filter.

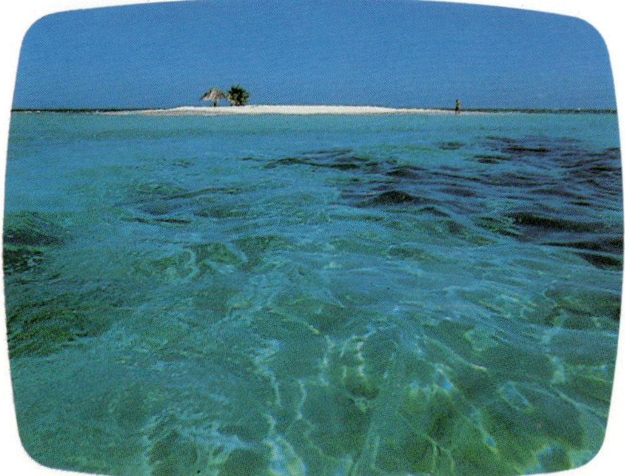

Polarizing filters can eliminate unwanted reflections from non-metallic surfaces such as water. Contrast is heightened and blue skies look darker. The filter can be rotated to vary the effect.

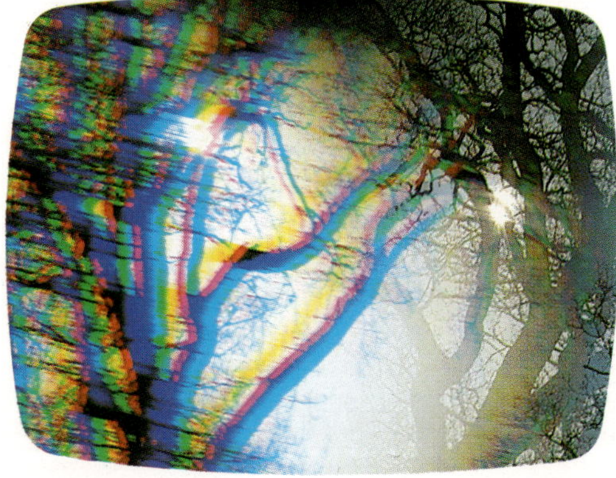

A diffraction filter gives its effect by splitting up the colours in the light falling on it. Some simply split the image into its spectral components; others can also multiply the image.

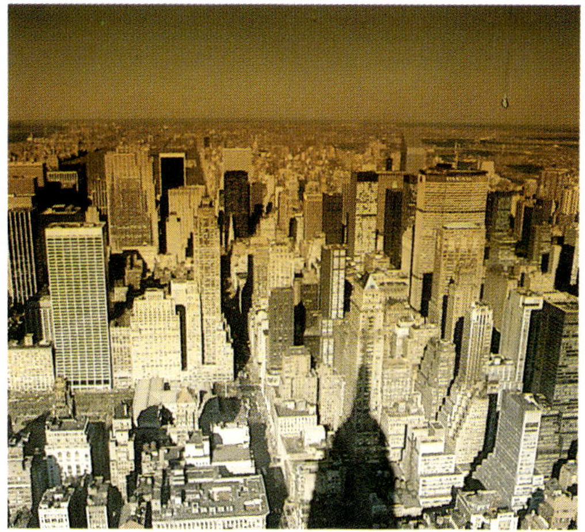

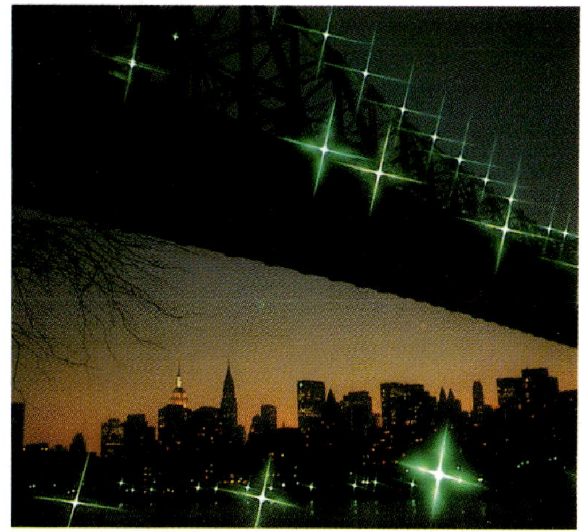

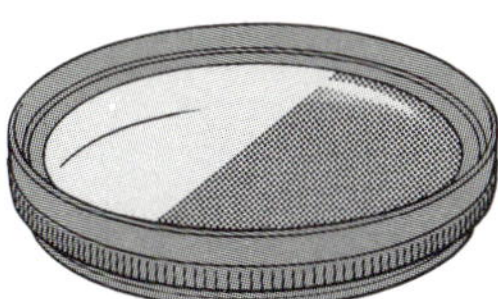

Half colour filters are half coloured and half clear. When mounted in front of the lens they create interesting colour effects. The foreground of a shot may look normal, but the sky may be an eerie red. The division between the coloured and clear halves may be abrupt or graduated. Half colour filters are commonly available in yellow, blue, red, violet, grey, pink and green. They can be used in conjunction with other filters (star burst, for example) for even stranger effects.

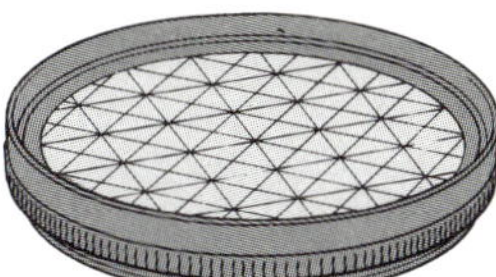

A cross screen is a clear filter with two sets of parallel lines scored on it at right angles to each other. It is useful for enhancing pin-point light sources. If a row of street lights is shot through a cross screen at night, four streaks of light spread out from each light, turning it into a four-pointed star. The same effect can be achieved from point reflections produced by metalwork or glassware under strong lighting. If more sets of lines are scored on the filter, six- or eight-pointed stars appear.

Kaleidoscope effect

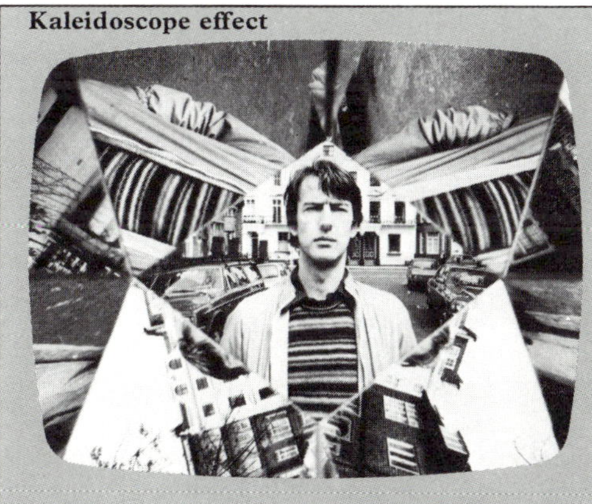

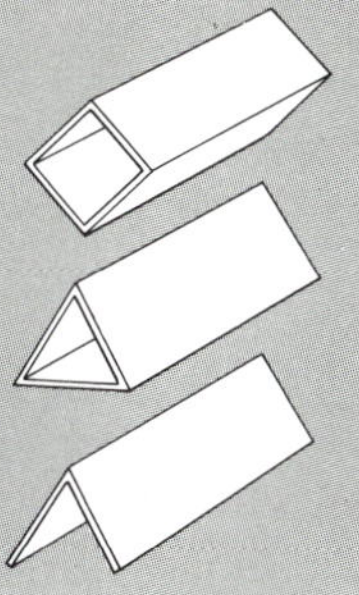

By arranging long rectangular mirrors of equal size into a square or triangular-section tube, and mounting this in front of the camera lens, you can produce multi-faceted semi-abstract shots as kaleidoscopic images of your subject.

Simple wipe effect

A simplified wipe (one picture replacing another by moving in from the side of the screen) can be made using a large mirror. Set up the second shot in front of the camera, and place a mirror at 45° to the camera, *above*. Then place the subject of the first shot so that the camera picks it up from the mirror at the same focus setting as the second shot.

Start recording, then pull the mirror smoothly sideways so the image of the first shot moves off camera to be replaced by the direct second shot.

A video backpack

Making a video movie with a home deck is limiting: the machine is too cumbersome to carry. Moreover, your radius of action is always governed by the length of the power cable and the position of the nearest electrical outlet. A portapack (a portable VCR powered by batteries) is designed to be used on location with a camera.

With a portapack you can make recordings almost anywhere. However, it has a few drawbacks which make video recording a more restricted art than cine filming. Even portapacks are heavy (although new models are progressively lighter) and you have to work within the limitations of your batteries.

Generally speaking, you can operate a camera with a fully charged portapack for between half an hour and an hour at a time, depending on how much power you use in energy-consuming movements, such as zooming, or rewinding the tape. When the batteries are exhausted, they can be replaced with another set, and later recharged with a special unit which can also provide electrical power when a supply is available.

There are relatively few portapacks on the market: most major manufactures offer only one and so there are more VHS and Betamax portapacks, but so far no Video 2000 version. If you have a VHS or Betamax VCR, you need a portapack of the same format for location recording, plus a power adaptor/battery recharger unit. Video 2000 users have to use a portapack of one of the other formats, and copy their recordings on Video 2000 cassettes to play them back on the home deck.

Because they need to be simple and compact, portapacks lack tuning and timing facilitities and, therefore, cannot be used for recording broadcast television off air. If you have no VCR, consider buying a portapack and a tuner-timer unit. This will add all the facilities of a home deck to your portable, and recharge batteries used on location.

Generally speaking, it is better to buy the portapack recommended for your camera; if you buy a camera from one manufacturer and a portapack from another, there may be compatibility problems. Picture quality and durability should be just as good with portables as with their home deck equivalents, but a monitor is still necessary for fine adjustments, and a portable monitor would be useful for regular location work.

Location check list

Before setting out on your first location assignment, make a list of all the equipment you are likely to need. When you return, add to it any items you missed and use it as a check list for subsequent trips. Then, on each assignment, follow these rules:–

1 Assemble and pack all the equipment on your check list.
2 Make sure that all batteries (including spares) are fully charged.
3 Check that enough tapes have been packed for the planned recording time and one more in case you overshoot.
4 Connect all the cables securely to their corresponding terminals and ensure that none is under tension or loose.
5 Clean the camera lens and set the controls to give correct picture and colour balance. Check the picture on a portable monitor, if possible.
6 Check the picture in the viewfinder and run a test shot to make sure that audio and video are being recorded properly.
7 After shooting, play back the recording through the camera viewfinder (or the portable monitor) to check all is well.
8 Label the recorded tapes as soon as possible.
9 Clean the camera and portapack before putting them away.

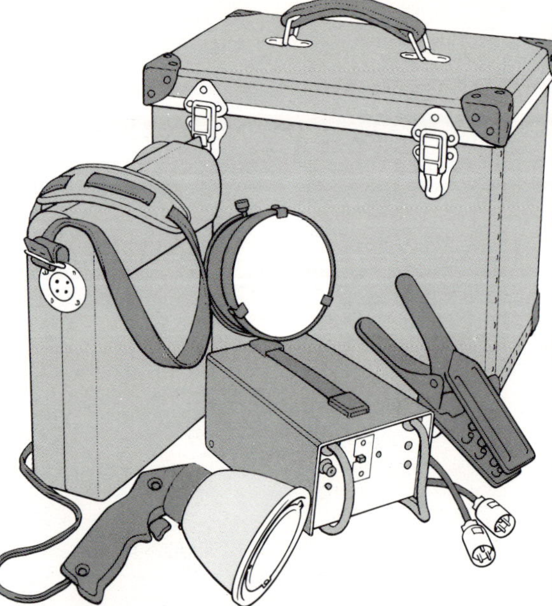

Location lighting is supplied in kits, *above.* A basic kit would comprise two hand-held lamps, with clamps to anchor them and a 12V 100W battery to power them for up to an hour; a battery charger; a daylight conversion filter to eliminate colour temperature compatibility, and a sturdy carrying case.

The transportation of equipment is always a headache for photographers and tape-makers. Everything needed for location shooting can be fitted into a strong, light, waterproof carrying bag or case with plenty of side pockets. The American backpack, *below*, designed for the BVU-110 U-matic portable VCR, provides a useful model.

The carrying straps must be capable of supporting the weight of the fully loaded case. This bag has extensible shoulder straps, **1**, fitted with a comfortable shoulder pad, **2**, to spread the load, and a mike holder, **3**. It also has hand straps, **4**, with reinforced grips.

White panels, 5, held in place by pins or tape, can be used to adjust the camera's white balance controls to suit the lighting conditions.

Loops, 6, can be used to hold a headset, or guide cables.

A flap, 7, protects the portapack and a hood, **8**, shields the camera input socket. The pockets, when filled, cushion the machine from knocks in transit.

The pockets, 9, and **11** are large enough for spare cassettes and batteries, as well as all the equipment ranged *below*. Rolls of tape, **12**, have many uses, from pinning a mike in place, to fashioning a reflector. A camera can be anchored with a clamp, **13**, or a tripod weighted with a string bag, **14**, filled with stones. Filters, **15**, are needed for colour correction and special effects. Pack a stocking, **16** as an extra soft-focus filter, and a roll of kitchen foil, **17**, in case you need a reflector.

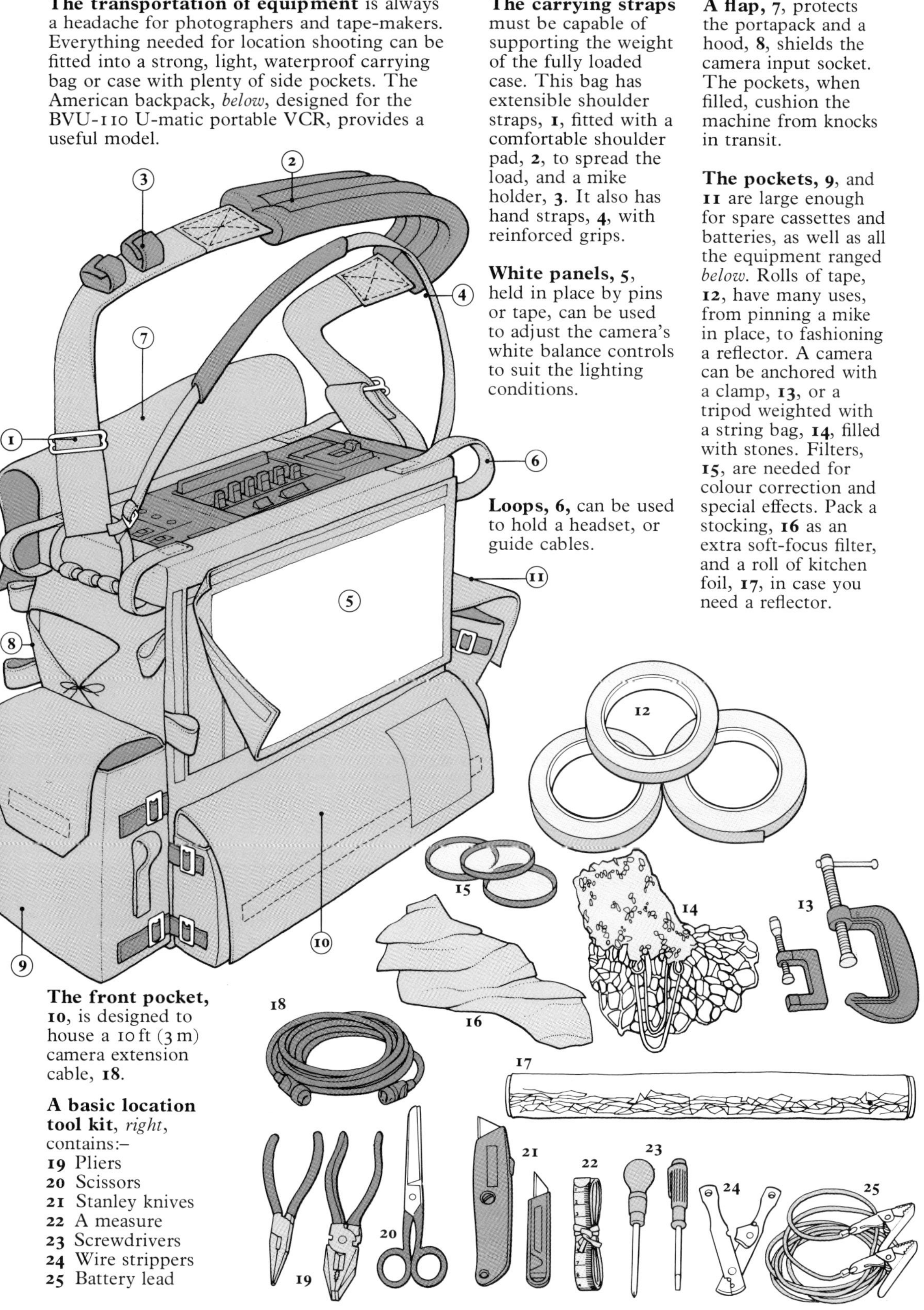

The front pocket, 10, is designed to house a 10 ft (3 m) camera extension cable, **18**.

A basic location tool kit, *right*, contains:–

19 Pliers
20 Scissors
21 Stanley knives
22 A measure
23 Screwdrivers
24 Wire strippers
25 Battery lead

Street scenes

The new freedom a portapack offers brings new problems for you to solve. Although your range and choice may be widened, you lose the tight control over events which you had inside the house. This means you have to learn to recognize good subjects and good shots as they happen. Your planning has to be much more exact, and your reactions much faster, if you are to seize opportunities as they arise.

Take things slowly at first. Move out into the garden and begin by recording any interesting pictures within range. Practise framing and zooming with a greater depth of field (bearing in mind that zooming uses up battery life faster than any other camera movement). Practise being selective: learn to recognize when a shot is not going to work, and be prepared to wait for a better one.

Next, move out into the street. You can try to capture people on tape, zooming in to pick out a face in a group, or following an individual. Eventually, try following a car, keeping it in frame as it approaches and swings past. Keep an eye open for the interesting and the unexpected.

At this stage there is no need to think about building up a logical sequence. You are practising shooting, and worrying about the place each shot might have in a sequence could be inhibiting. In time, you should find you are developing an eye for a good shot.

Look out for useful aids to a more interesting shot; something to shoot a subject through. Try shooting a subject past or through foliage or a fence or, alternatively, focus on the leaves of a tree in front of the subject and then change focus, so that the sharp picture of the leaves dissolves and the subject comes unexpectedly into focus. Done well, this can be a very attractive dissolve between two images.

To make obvious but necessary shots more interesting, try and treat them in a novel way. You can experiment with different camera angles and viewpoints. Shoot from bridges or elevated walkways to give a bird's eye view of a scene; or shoot from subway steps to give a low-angle perspective. Shoot across a busy road at faces on the opposite side, so that passing traffic provides an out-of-focus punctuation to your shot. Learn to look at the apparently unpromising location in terms of the possibilities it presents, and the kinds of shots you can find in it.

Shooting in daylight
Before you venture outside with your camera and portapack, make sure the camera's colour controls are set for the prevailing lighting conditions. You will be operating without a monitor, and so no longer able to make quick visual checks of the colour balance in the picture.

Shooting outdoors means that you can rely on sunlight for most of your lighting requirements, but beware of shooting toward the sun. Even if you are sure that there is no danger of direct sunlight entering the camera lens and damaging the tube, the details of your subjects, particularly people's faces, will be in deep shadow and will tend to look black against the bright surroundings. You can learn to recognize when this is happening, even in a black and white viewfinder, but where the shots are really important, the only insurance is to use a portable, battery-operated monitor.

If you have to follow a subject from a sunny to a shady spot, move the camera slowly. An auto iris will react, but not instantly: there will be a momentary time-lag before it opens enough to compensate. Alternatively, you can fade down, or stop, as the subject walks into the shade. Then, move to a new shooting point, frame up a new shot in the darkened area and begin recording again. This way the changing lighting conditions, and change in shot, can be used to dramatic effect.

Yet another alternative is to lift the patch of shadow, much as you would with a fill light in the studio: a professional might have an assistant aim a hand-held, battery-operated light on the subject from a position at an angle to that of the sun's rays.

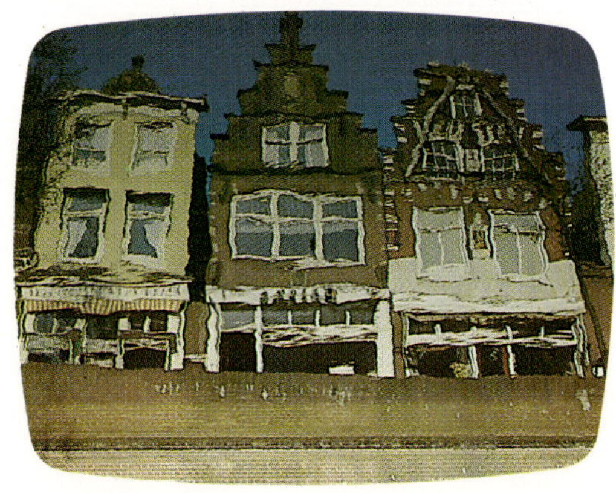

Look for out-of-the-ordinary shots. Try shooting a subject reflected in water, or in a shop window, then zoom out to show that it is a reflection. A direct shot may then seem more interesting.

Use groups of people talking to experiment with framing. Shoot across a busy road at faces opposite so that passing traffic provides an out-of-focus punctuation to your shot. Take close-ups of inanimate objects for interesting cutaways.

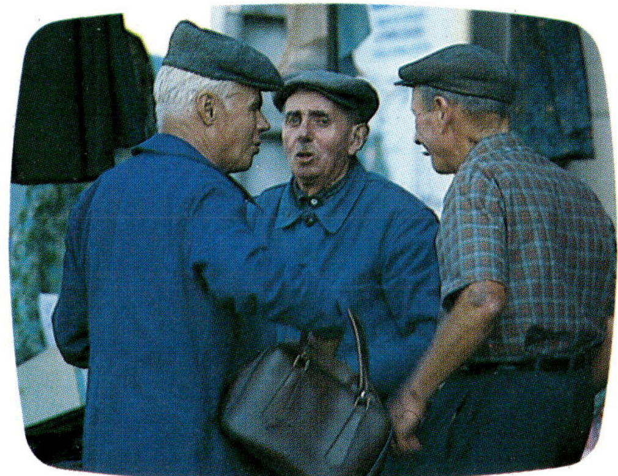

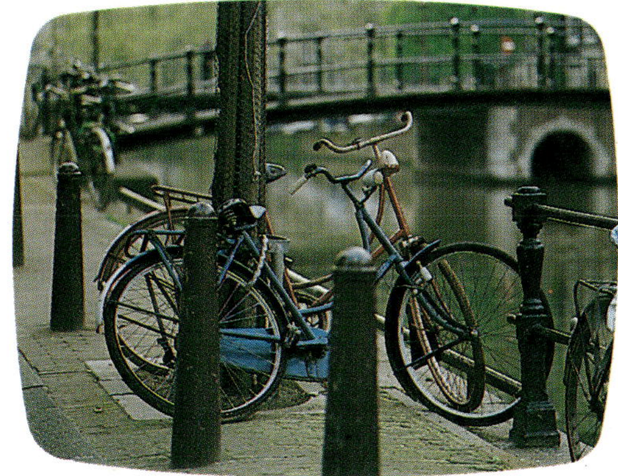

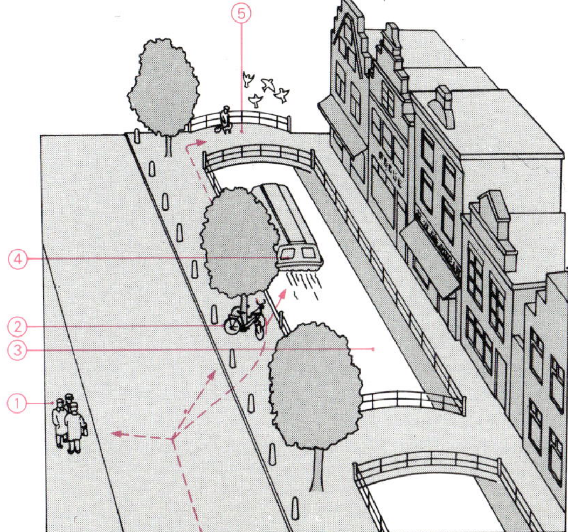

Taking the camera into the street will provide the video novice with a wealth of subjects for essential shooting practice. Use the viewfinder to assess the potential of every view along and across the street. Pick out individuals in a crowd by accurate zooming, **1**, *left*. Look out for shots that summarize the atmosphere of a place, **2**. They often make good opening shots because they establish the location in the mind of the viewer: houses with stepped gables, **3**, say 'Amsterdam' to most people.

Practise following moving objects, and keeping them in frame. Begin with something slow, such as a boat, **4**, or people walking along a road, then progress to faster objects, such as moving bicycles. Learn to take advantage of sudden events, such as a flock of birds taking off from the street, **5**. It demands fast reactions.

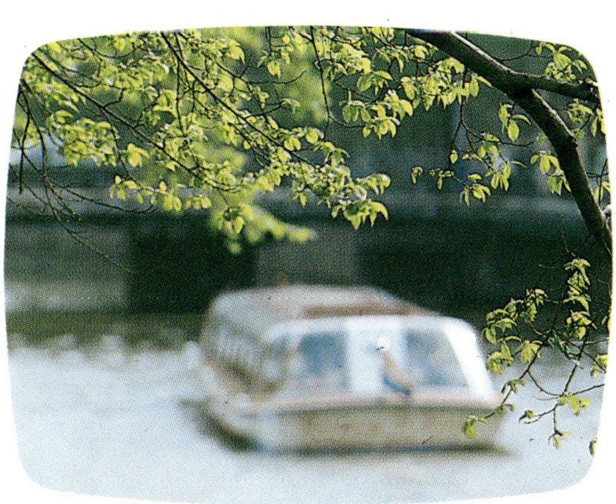

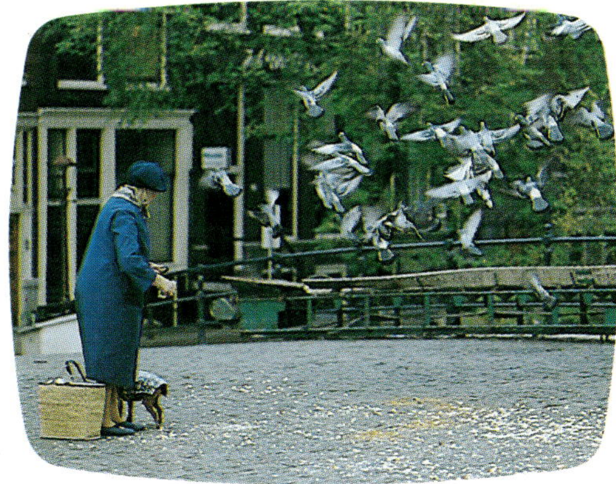

Create dramatic interest by opening on a moving subject out of focus: it will appear as a moving blur. Bring some bankside foliage into sharp detail, then gradually bring the boat into focus.

Make the most of every shot by using the zoom to make two shots out of one. Here, you could open with a long shot and then keep still, or follow a pigeon as it flies, and finish by zooming out again.

City walkabout

Having discovered the freedom offered by a portapack, the next step is to make full use of it by putting together a sequence on location. Try not to be too ambitious at this stage; it is much easier to cope with problems in a familiar location than on the top of a mountain, or on the deck of a boat.

Begin by going into town on a Saturday morning. Shoot scenes with plenty of colour and movement: crowds of shoppers or children playing in the park. Concentrate on looking for interesting shots.

There are two main points to remember. First, video equipment is bound to be noticed, so learn to ignore the attention you will inevitably attract. Second, even portable video equipment is heavy, so keep the walking and equipment-carrying to a minimum by planning out the shooting sequence in advance. Make time to tour the location ahead of the day you plan to shoot, and work out a rough sequence of shots. Try sketching out on a piece of paper how the shots will look. Afterwards, your extempore storyboard can be rearranged in a more logical order, to cut out unnecessary moves.

There is no need for a detailed written script, a natural sound-track is often more appropriate than an elaborate commentary. For instance, a shopping trip might open with a shot of people in line at a bus stop. Then you need a shot of a bus approaching. You might want a quick shot of people looking up the road, the bus halting and the people climbing on board. You might then cut to a shot of a bus arriving in town, as a handy way of shifting the location, and follow this with a shot of a store, and so on.

It is important to keep the storyboard simple and flexible, because the shots you record may have to be quite different from those planned. If it is impossible to make a shot planned on the storyboard, keep looking for something similar, or some other shot which will make the same point. Something entirely unexpected might happen – a parade, a ceremony, or an argument in the street. Events like these can be a gold mine to the aspiring video director, but only if they can be fitted into the intended sequence.

Even if your storyboard is no more than a list of shots giving clues to an intended sequence of visual ideas, having a definite plan to work to makes it much easier to take advantage of unexpected opportunities.

Editing in camera

Every time you switch the camera on, the picture takes a few seconds to build up; and every time you stop it, it breaks up again. This means that when you play back your recording, the picture may break up between shots.

If you press the 'Pause' or 'Stop/Start' button on the camera, instead of switching off at the VCR, the recording circuits run back a little before recording begins, and the picture has time to stabilize before new material appears. This is known as backspace editing, and it means that a sequence of several shots can be built up without the picture breaking up between each one; but as soon as you stop recording, the problem of picture break-up recurs.

You can avoid this by fading the picture out. Close down the iris, so that the image darkens. When you are ready to record again, open up the iris, focus and frame, then close the iris down again, start recording and bring the picture up by opening the iris. The picture break-up will happen, but when the screen is dark.

The fade-out-fade-in technique is often used by professionals to show the passing of time between scenes, which is when you are most likely to need to switch the machine off. It can be a problem, however, where one shot is to be linked with the next on a different location. The answer is a device like that on the Hitachi 6500 which, even when switched off, can store in its memory for several days the synchronization of the last shot recorded. The next time any material is recorded, the VCR automatically matches it up to the previous material.

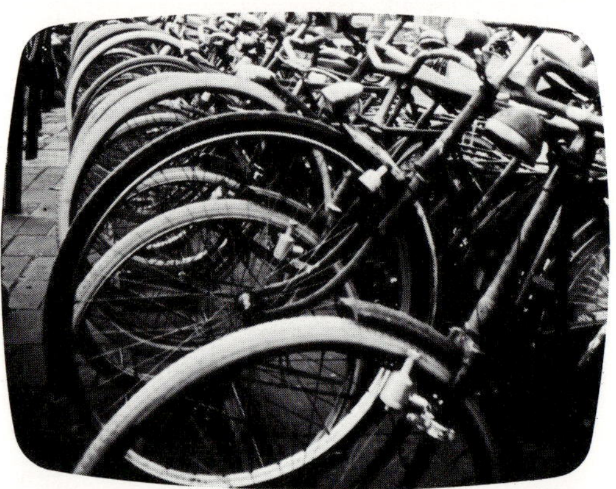

The opening shot in the sequence might be a view along the bicycles in their stand, *above*. Although there is no movement in this shot, natural sounds such as traffic, and the voices of the riders approaching, will create an appropriate atmosphere for the subject.

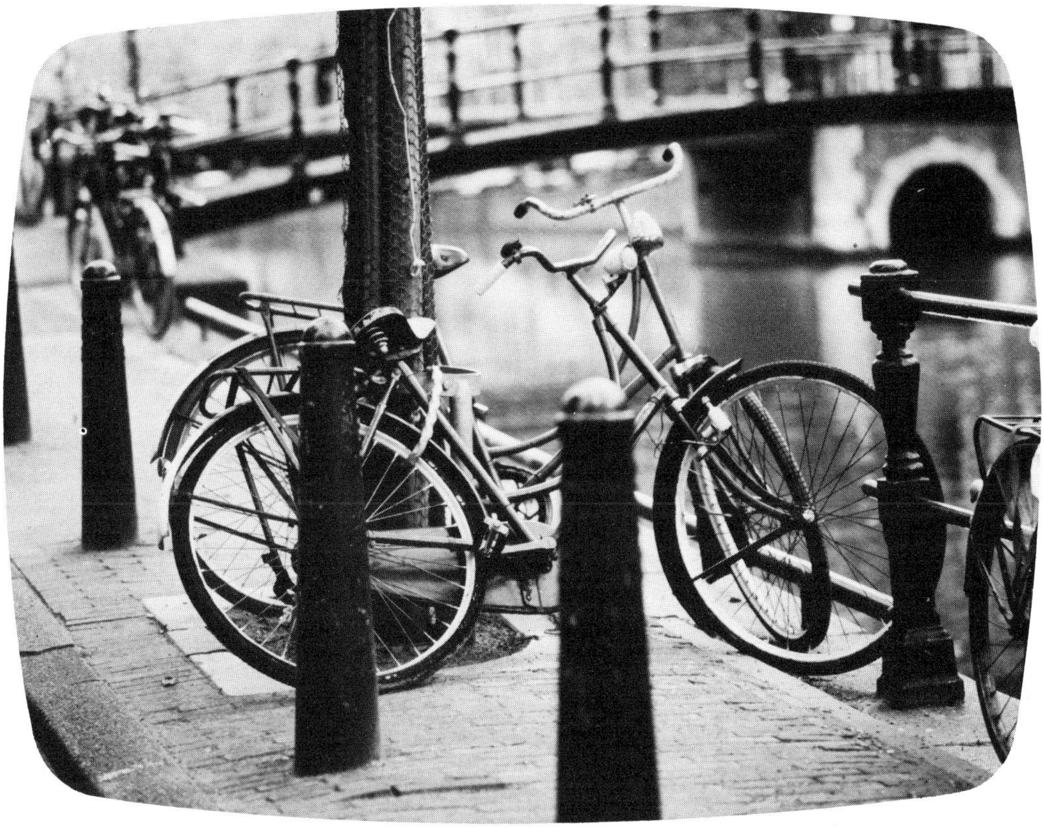

You should by now be thinking in terms of a series of images combined to put across an idea. Look out for something different: a sequence about a bicycle trip, for instance, might take bicycles, not their riders, as its subject, and follow their progress from bicycle shed or stand to destination.

The subject of the main shot in the sequence, *above*, is undoubtedly a pair of bicycles, yet they are shot through a line of bollards and against a canal back-drop. However, careful framing and selective focusing leave the viewer in no doubt about the subject.

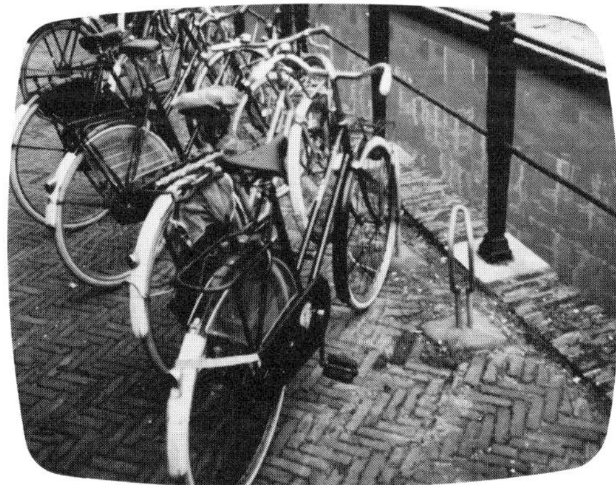

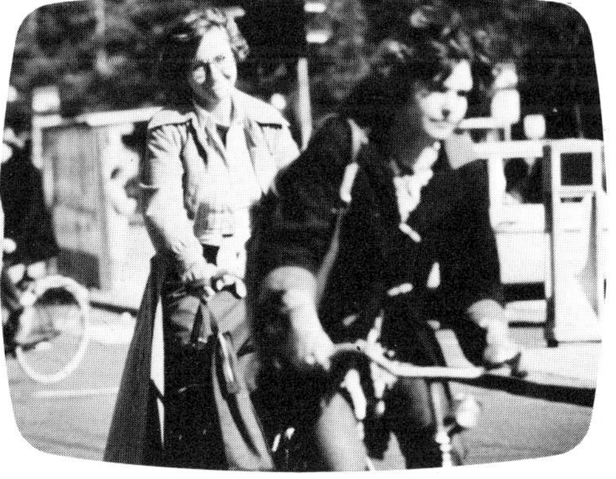

Experiment with different viewpoints before making the shot planned on the storyboard. If the planned shot is not available for some reason, knowing how it is to relate to the others in the sequence will help you think of some alternative, such as a different angle, *above*.

A following shot will make any action sequence more interesting to watch. A camera operator could follow cyclists in a car, and even pass them, *above*, swinging the camera round until they appear to be cycling towards it. You can then shoot them as they dismount.

Making a documentary

During the last few years, the portable video recorder has revolutionized video production. Before that it was necessary to carry heavy mains machines, and miles of extension cable, to try to reach outdoor locations which were still within reasonably easy reach of an electrical power supply. Now, domestic video is making the same move that broadcast television made during the sixties: recording outside the confines of the studio in every kind of location, on every subject.

This means that the immediacy and audience involvement of video can be brought to bear on subjects which would previously have been impossible: from a narrow boat on a canal, to a combine harvester working deep in the countryside. The whole world is there for the recording.

Although your equipment costs and budget will be a small fraction of those of broadcast television, you will encounter the same benefits and the same difficulties as the professional. To begin with, the biggest difference between location work and shooting scenes at home is that both events and subjects are less easily controlled. As a result you can only try to capture on tape parts of the action which seem to you to present an image and give the overall feeling of the subject.

For instance, a horse-racing event, or a canal-boat trip, cannot be shot from start to finish; that would use too much tape and be boring to watch. To make an interesting documentary of such an event, you would have to try, first of all, to isolate the quality the documentary is to put over: the excitement of a horse-race, or the sense of tranquillity of a canal trip, and then plan what kind of highlights are necessary to create that quality.

Keep the planning of a documentary simple and flexible; but better a list of possible shots scrawled on the back of an envelope than no plan at all. Do not plan a sound-track at this stage, but be sure to record enough atmosphere through the camera microphone to give the finished production that extra, lifelike quality. When covering a subject which allows you plenty of time to set things up, try to be slightly more ambitious. On a canal boat, for example, make a point of including all the sounds of the lock machinery being worked, to go with the picture.

The grammar of shooting
The easiest way of working out the relation of each shot to the others in your plan is to think of the whole sequence as a paragraph of text. Each shot is equivalent to a word, and a sequence of shots makes up a sentence. Shots, like words, have to fit together logically, so that the audience can follow the sequence; a shot which is not logical will only confuse the viewer.

Cutaway shots are needed to bridge gaps in the continuity of the main shot. For example, a sequence showing a boat moving into a canal lock, being raised to the next level, and then leaving the lock, would take too long to record. You might try showing brief shots of each phase of the sequence, but when you played back the three shots in order, they would look ridiculous: the boat would seem to jump about on the screen.

Instead, you could separate them with cutaways. After the boat enters the lock, show a close-up of the gates being closed, another of the winding gear being turned and then the boat rising with the water. Then take a shot of the gates opening, one of the steersman opening the throttle, and finally shoot the boat leaving the lock.

Style is as important as continuity. Try to vary the images reaching the viewer by including close-ups in between long shots. Remember to punctuate the end of each sequence visually in some way. You could use a long shot, then cut to another long shot to emphasize a new place and a new time for the next sequence. If a longer pause is required, remember the fade-down-fade-up technique. This can also be used to suggest that a fairly long time has passed between two views of the same subject. It will not matter if there are differences in the details of the two pictures.

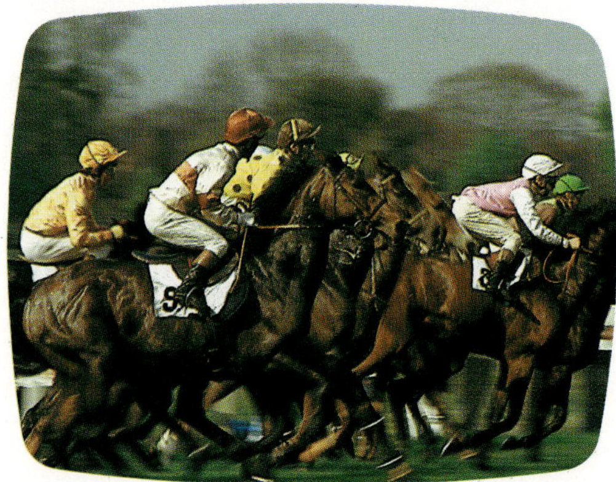

It is impossible to shoot an entire race, but you can build up a sequence by shooting the start of one race, *above*, and the middle and end of subsequent races.

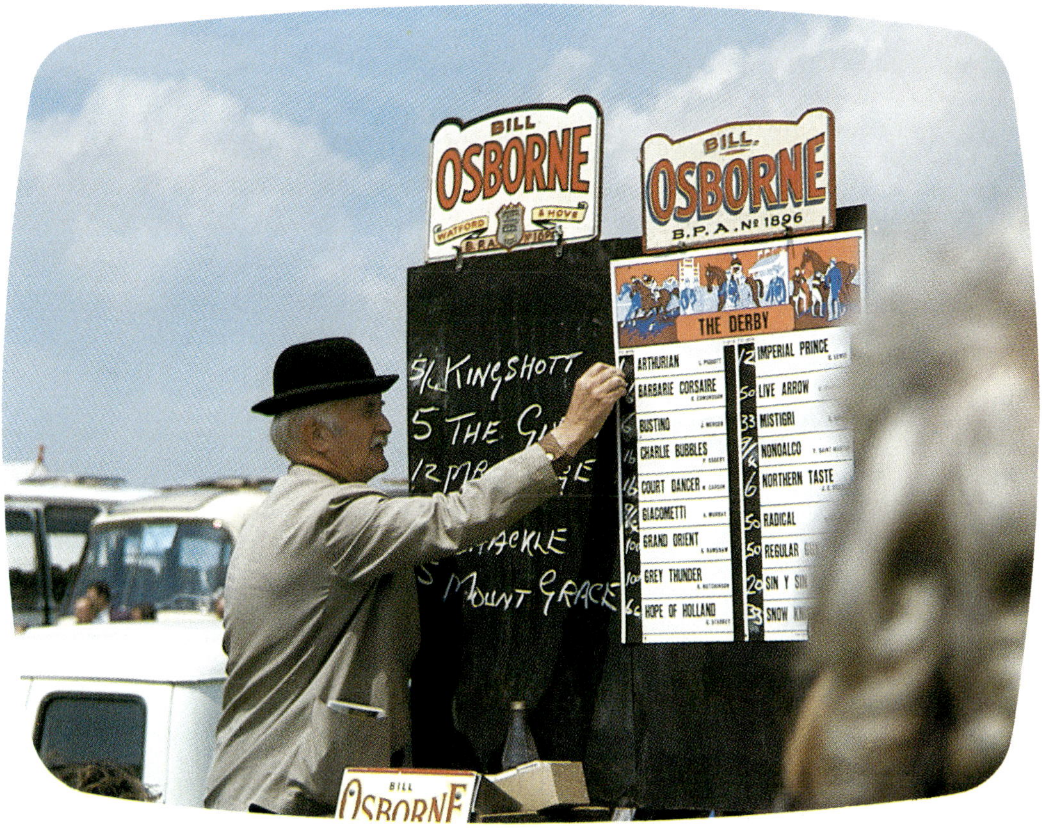

Making a documentary about horse-racing presents problems. It is difficult to record a single race from start to finish, unless you have a particularly good viewpoint. It might be more practical, and certainly more effective and entertaining, to try to convey the atmosphere of a horse-racing event, to which the race is only one contributor.

There is a lot of off-course activity: pick out the lead pony on an American course or, in Britain, a bookie, *above*, by careful framing and zooming. Let the mike pick up the reactions and comments of the spectators.

Off-course scenes for cutaways could include shots of grooms leading horses in and out of stables, types and characters around the course, and anxious or triumphant faces in the crowd.

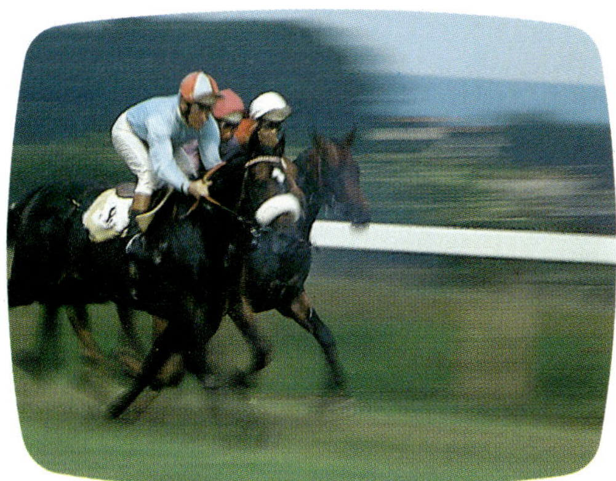

Practise following groups of galloping horses by panning. Make sure you can keep them in sharp focus all the way through the shot, before recording.

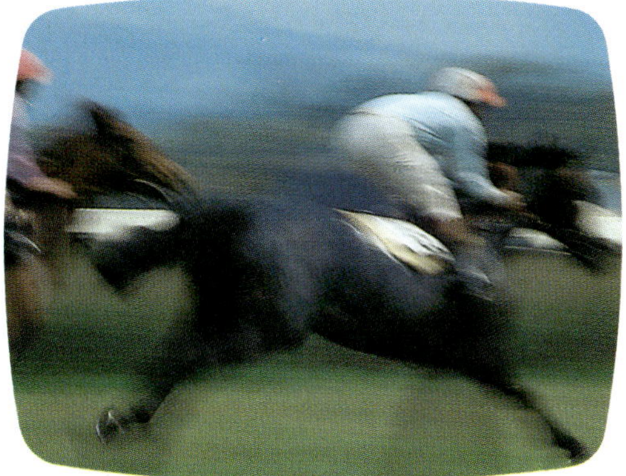

A fast pan (called a swish, whip, zip or blur pan) can make a spectacular connecting shot – perhaps between a lagger and the leader during the run up to the finish.

Shooting team games

If your family and friends are as intrigued as you are by the demands and the possibilities of video, involving them in the production process will be more rewarding for them than always watching, and reacting to your finished efforts. One person can rarely produce a video recording alone. Although one's inspirations, vision and control may be essential to transfer an idea on to video tape, the help of an inexperienced assistant can make a noticeable improvement.

Planning and shooting even a simple sequence can be surprisingly demanding; it is the thinking and hard concentration that makes video such an absorbing hobby. However, video equipment is still very cumbersome, a fact you appreciate when you have to move around on location. When you add to all the mental effort the physical demands of carrying around and looking after the location kit, the result is usually fatigue.

If you are tired when you are shooting on location, there is naturally an irresistible temptation to stay in one place, rather than to move to a different viewpoint, which might bring you a better, or more dramatic shot. The final results will usually disappoint you, for although no one else may be aware of the shots you missed, you will invariably be your own severest critic. A keen and willing assistant can therefore be a help on the simplest level: by relieving you of some of the physical effort of video production.

An extra person (or more, since there is no reason to limit yourself to just one assistant, if you have more help available) can be particularly useful in helping you to speed up a move between one camera position and the next. This is essential when you are having to react to something which is happening outside your control.

A sporting event, such as a football match, is a good example. Rugby, and the football games played in North America, involve frequent intense battles over a small section of the field at a time. If you are shooting from one position throughout, no matter how dramatic the action, you will miss the opportunities for the thrilling close-ups which make the match visually exciting for the viewer. A tape covering the whole match could be tedious and costly, but your success in catching the highlights depends on your ability to spot trends, and reposition yourself to take advantage of them.

Video camera lenses

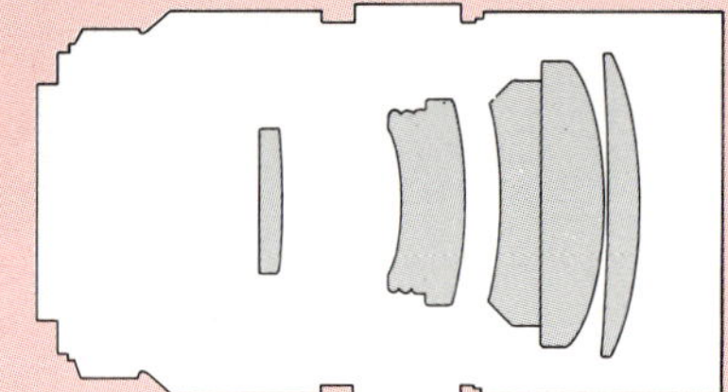

With a telephoto lens on your camera you will be able to cover more of the action. Many video cameras now offer a 6:1 zoom lens, some have 10:1 lenses, and JVC offers a 14:1 zoom lens as an optional extra on its 3-tube KY-1900. You can extend the range of your camera lens by fitting a telephoto adaptor ring in combination with the lens, but unless the optics are precise, you may lose definition in extreme close-ups.

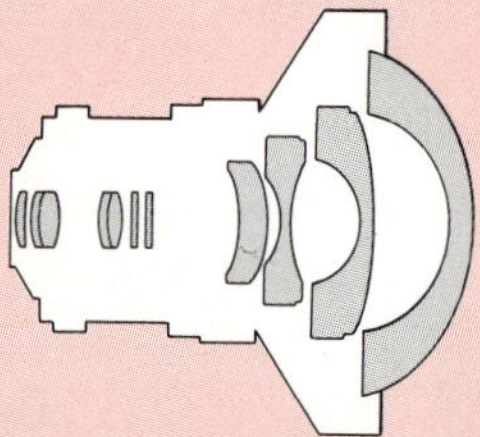

Video cameras are usually made to be used with the lens that comes with them, but if your camera has a standard lens fitting, a film camera lens – perhaps a telephoto lens with an extra zoom, or longer focus, a wide-angle lens or even a fish eye lens, *above* – will fit. Make sure you buy it from an expert salesperson: the geometry of a film camera lens may not adjust to the optics of a video camera without an adaptor.

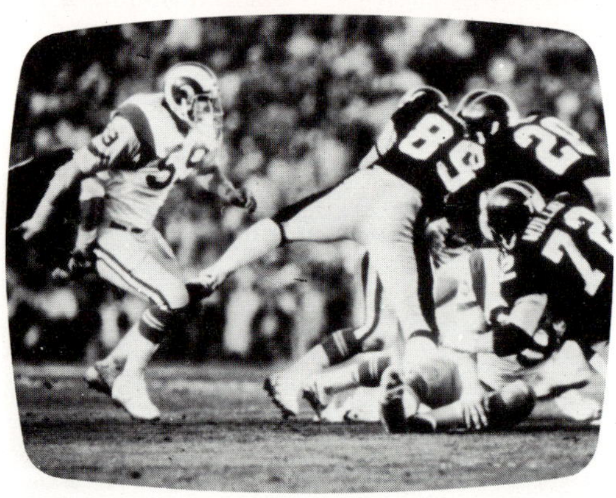

For the viewer's sake, an action shot or foul should be in close-up. At full zoom, steady the camera against something (telephoto lenses magnify shakes) and be ready to zoom out with the ball.

The distorting effect of a wide-angle or a fish eye lens can make an effective opening shot, *right*. Here, a camera, angled from a crouching, kneeling or prone position, emphasizes the players' power and strength. They are not facing the camera, yet the lens transmits the threatening atmosphere.

Do not cross the pitch after you have established your shooting position. A shot from the other side will confuse the viewer over which way the teams are playing.

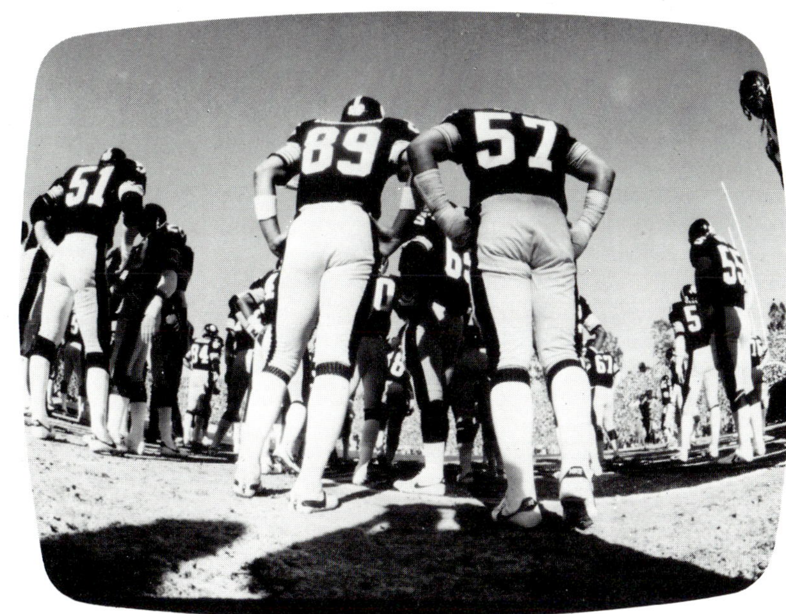

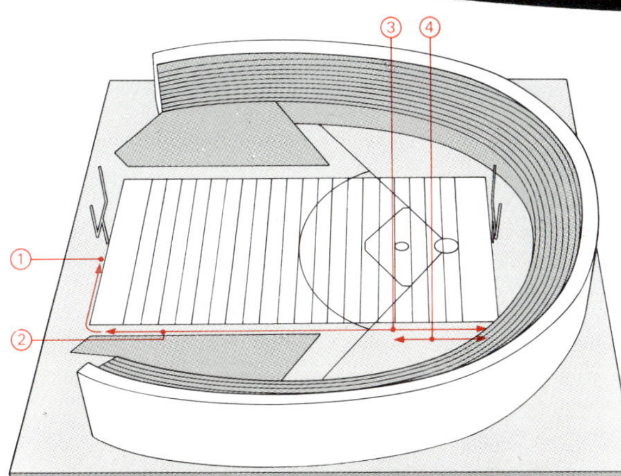

Your freedom of movement will determine the way you cover a sporting event. Set up near the half-way line if you have to stick to one location for the whole match. Position **4**, *left*, gives you a view of the start, and from **3** you have a clear view of the action in the centre of the pitch, **2**.
Your assistant can help you set up, and work the VCR.

If you can move around, try and stay with the team that seems to be gaining advantage. Your assistant's job will be to spot trends as they develop, warn you and help you get from place to place in time. Risk setting up at the goal line, **2**, to wait for a dramatic touch-down, then move round the corner, **1**, to shoot the conversion.

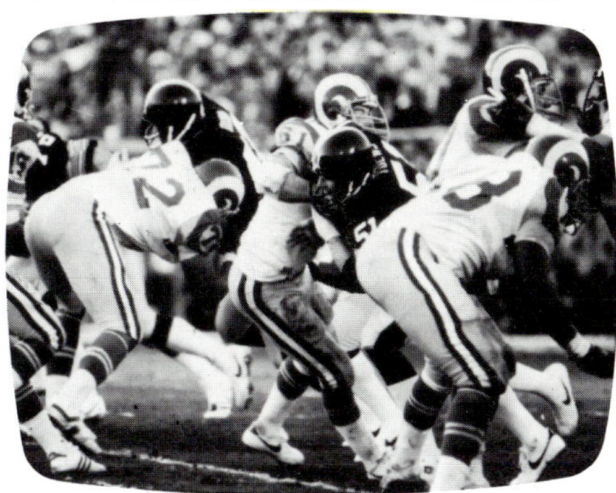

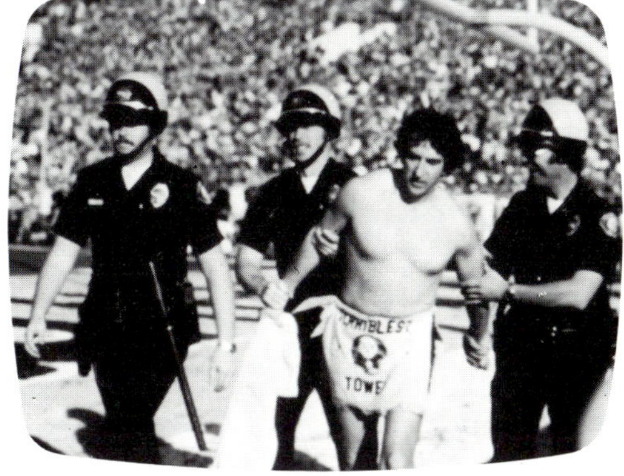

Pan to follow action moving across the field of vision. It needs a steady hand and tripod, but the camera stays in focus longer than when zooming. Move out at intervals to review the game's progress.

Take cutaway shots of spectators, and of players watching the action at the far end of the field. Keep the camera running after the final whistle goes: the crowd's behaviour could make a good end to your tape.

Fast-action shooting

An extra pair of helping hands will not only cut down your response time to developments in the progress of a team sport, it will extend your range of activities. With an assistant you can try shooting sports such as canoeing or slalom skiing, whose locations are invariably harder to reach than a nearby football stadium.

Canoeing, for example, calls for good close-ups at the fiercest rapids, which may mean a long and detailed reconnaissance up and down the course. With an assistant, you can search for unusual viewpoints from which to start and finish the recording, and to provide visual variety. By carrying the equipment up a nearby hill, or on to a bridge overlooking the canoeists' course, you might be able to record a really dramatic bird's eye view. However, have your assistant guard the video kit while you scout; if you both go you double the effort.

Video taping skiing calls for even greater effort, since you have to move through snow. For the most dramatic shots of such fast-action sports you need to know in advance when each competitor is about to appear in camera range. You have to frame the shot and have the camera running, so that in the recording the action bursts upon the viewer. However, in a hilly location you may not be able to see the competitors approaching the spot where you have decided to begin.

An assistant is invaluable in such a situation as this. Work out in advance the spot the competitor will reach at the moment you start the camera, and position your assistant in a place that gives a good view of both you and the oncoming competitor. Then arrange a cue: a wave of the hand will not appear on the sound-track. If there are several competitors, try one or two dry runs with the camera switched off so that you know you can react quickly enough to the cue to begin recording at exactly the right moment. Cueing is essential if the success of a shot depends on timing, especially when you have no editing facilities.

Fast water sports such as sailing or water skiing are difficult to capture dramatically unless you can become part of the action. Look for a place to moor a rented or borrowed boat out of the spectators' line of view, but close to the competitors. Your assistant can help steady you for awkward shots which call for a well-braced camera position.

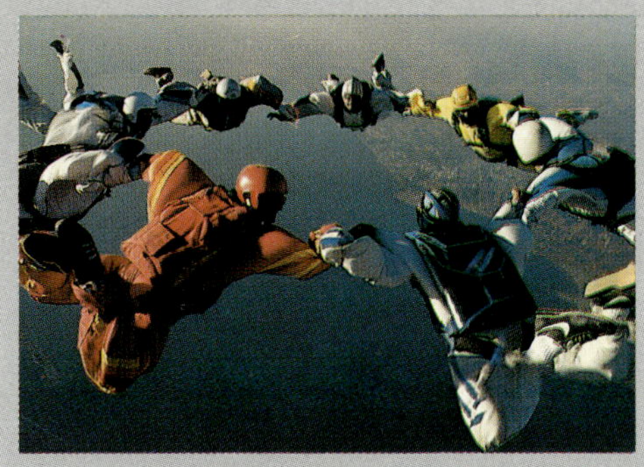

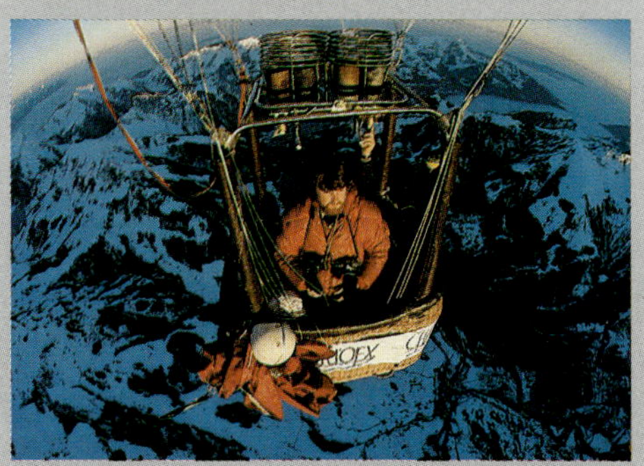

Shooting on the move

The best location vehicle is a hot-air balloon. It is entirely smooth and gives an open view of the ground. If you use one, keep your scripting flexible: you have no control over the direction in which you fly.

Airliners offer only restricted window space and from the centre of the fuselage you see only an expanse of wing. A high-wing single-engine aircraft gives you a better view. In aircraft or automobile, try mounting the camera on a folded tripod and bracing it against the seat to steady it.

In small boats it is difficult to keep equipment dry, so wrap it in tough polythene bags. On a larger yacht, the sails blot out much of the view, but with an assistant to hold you, there are spectacular shots to be had from the bows.

Trains can offer a good viewpoint for passing shots of town and country. Try using a tripod if the track seems smooth. Never, whether in train, car or aircraft, steady the camera against the window: the vibration will be transmitted directly to the recorded picture. Instead, cradle the camera in your arms and angle the viewfinder so that you can see the picture.

Remember, however, that when shooting from a moving vehicle is part of the story, movement and background noise establish the immediacy of the action. The gentle rise and fall of the sea, for example, reminds the viewer of where the action is taking place.

When shooting from above, do not waste picture space on mountains and sky if there is any action down below. A medium shot, *right*, cuts out the context in which the sport is taking place, but is wide enough to show what is happening. Remember that zoom and telephoto lenses will magnify any unsteadiness in your hold on the camera, and following a moving object smoothly calls for a great deal of practice.

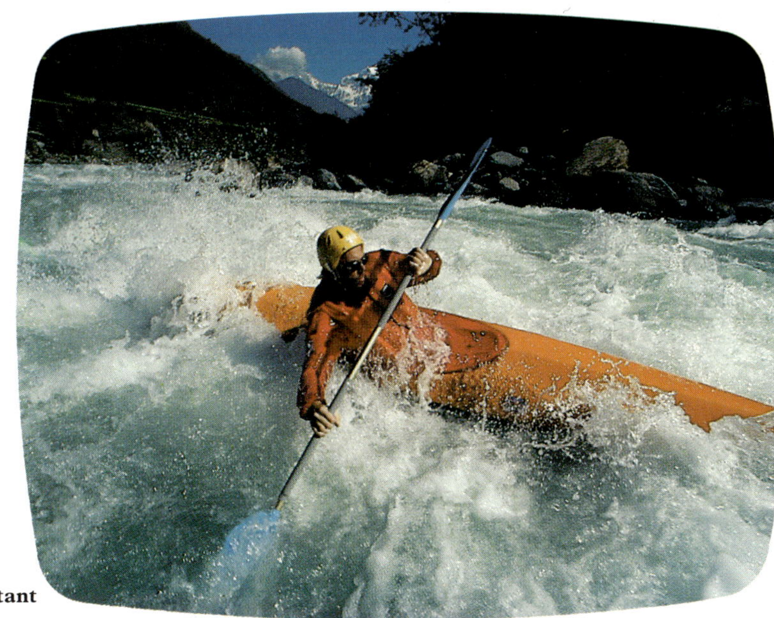

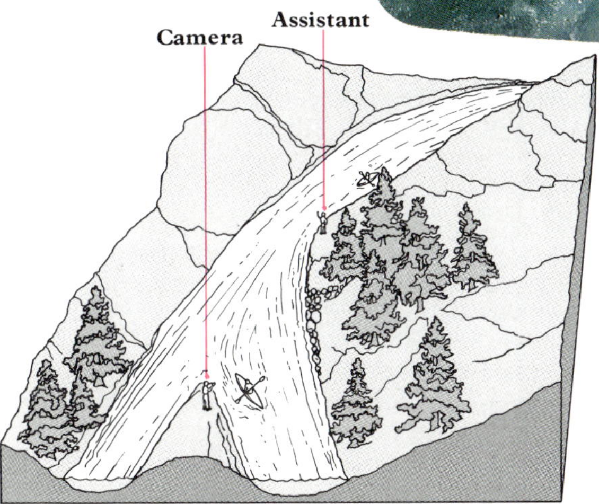

A good vantage point is essential when shooting any sport. Set your camera up in such a position that most of the action takes place at a constant distance. With canoeing this may have to be at the top of a gorge, or on top of a hill, *left*. Station your assistant in a position from which to signal you just before each canoeist is in view.

Always begin with a wide shot so as to establish the scene in the mind of the viewer. Shots from high up make spectacular openers: you could tilt down from the sky and hills to the river below, and zoom out to show the setting from time to time. To close, zoom out, tilt up to the hilltops, and fade out.

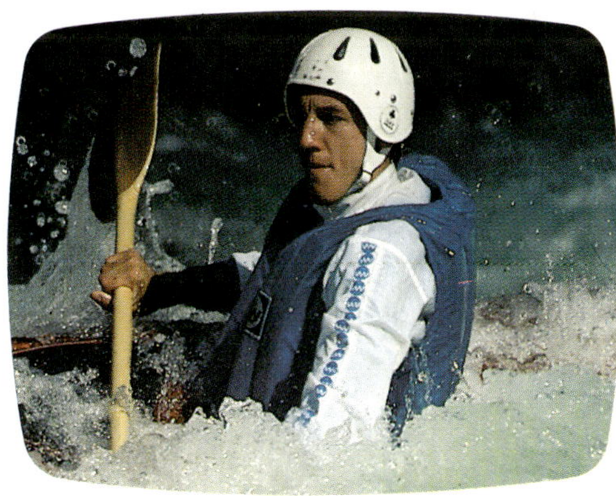

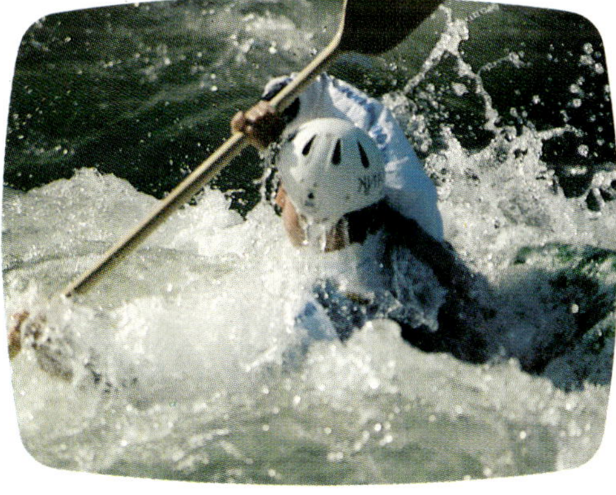

Canoeists all look alike in a long or medium shot. When you set up the camera, pre-set it for close-ups and zoom in wherever possible, to show the concentration on the canoeists' faces.

A close-up gives the viewer far less information than a long or a medium shot. Before you zoom in, make sure the preceding shots have made the viewer clear about who or what you are shooting.

The TV interview

By helping in front of the camera, as well as behind it, your assistant can add a new dimension to your recordings. With another person presenting, introducing, commenting, acting as a television reporter, or interviewing other people, you can try out some entirely different ideas. Should your assistant be camera-shy, teach him or her to operate the camera while you do the presenting. Once they see how painless and simple it is they may feel emboldened to try it out.

Good presention is an art which can be learned: acting schools run training courses designed to teach people to project their personalities under the stare of the camera lens and the glare of the studio lights. For the amateur, it should not be so difficult. Professional television is, to a great extent, a prisoner of its complexity; people tend to be intimidated by the awesome bustle of a large broadcasting studio. Provided you keep the atmosphere informal, and rehearse often, the ease with which most retiring people can learn to feel relaxed and appear professional in front of your camera will surprise you.

Begin by trying it for yourself, so that you understand the problems. Sit in a chair with lights positioned, and the camera set to record you as a straight to-camera presenter (like a newsreader, for example), and deliver a simple statement. You will probably find it easier to make up what you say as you go along, following brief notes, than to commit a speech to memeory. Unless you are a natural actor, the strain of remembering learned lines will show in the expression on your face and in the intonation of your speech. Try and treat the camera as you would an old friend to whom you are telling an interesting story.

Split up the story into parts which are neither too long nor too complex to remember. The camera operator can then maintain visual interest by varying the camera angle, or the framing of the shot, for each part. Timing will be important. You should begin speaking just long enough after the beginning of the shot for the viewer to have taken in the fact that you are on camera and about to say something; but not so long after the beginning of the shot that your viewer is aware of a laboured pause. Practise the timing, until it becomes second nature. You will need a cue from your camera operator so that you know when the recording is about to begin: it should be a signal you can see without seeming to look away from the camera, such as an obvious wave of the hand, or a nod of the head.

Finishing needs good timing, too. When you deliver the final words, continue to look at the camera; do not drop your gaze. Let your assistant know what the final words will be (professionals call it an out-cue) so that the camera can be stopped after a pause of a second or so. Remember your out-cue, so that you don't add something afterwards. You might be cut off in mid-sentence.

Almost everyone is shocked when they first see themselves on screen, just as they are when first hearing their voice on a tape recorder. The reason is partly due to everyone's self-image being far removed from the self they see on the screen. Another cause for alarm is that the personality emerging for the first time through the camera is one reacting self-consciously to a daunting and unfamilar situation, and at first is nothing like the relaxed, everyday self.

Finally, remember that the television screen allows you to study yourself, and anyone else, far more closely than you would be able to in ordinary circumstances. You are likely to be over-conscious of mannerisms or characteristics which seem uncomfortably conspicuous. The viewer, however, will probably not notice them.

In time you will find yourself relaxing in front of the camera. As this happens, you will grow more accustomed to, and more pleased with, the impression you give on the screen. As you learn to think more freely in front of the camera and to cope with more complex cues and routines, you will find a growing pride in your mastery of a difficult craft.

A good presenter should be able to persuade other people to talk to the camera. Your evident ease is the first requirement; it will help other people to respond in kind.

Begin with the easiest possible interview assignment, with someone who knows you well, in informal surroundings, then try asking for an interview with someone you know, but not so well, such as the captain of the local sports club. It may help in both cases to talk through the questions you are going to ask, and the answers you are looking for, before you do the live interview. It will avoid confusion, but do not go over it in too much detail, or the interview may look over-rehearsed.

LS street with crowds. Slow ZOOM in to MCU PRESENTER. Presenter turns to camera.
Presenter: *'Only three days before the elections . . .'*

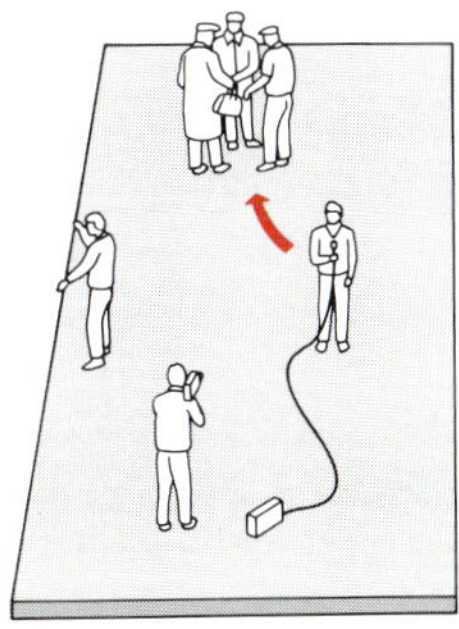

The 'vox pop' interview, where a reporter asks people in the street their opinion on some issue, is more difficult than an interview with a friend or associate. You have to introduce yourself, ask your question as briefly as possible, and hope the people respond. It can be disconcerting if they cross the street when they see you approach with a mike.

You need a long mike cable, and a cue: you must turn to the camera just after the zoom.

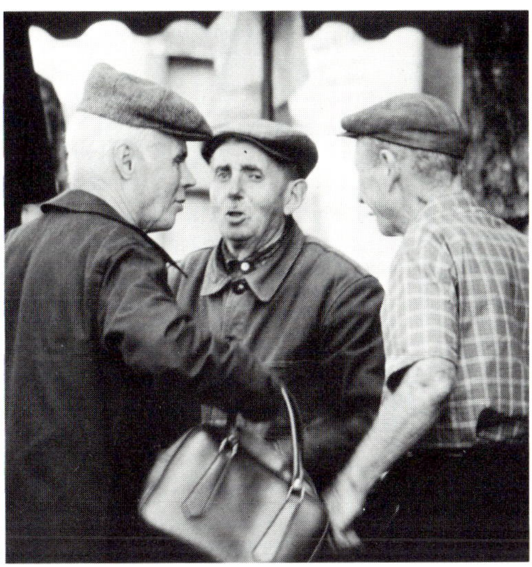

ELS beach. Slow ZOOM in to reveal PRESENTER on shore. PRESENTER turns to camera.
Presenter: *'Here we are on the spot . . .'*

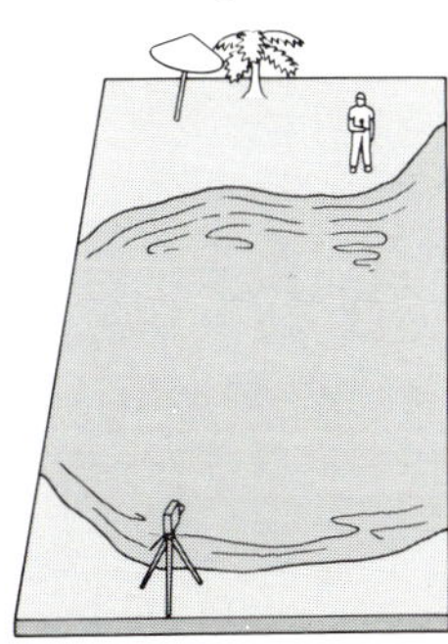

Instead of simply cutting to a presenter in time for the first words, try some professional tricks for bringing him into shot. Here, he is too far away to reach with a long mike cable. Signal him to begin speaking with a prearranged gesture when you reach a certain point on the zoom. After shooting, wind back to the cue point, press the 'Audio dub' button on the VCR and record the voice over on playback.

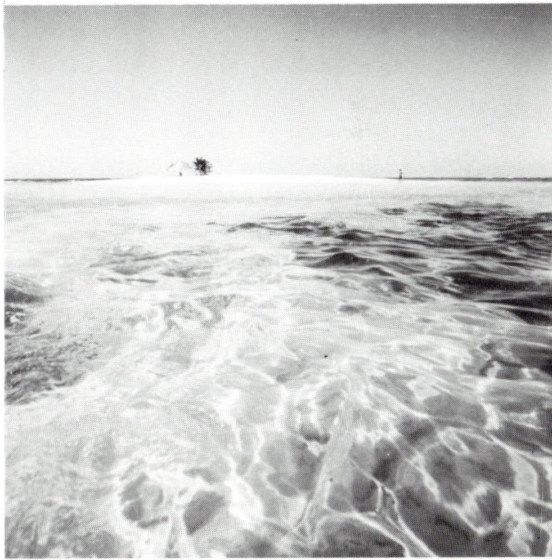

WS looking down valley. Slow PAN to reveal PRESENTER, who turns to camera.
Presenter: '. . . *will we know whether the freeway will be here.'*
PRESENTER turns away. PAN away and fade.

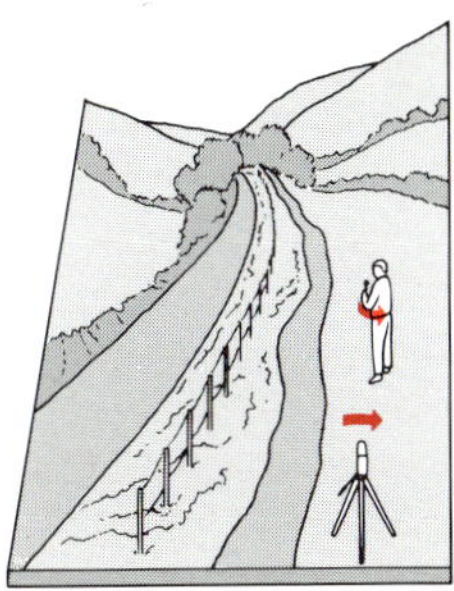

This is one way of ending a documentary and combining a wide shot of scenery which can tend to lack punch when backed only with a voice-over commentary as the final shot. Instead, pan from background to presenter, zoom in during the commentary, then pan away to the scenery before the slow fade out. You will have to cue the presenter to begin speaking, and the presenter will have to establish a cue-out for you to pan away.

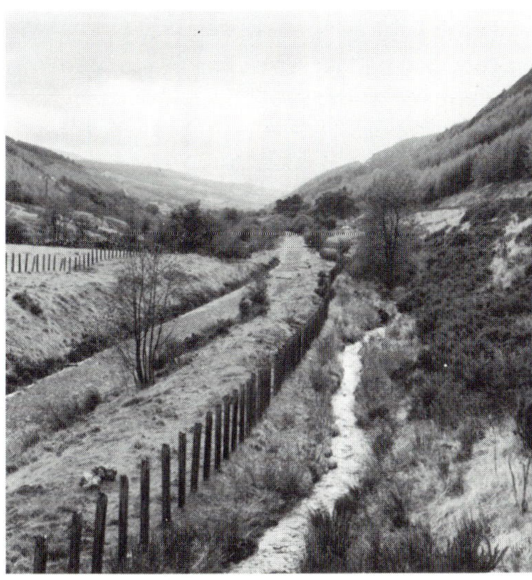

Typography for a TV screen

As soon as you start using a VCR, you will find titling useful in keeping track of all the shows and excerpts you record off air. When you start making your own video shows you will realize that not only do titles do an essential job at the beginning and end of each recording, but that titles and graphics can be used together to emphasize or explain points in your recordings, and to make them more interesting to watch.

Titles can be simple or ambitious, but they must suit the medium. Video pictures are not as crisp and detailed as cine film or slides, so titles must be kept simple, and the letters have to be clear and large enough to show up clearly on the screen. The shape of the screen, and its relation to the area seen by the camera, restricts the design of each frame; words should not be positioned too close to the edges of the screen area or they may be distorted, or disappear, when the titles are screened.

These factors limit the amount of information that can appear on a small screen. If you have a lot to say, it is better to spread the information over several frames.

Informal titles are the easiest to make. The words can be written on a blackboard in chalk, or on paper with a paint brush or broad felt-tipped pen. As a rule, it is better to avoid harsh contrasts, such as blue and green on the screen, and to avoid bright colours, especially deep reds, which can cause flares and echoes. Pale buff or light gray paper will look white on the screen without glaring reflections, and pastel colours work better in video.

By using your ingenuity, you can dispense with drawing out titles on card, paper or boards, and at the same time link them more effectively with the theme of the recording. A tape shot at the beach could have the titles drawn in the sand with a stick, for instance. When titles are shot out of doors under natural daylight, reflections and contrast are less important than they would be under indoor lighting.

More formal titles need laying out so that the letters follow the same style and the words appear in the same part of the picture in each shot. The easiest way of ensuring this is to use a set of standard cards marked out to the same measurements, and make up the words from a suitable typeface in one of the ranges of rub-on lettering.

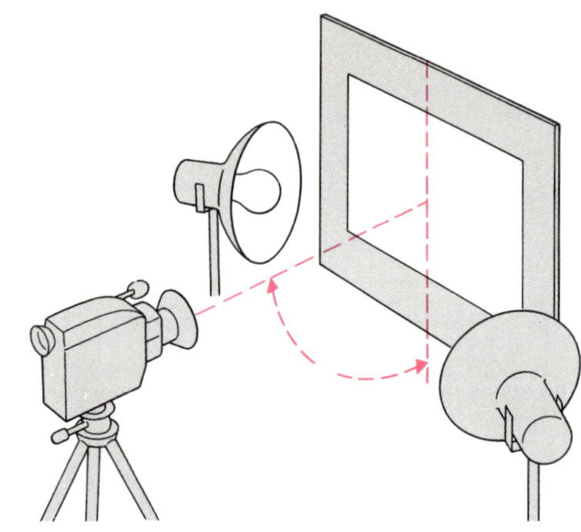

The easiest way of shooting titles is to fix them to a vertical surface, *above*. The camera must be positioned square-on to the surface, or the titles will look distorted. Position a lamp on either side of the camera, angled at about 45° to the card, exactly the same distance away from it, and slightly above the camera.

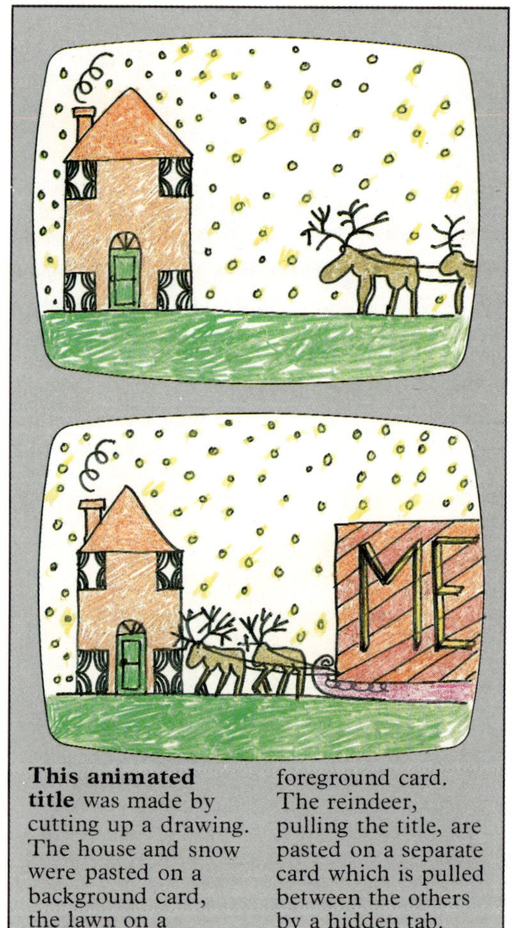

This animated title was made by cutting up a drawing. The house and snow were pasted on a background card, the lawn on a foreground card. The reindeer, pulling the title, are pasted on a separate card which is pulled between the others by a hidden tab.

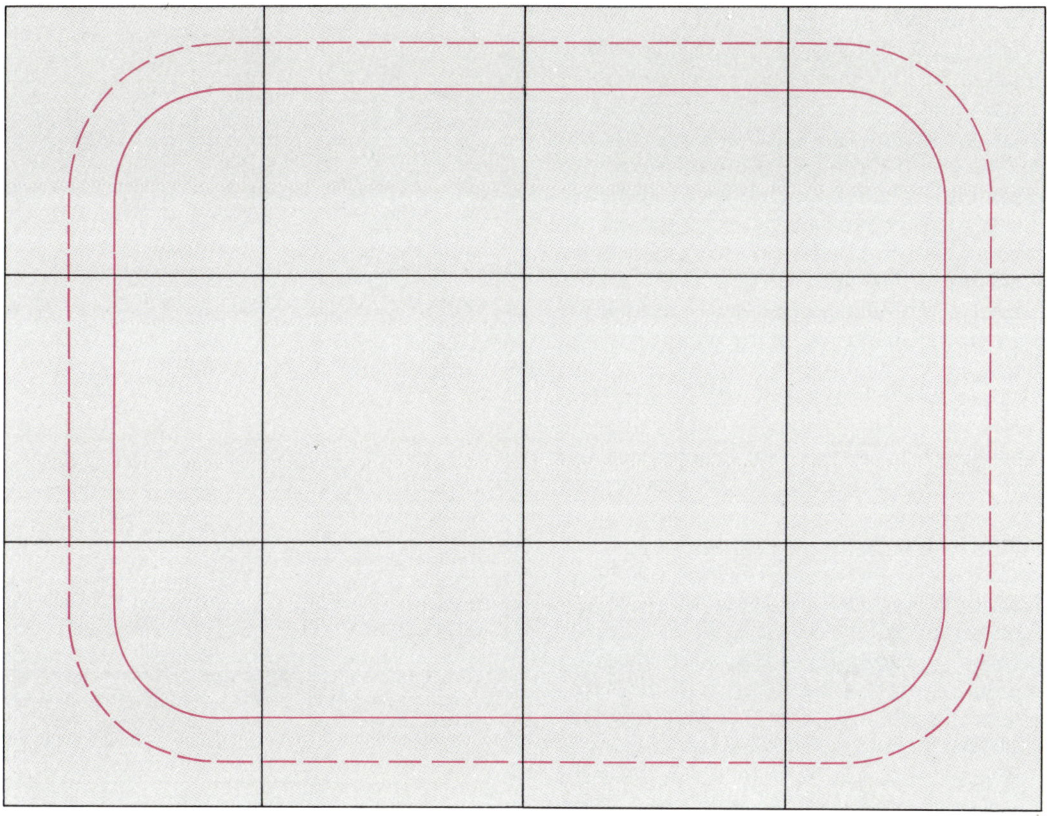

The TV screen is 4 units wide and 3 units deep, and titles have to fit within its shape and size. Titles are best drawn on cards measuring about 9 in (22.5 cm) deep by 12 in (30 cm) wide. The basic grid, *above*, or an enlarged version, may be used as a template for cutting out titling cards. It is best to limit the words or graphics to the safe area, that is, within a border of about $1\frac{1}{2}$ in (3.75 cm) all round. Lettering which extends into this border may vanish during playback.

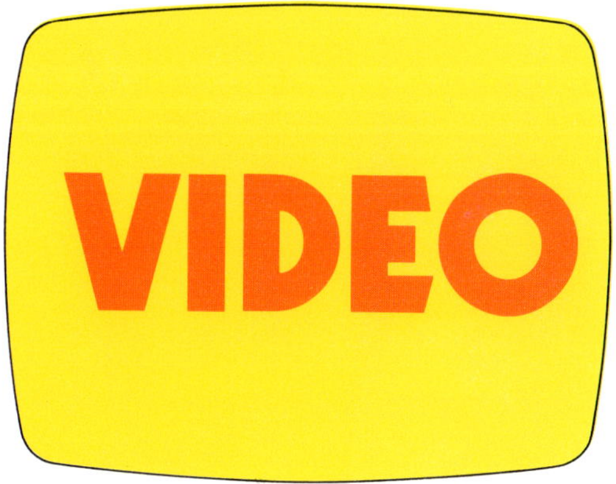

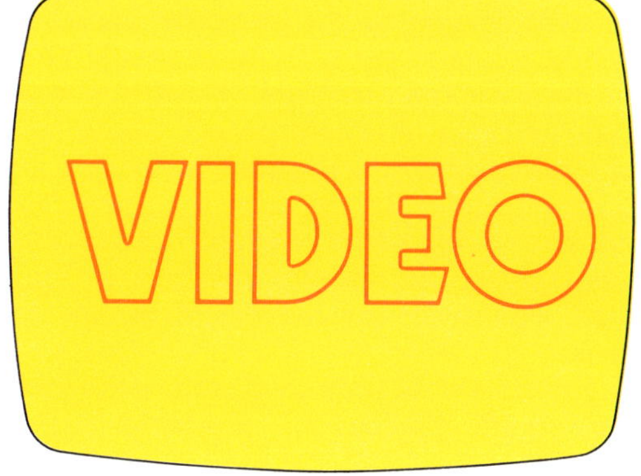

Titles should be brief, to be effective. Keep the number of words to a minimum, and change the cards quickly.

Lettering should be no smaller than one-tenth of the picture height and should be solid (filled in) *above left*, or a bold outline. Avoid letters with fine lines and serifs (decorative lines extending from the letters of some typefaces): they may disappear altogether on camera.

Avoid using red in titles. The colour contrast in the title, *above left*, will make the letters show up on the screen, but video cameras have problems with the red end of the spectrum and the version, *above right*, will blur into orange.

Ways of shooting words

Title cards tell your viewers what you want them to know, but they are essentially a static medium in the middle of a sequence of moving pictures. Perhaps the easiest way of bringing movement into titles is to write the title on a card large enough to allow you to zoom slowly in on it, giving the audience time to read the message before stopping the zoom and cutting to the next shot. However, if several titles are needed, this can be distracting and it might be better to look for straight cuts from one static picture to the next.

Alternatively, the cards can be mounted on a pair of rings on board, rather like a ring-file notebook. A hand, on or off camera, can flip each card out of the way at the end of each successive shot.

Either keep the VCR running so that the whole title sequence is treated as a single shot, (time each card change carefully to suit the time taken for the audience to read the message), or stop the VCR, and restart it after changing the card; this will make the sequence look just like the first method, where each card is shot separately.

Several titles can be built up in different places on the same large sheet of paper. This is particularly useful for setting lettered titles against a background drawing. Start the sequence with the camera aimed at the initial title, give the audience time to read it, and then pan across the picture to the next title, and so on.

The same blend of title and pictures can be achieved by placing the title somewhere in the middle, or to one side, of the scene. You can then cut from the title 'The Christmas Party', written on a card propped up in an armchair, perhaps, to the party happening in the background. In the same way, you can pan across from a shot of people leaving the course at the end of a race meeting, to 'The End', written on any convenient surface in chalk.

If other titles, such as credits (who did what in the production) or acknowledgements (thanks to people who helped) are needed, they can be set up in different places and cut to as a way of building a credits sequence into the last few shots of the recording. All of these techniques attempt to bring a bit of extra interest to a sequence, which might otherwise not hold the audience's attention.

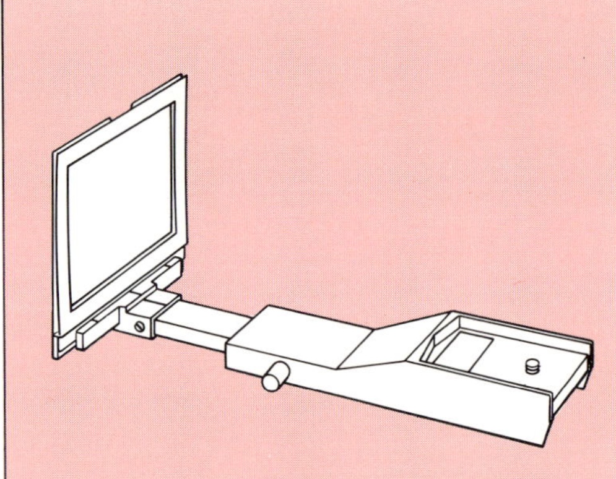

Small title cards can be used close to the camera, but they must be positioned carefully. Sony make a handy camera titler, the HTV 2100, *above*, to hold title cards (which can be as small as a credit card) square-on and at a fixed distance relative to the camera. The macro adjustment on the lens is used to focus on these small cards, and also on colour slides or sections of pictures. You can even use typewriter lettering, provided it is clear, evenly printed and correctly aligned. It is ideal for routine titles.

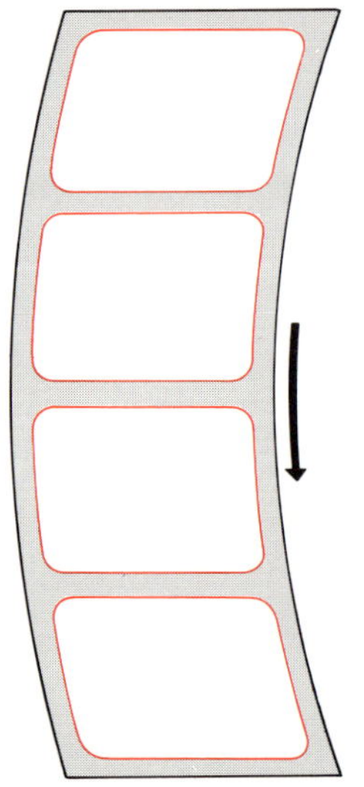

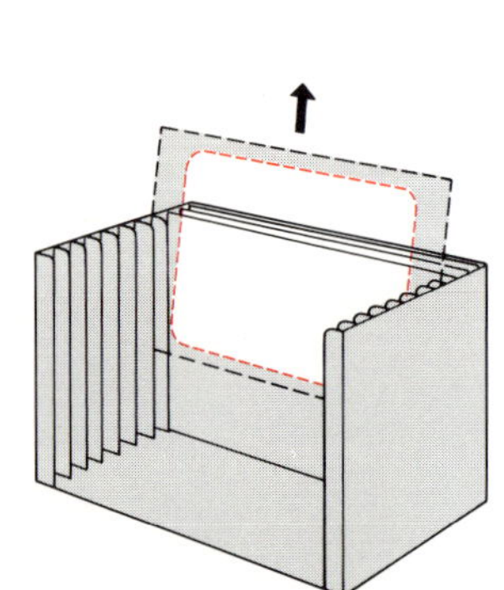

Several titles can be displayed as a moving sequence by mounting them on a long card, *above left*, and tilting the camera down them. A title card box, *above right*, can be used to hold the cards between grooves on either side. When each title has been shown, the card is lifted out of the box to show the next title. Keep each title on camera long enough for the viewers to be able to read it at a reasonable rate.

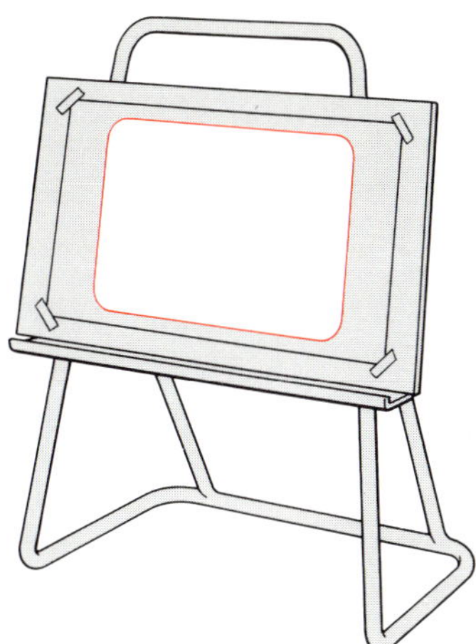

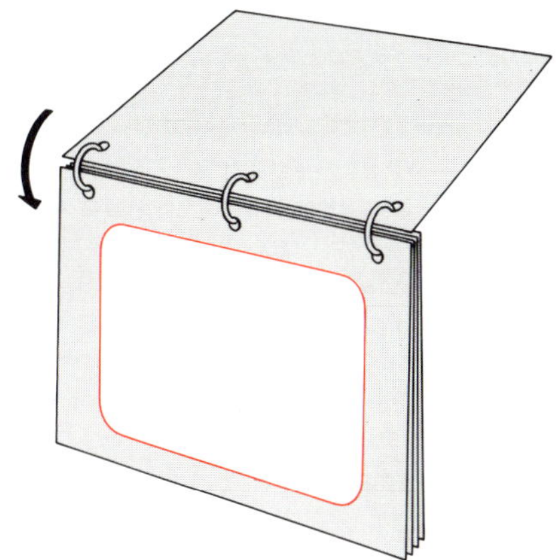

Titles can be drawn out on standard-sized cards and mounted in order on an easel, with the last card to appear on the screen placed at the back. Each card is shown for as long as it takes to read it, and then removed to show the next one. The lines and words should be in roughly the same place on each card, so that the titles do not appear to jump around the screen as the cards are changed.

Clipping the title cards into a ring binder greatly simplifies the card-changing procedure. It is then merely a matter of turning the title pages so that each one appears before the camera in succession. Position the binder so that cards drop in front of the camera in order, or they can be lifted out of sight one by one, to reveal the card underneath. Start and stop the camera from the VCR each time.

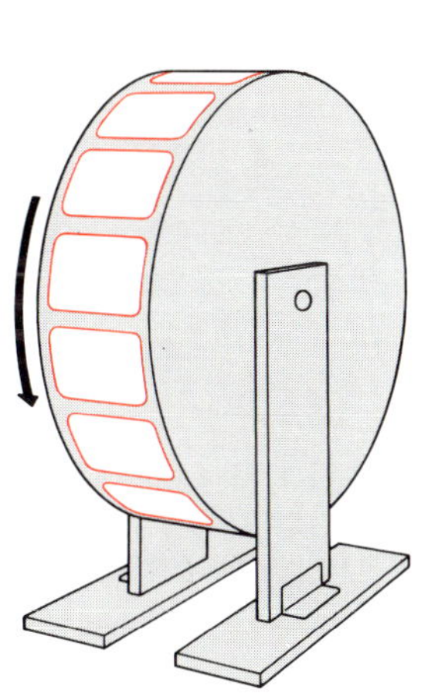

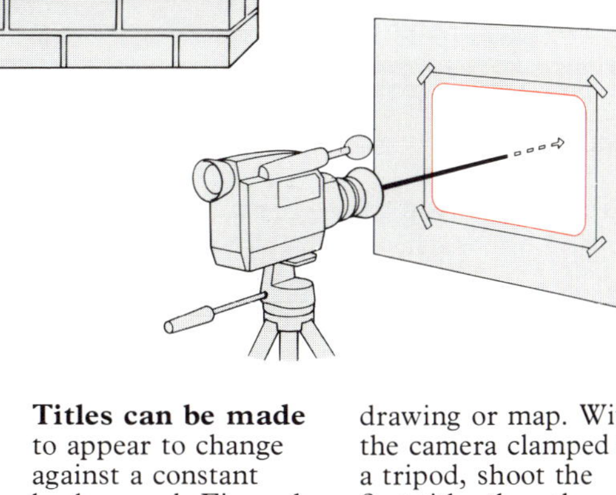

A series of cards that is too long to fit on a concave sheet can be fixed to a drum, *above left*, which is rotated in front of the camera.

Titles can be made to grow, for instance on a wall, *above right*. Stop the camera, add a letter or a word to the title, then restart the camera. The rate of growth can be matched to the viewers' reading rate.

Titles can be made to appear to change against a constant background. Fix each title or credit to a sheet of clear plastic, and attach the first one as an overlay on a background picture: perhaps a photograph, drawing or map. With the camera clamped to a tripod, shoot the first title, then the others in order, stopping and restarting at the VCR after each one.

Animating graphics

Even a relatively cheap cine camera can be made to shoot frame by frame for cartoon and time-lapse effects. However, with a video camera it is not possible to shoot frames individually, so this kind of animation is not possible.

The shortest sequences that can be shot separately with a video camera vary with the format and the machine, but with a single home deck you can only build up a sequence with a series of shots lasting for two or three seconds at a time. This gives the animation a jerky quality – rather than the smoothly flowing movement possible with film – yet it can be surprisingly successful as an occasional special effect. When the length of time between each of the pauses is just right, the result looks like a fast succession of stills.

All kinds of video animation can be built up on the basis of a simple titling technique. You can draw a graph, or add arrows to a map, by breaking down the material to be added to the background into small components which can be superimposed one by one. Fix the camera securely on a tripod, fix the background card on a wall or easel and draw the first part of the line, or add the first arrow. Run the VCR for a couple of seconds, then press the 'Pause' button; add another centimetre or so to the line, or another arrow to the map, release the 'Pause' button for a second or two, and so on. You may need an assistant to make the additions to the background while you operate the camera.

There are other ways of producing smooth animation with a video camera. A method of producing moving diagrams, used since the earliest days of television, is to construct what are known as pull-outs or reveals. There can be two or more layers of card; the first graphic or title to appear on the screen is written on the top card – the first layer – and the graphic or title which follows it is drawn on the second. The top card is then pulled aside, or a hole is cut in it, and the card behind pulled aside in turn – so that the graphic seems to change, or names appear, in real time, (that is, while the camera is running). A series of titles can be produced in this way as an animation, with the camera running from start to finish.

Pick background music with an appropriate pace and rhythm to accompany your video animation. It is even possible with practice to time the frame changes to music.

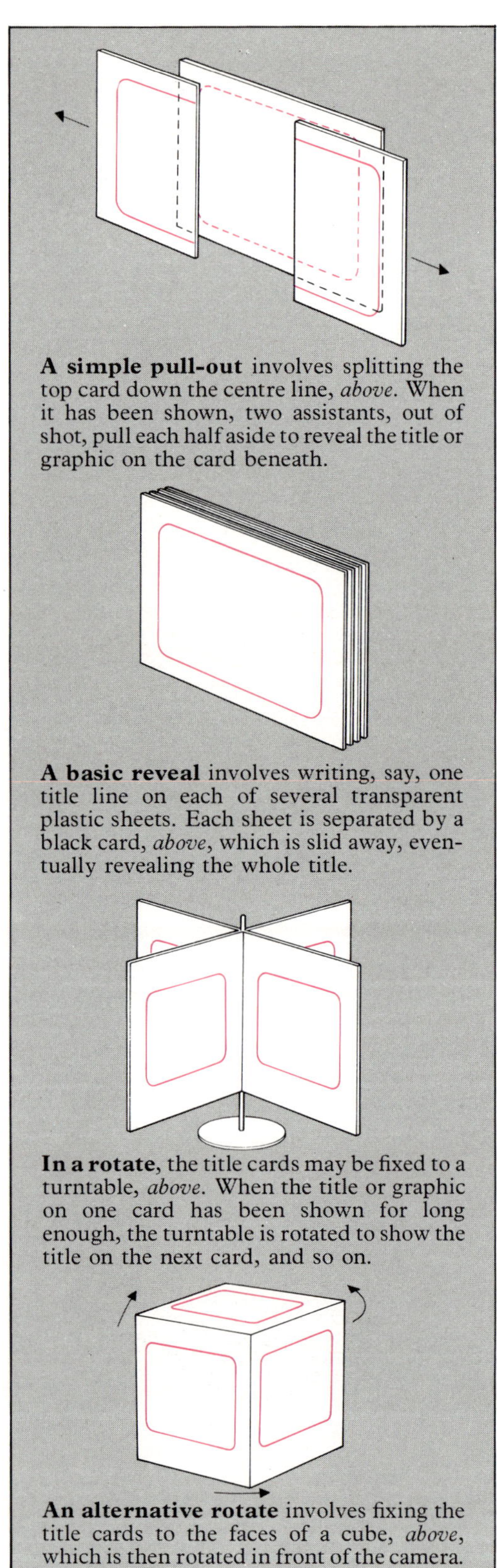

A simple pull-out involves splitting the top card down the centre line, *above*. When it has been shown, two assistants, out of shot, pull each half aside to reveal the title or graphic on the card beneath.

A basic reveal involves writing, say, one title line on each of several transparent plastic sheets. Each sheet is separated by a black card, *above*, which is slid away, eventually revealing the whole title.

In a rotate, the title cards may be fixed to a turntable, *above*. When the title or graphic on one card has been shown for long enough, the turntable is rotated to show the title on the next card, and so on.

An alternative rotate involves fixing the title cards to the faces of a cube, *above*, which is then rotated in front of the camera.

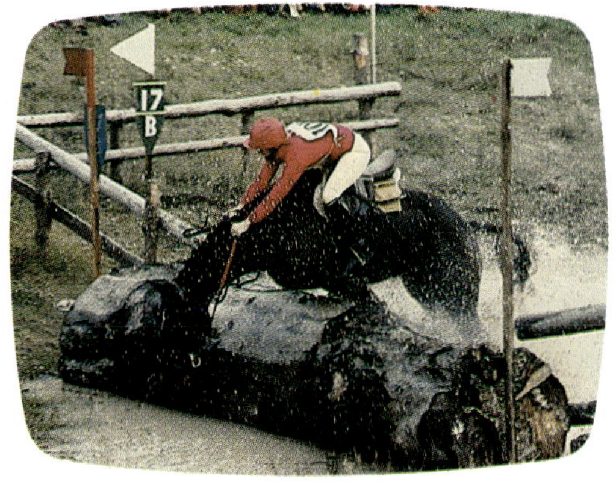

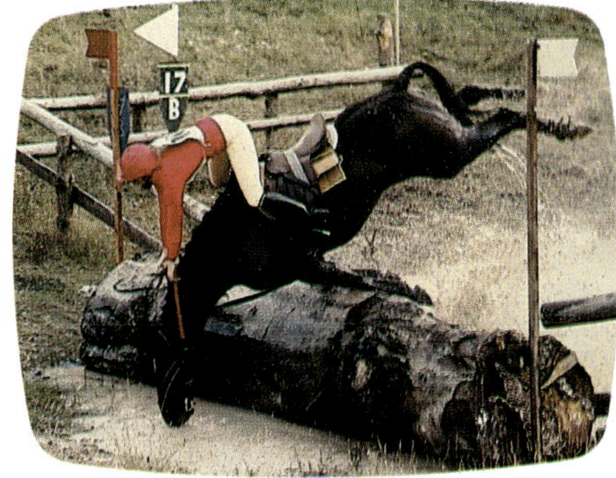

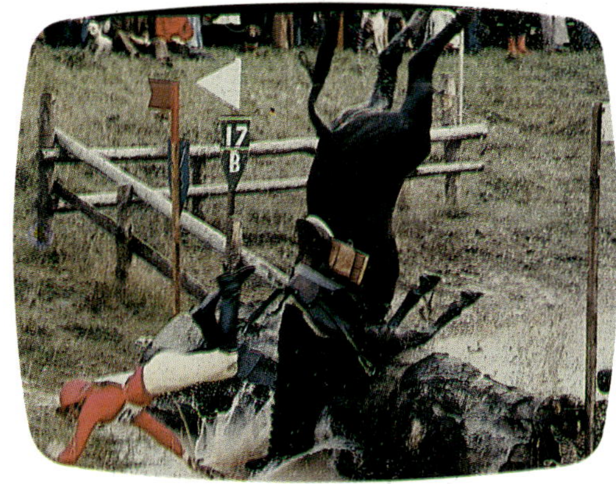

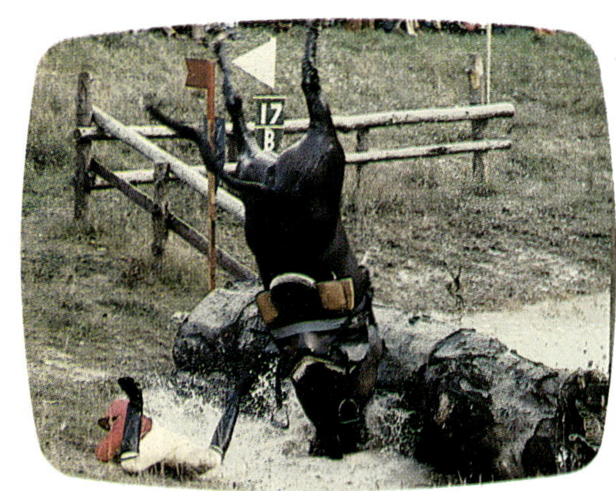

You can transfer slides (and also cine film) to video tape by projecting them on a screen and recording the picture with a video camera. This should be positioned next to the projector and framed on the projected picture. Make sure that the screen provides a bright enough image for the camera without flaring on the highlights. If this happens you may need a less reflective surface, or a telecine adaptor, *below*. The slide is projected through a ground glass screen which brightens the picture and reflects it on a mirror angled at 45° to it. Remember, the mirror will reverse the image.
The click of a stills camera shutter would be an effective alternative to music for the sound-track of the slides, *above*.

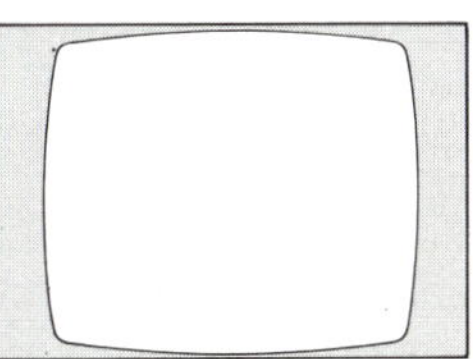

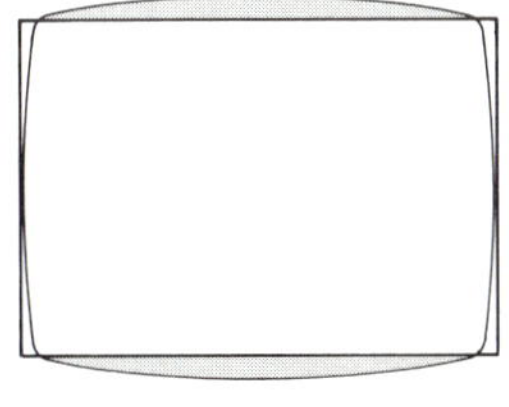

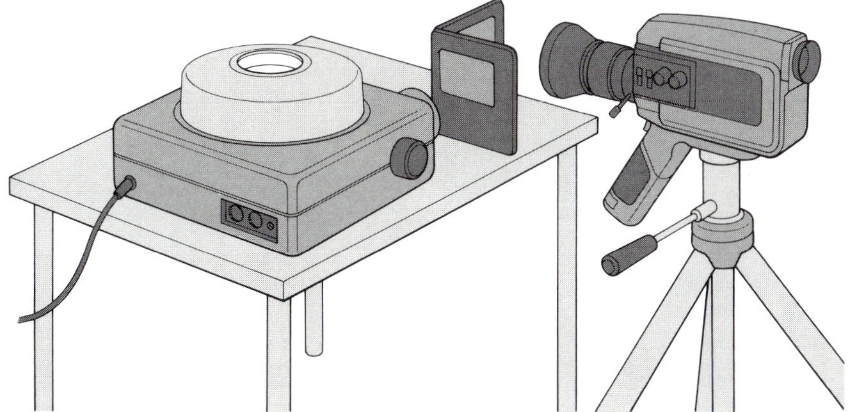

Slides and 8 mm frames are 3 units wide and 2 units deep, *top*, wider than the 4:3 aspect ratio of TV screens, *above*. When converting to video, connect the camera to a monitor to ensure that all the important material is framed by the lens.

Professional effects

Titles on broadcast television rarely stand alone on a blank white background. They appear superimposed on coloured grounds and mixed in with pictures, but in such a way that the essential information comes over with a minimum of distraction.

To colour a background you have merely to reset the colour balance on the camera, checking the result on the monitor, until you find the hue you want. With similar ingenuity it is possible to advance simple titling techniques to create more exciting effects.

Mount a camera securely on a tripod and focus it on a vertical surface, such as a wall or easel. Eliminate all unnecessary lighting. Now position a slide projector so that the image of a slide is projected on the wall or easel. This requires practice. To avoid distortion, the camera and projector should be side by side, and as far from the screen as is possible without the projected image losing intensity. You may also have to adjust the lighting and colour balance to achieve the most lifelike results with the slide.

Record a few seconds of the slide, then press the 'Pause' button on the VCR. Take a line of the title on a white card and fix it to the wall or easel with pins or tape, positioning it so that it shows up clearly. Make sure the camera does not pick up any tell-tale edges. Release the 'Pause' button and record the shot. The title will seem to superimpose itself on the image.

You can go on to use a succession of title cards with the same slide, or change to a different slide for each card. You can cut from slide to slide with a single projector; but if you have access to two projectors and a dissolve unit, you can mix from picture to picture, printing up a new title once each new picture has been seen. Since the slides do not move, you can stop the camera once the mix from one slide to the next has been recorded, fix the title card, stop the camera again while you remove the card ready for the next dissolve, start it again to record the dissolve, then stop it to put up the next title, and so on.

From there you can progress in whichever direction your imagination takes you. Try shooting some action which is taking place relatively slowly, such as the approach of a slow-moving train, as a series of stills. If you shoot the sequence of one slide dissolving smoothly into the next with your video camera, the result can be sensational.

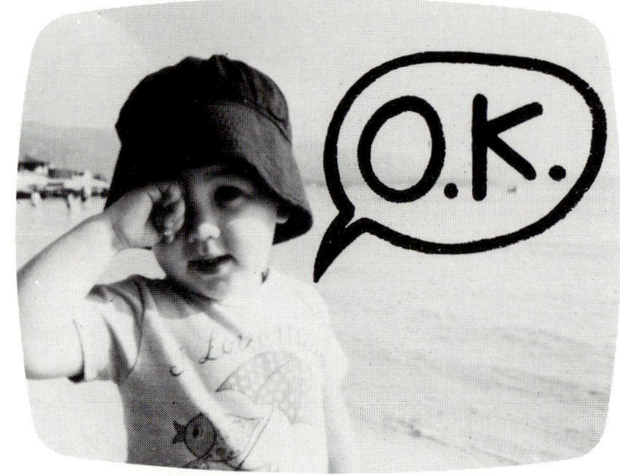

An SEG can superimpose lettering from a titling camera on a blank screen, or on the image from a colour camera.

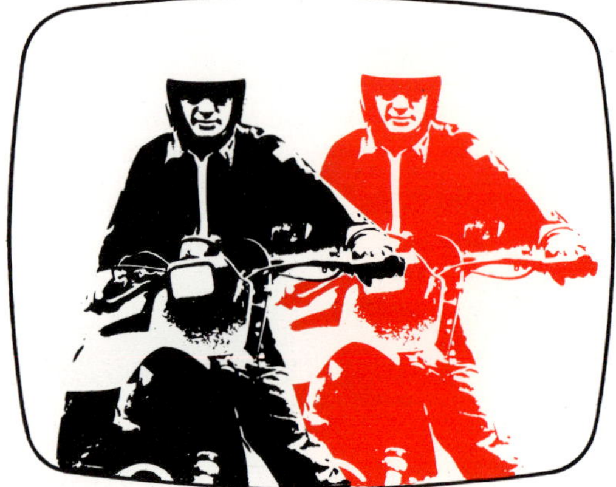

The images from the titling and the colour camera may be reversed so that the two motorcyclists change colour.

Cine film can be used, in the same way as slides, as an effective background to titles. Also, where it is easier to record on film: in animation, slow-motion, speeded-up special effects, and so on, you can copy the film sequence, (run at normal speed) on video tape at the right place in the recording.

The transfer of film or slides to tape is called telecine conversion. You can do it by projecting the film on a screen and recording it with a video camera. This is positioned next to the film projector and focused on the screen. Just as when copying slides, a telecine adaptor in folder or box form, *below*, can be used to brighten the projected image and reflect it to the video camera.

Film projector

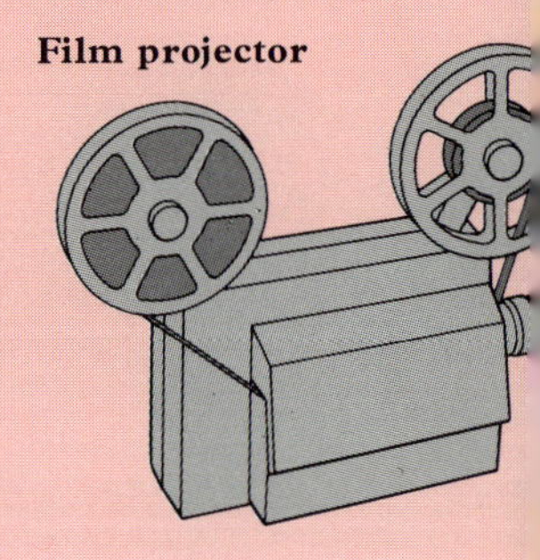

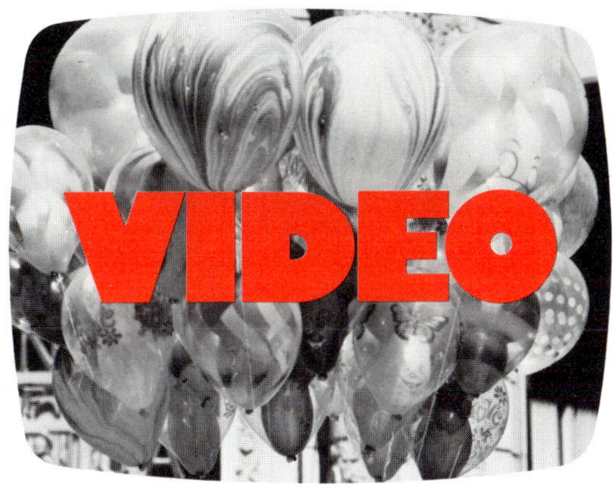

Superimposed titles can be generated by the SEG in monochrome, yellow, magenta, red, cyan, green or blue.

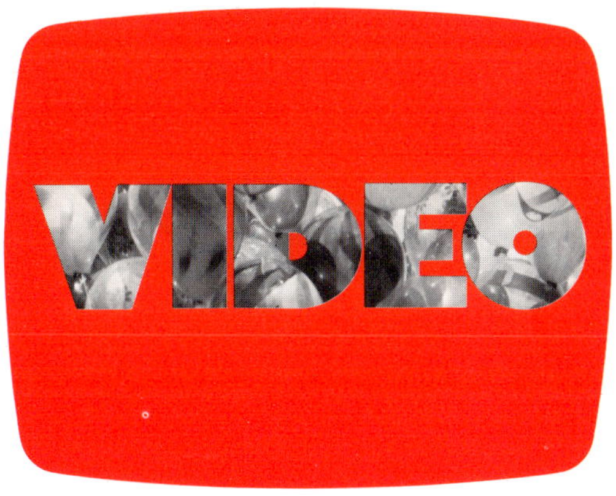

The lettering can become a window, through which the image from the colour camera appears, and may then expand.

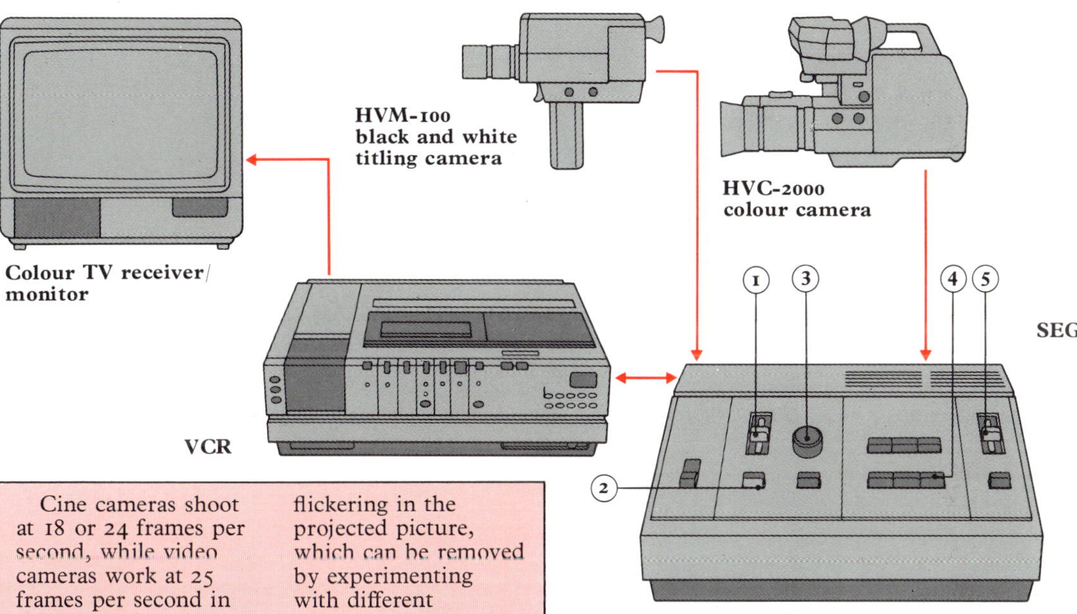

Cine cameras shoot at 18 or 24 frames per second, while video cameras work at 25 frames per second in the UK and 30 frames per second in the USA. This difference can produce lines or flickering in the projected picture, which can be removed by experimenting with different projector speeds. However, 16 mm cameras have a 25-frame setting.

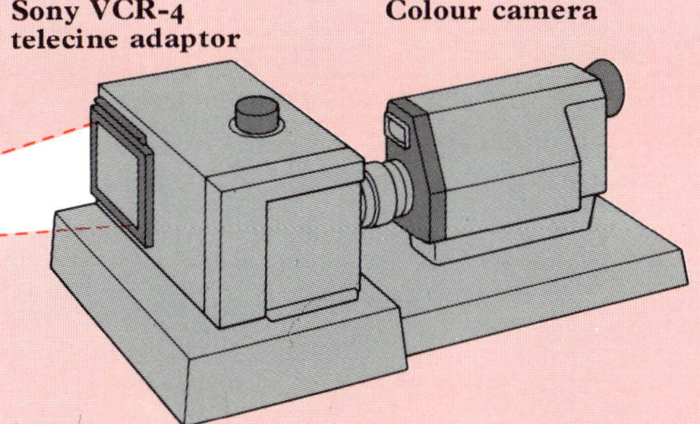

Professional effects equipment is now on the market at prices the home enthusiast can afford. A black and white titling camera costs a fraction of the price of a colour camera. It is used to blend titles into a main picture sequence through a special effects generator (SEG), such as the Vel Minimix or the Sony HVS 2000, *above*. This works by switching between two cameras to produce split-screen effects, to wipe one picture off the screen with another, or to combine them in various patterns according to pre-set instructions:

1 Key level (outline) control
2 To reverse superimposed picture
3 Colour adjustment control
4 To remove superimposed picture immediately
5 To remove superimposed picture gradually

Recording sound

Sound is the trump card in the video producer's hand. Whatever the quality and clarity of moving pictures, sound gives them an arresting, life-like immediacy they would never otherwise achieve. Most video cameras have microphones built in, but for a more professional sound, it is better to plug a microphone into the 'Mic in' socket of your VCR, to replace the camera microphone.

Microphones work by converting sound waves in air into electrical signals which vary in exactly the same way. The signal has to be amplified before it can be recorded on magnetic video tape, and when the tape is played back, it has to be amplified again before it can be fed into the loudspeaker of the television.

All the many different microphones useful for video recording can be classified as dynamic or electret condenser in type. The advantage of electret condenser microphones is that, although more expensive, they are smaller and lighter than dynamic ones; they generally give a better, more even response to sounds of different frequencies, and they produce a stronger signal. Their greatest disadvantage, however, is that they run on batteries which must be checked regularly.

According to their design, both dynamic and electret condenser microphones respond differently to the sounds around them. An omnidirectional microphone picks up the sound equally from all around while a unidirectional one picks up sound only from the direction in which it is pointing. Although it picks up some sound from areas at the sides, the unidirectional or cardioid microphone is unable to pick up sound from behind. A unidirectional microphone with a very narrow pick-up range is described as superdirectional. A microphone able to pick up sound from two directions at once, but not from all around, is known as bidirectional.

The type of microphone you choose, its shape and mounting, will depend on the job you want it to perform, but it is worth remembering that a unidirectional microphone can usually be used from farther away than an omnidirectional one.

The more trouble you take to record high-quality sound, the better playback facilities it deserves. Most television sets have small, inefficient loudspeakers, but connecting the 'Audio out' of your VCR to the input of a hi-fi amplifier will give you immeasurably better sound quality.

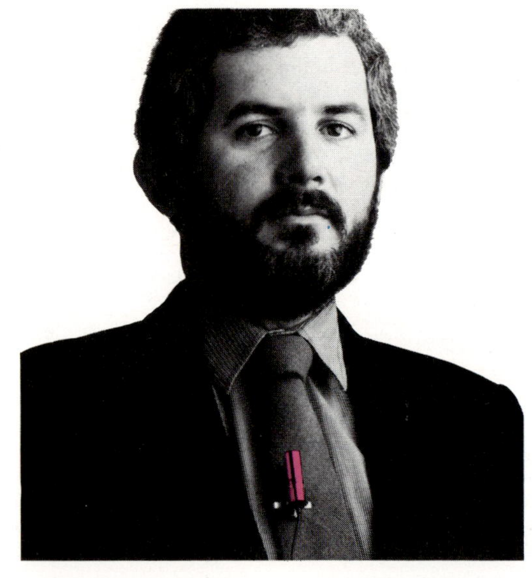

The unobtrusive necktie or lavalier mike is clipped on to clothes or worn on a cord round the neck. For best results it should be omnidirectional.

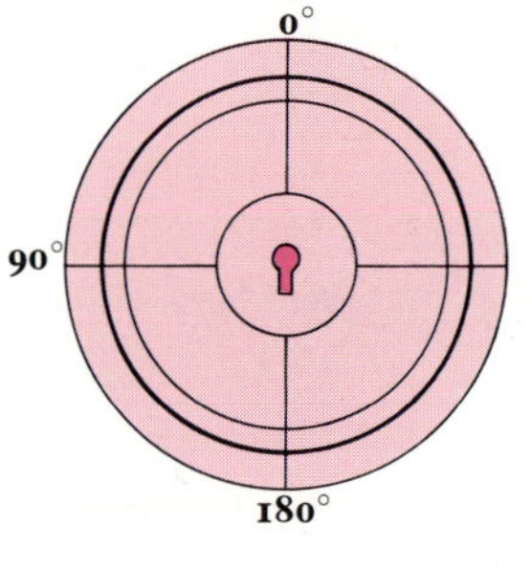

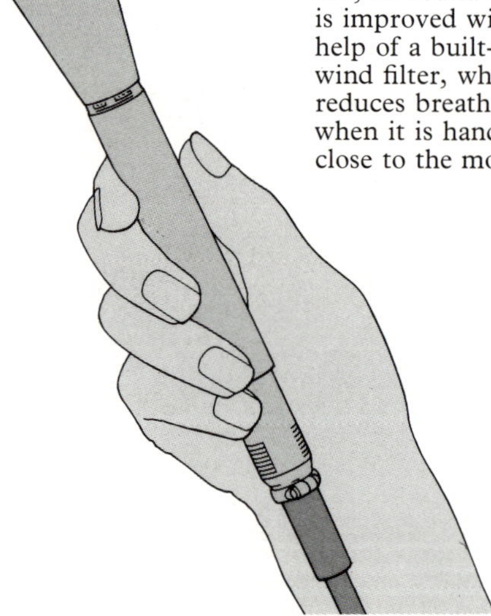

An omnidirectional mike, *below left*, as its name suggests, has virtually equal sensitivity to sounds coming from all directions, as shown by its typical pick-up or polar pattern, *above left*. This mike is dynamic in type and can be hand-held or mounted on a stand for greater versatility. Useful indoors and out, its sound quality is improved with the help of a built-in wind filter, which reduces breath noise when it is hand-held close to the mouth.

The JVC zoom mike has been designed especially with video recording in mind. The mike is mounted on top of the video camera, and is linked up to the camera lens so that both sound and vision can be varied in concert. This connection means that as the zoom lens of the camera automatically alters in function from wide angle to telephoto, so the pick-up capacity of the mike varies in unison from all round or omnidirectional to unidirectional and finally to superdirectional to give accurate pinpointing of sound.

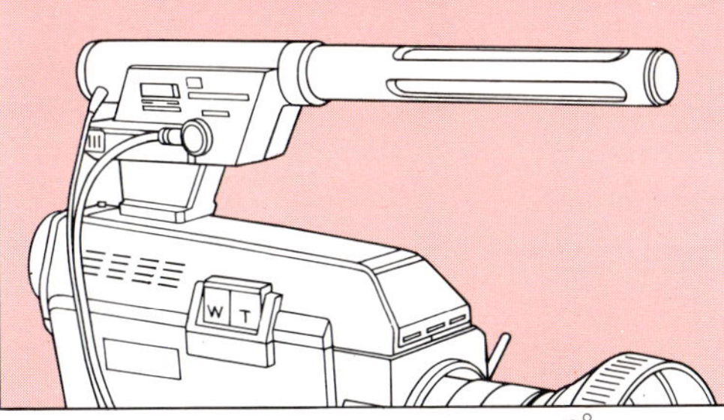

Mikes used for video recording are classified as either dynamic, *below*, or electret condenser, *bottom*. In the dynamic mike, the diaphragm, **1**, which is placed in a magnetic field, works like an audio speaker in reverse and produces a series of electrical signals as it vibrates in response to the sounds reaching it. In the electret condenser mike, the vibrations of the diaphragm, **1**, are converted into signals by means of a closely attached plate which is charged by a battery, **2**, placed inside the mike's head.

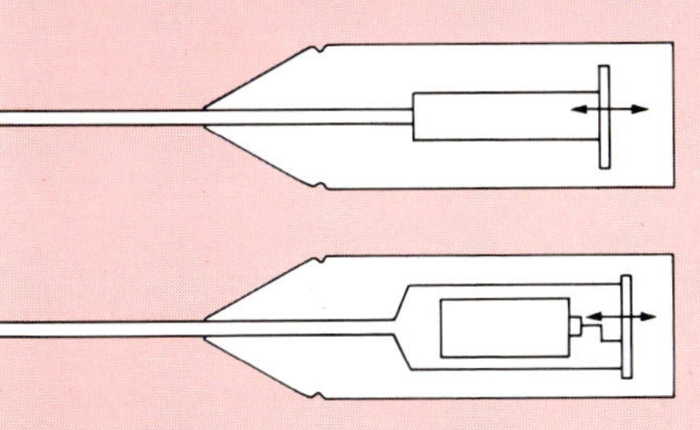

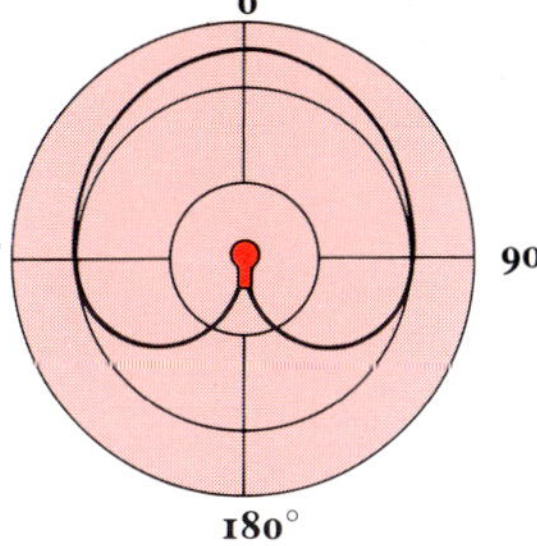

The unidirectional mike, *left*, picks up sound from one direction only, but over a wide area, as its polar pattern, *above*, shows. This heart-shaped trace also gives the mike its alternative name of cardioid.

A mike which is mounted on a boom, *above*, is ideal for picking up sounds in a particular area while, at the same time, keeping out of view of the camera. To make use of this particular mike, it is not essential to have an assistant because the mike extends like a telescope.

Choose a boom mike according to the type of recording you are making. An interview, for example, would be best with a unidirectional mike.

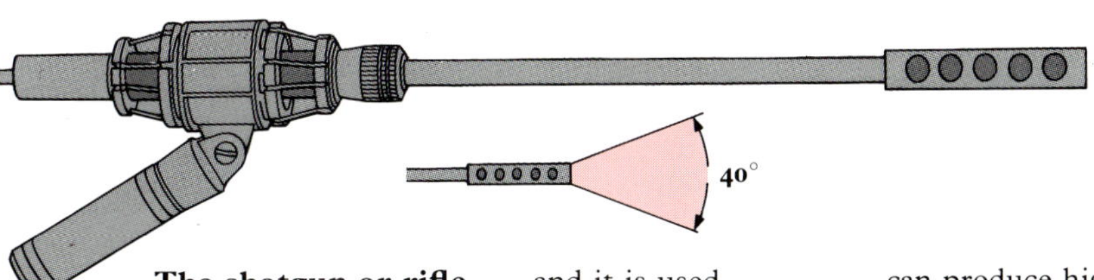

The shotgun or rifle mike is shaped as its name implies, and can be up to 6 ft (1.8 m) long. This is a mike for the specialist, and it is used held over the shoulder. It gives selective pick-up of sound over a narrow angle of only 40° and can produce high-quality recordings of sound from distances of dozens of feet. It is, however, heavy to use and expensive.

Using sound

Built-in video camera microphones are useful for recording general sound atmosphere – the hum of traffic in a busy street or the chatter of a children's party; but when you want to record specific sounds, such as the outburst of delight as your child opens a Christmas present, you need the extra clarity provided by a separate microphone.

Before recording, you must first decide what sort of sound you want to create, then select and set up the correct microphone for the job. If necessary it can be mounted on a stand or boom, but check that it can be easily removed, if required. For recording music, interviews and stage shows, choose an omnidirectional microphone.

Always make sure the microphone is really in the thick of the action, not to one side of it. Use the 'Bass cut-off' switch on the microphone to select the best kind of frequency response for the sound you are recording.

In many situations you will find the miniature, lavalier-type microphone most useful. This is an omnidirectional type used attached to clothes or hung round the neck. It can be cushioned on foam rubber and taped inside a musical instrument, such as a guitar, to give really high-quality sound recordings.

The omnidirectional microphone likely to be of least use to you is the one on your camera, and it is best replaced, if possible, by a directional boom microphone or, even better, the zoom microphone which acts as the audio equivalent of the zoom lens. Any directional microphone fixed to the camera will be much more sensitive than an omnidirectional type to sounds coming from the area toward which it, and the camera, are aimed. As a result, you are much less likely to pick up unwanted background noise.

To pick up only the sound you want, especially sound coming from a single source, use a carefully aimed unidirectional or cardioid microphone. You can also use different parts of the microphone's response pattern to balance two or more sounds blended through a single microphone, for example, a lead singer and backing group. The chief disadvantage of the cardioid microphone is the unwanted echoes it creates.

Most precise of all are the shotgun, or rifle microphone, which can be placed far enough from a speaker to stay out of shot, and the parabolic microphone, which uses a reflector to channel sounds from its target.

Sound-recording check list

1 Do be careful how you position audio cables, particularly when they are used with mikes attached to people's clothes.
2 Do add an extra mike to a setting to give extra atmosphere.
3 Do insulate the base of a desk mike to prevent extraneous noise.
4 Do hide the cord of a lavalier mike. It can be concealed by the folds of a dress, or threaded up a trouser leg.
5 Do place a mike in the correct position for someone who is going to make a speech, and adjust it beforehand.
6 Do remember that if you stop the camera and there are breaks in the sound, you must fade down the sound on camera as quickly as you can, then fade it up again, making sure that you return to the same level. Alternatively, fade down the sound and picture together and change the shot.
7 Do not let subjects handle the mike.
8 Do not leave the mike exposed to view except for interviews or concerts.

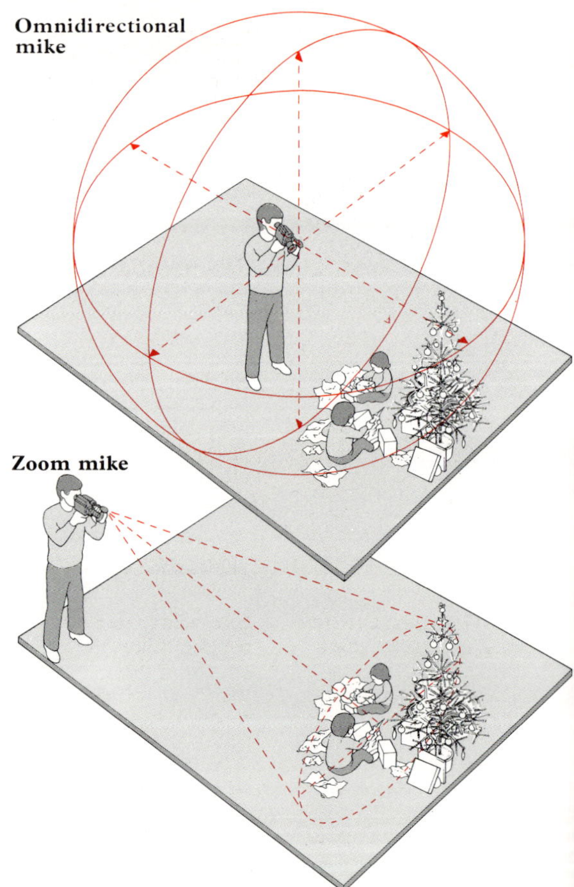

The distance between mouth and mike is crucial to the quality of your finished audio recording. When using a fixed, unidirectional mike to record the spoken human voice, the ideal distance is about 1 ft (30 cm). A subject whose lips are too near the mike will produce sounds with explosive 'pops' on the letters 'p', 'b' and 't'. If the subject is too far away, the voice will come over weak and indistinct, and this problem will be made worse by the intrusion of background noise from other parts of the room.

To record an interview in a studio-type setting, use a fixed omnidirectional mike, which produces a sound pick-up area shaped like half an orange. The two people should sit with their knees about 3 ft (1 m) apart. For the best mix of sound and vision and the best variety of visual angles, the scene should be shot over the shoulder and diagonally from the right- or the left-hand corner of the set. Getting a good interview also depends on having a relaxed subject and confident interviewer, who has taken the trouble to do some homework.

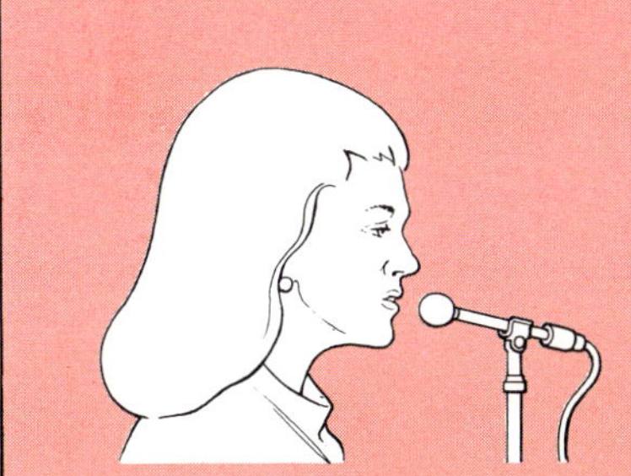

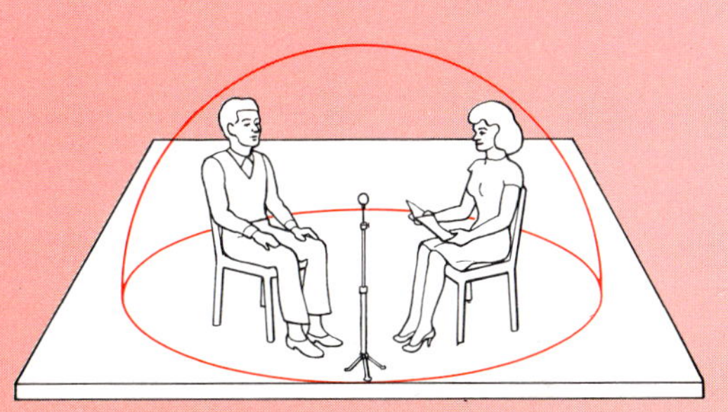

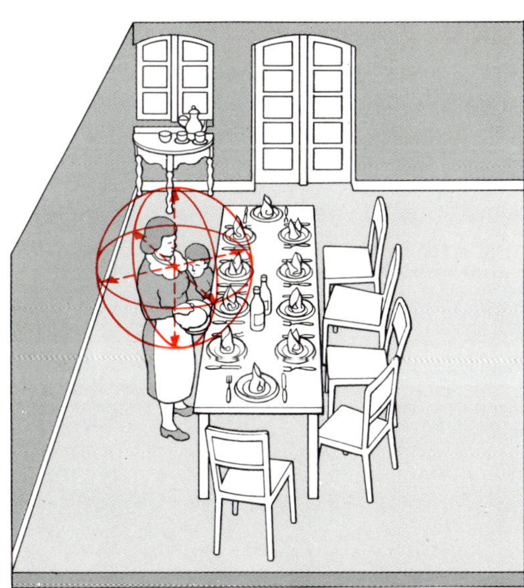

An omnidirectional mike, *opposite top*, picks up all the sounds that go to create an atmosphere, but also extraneous noise, for example from behind. To pick up only the sounds of the children, use a unidirectional zoom mike, *opposite bottom*.

Use a lavalier or necktie mike for filming a subject on the move, *above*. The mike picks up sounds from the subject but not extraneous noise. The only problem is the cable, which may trip the subject up, so it is wise to have a rehearsal.

A unidirectional mike is useful for picking up dinner-table talk, but if you stand it directly on the table it will pick up vibrations. A table-cloth deadens the clattering of plates and cutlery, and a foam pad beneath the stand will help to insulate the mike.

Position the mike nearest to the main subjects of the recording. Here it is concealed inside a vase of flowers. Brief everyone beforehand not to bang the table or clatter their cutlery while speeches are in progress.

Sound on location

Once you have chosen the correct microphone, good sound recording demands, in addition, care, concentration and plenty of advance planning. The problems you come up against will depend on your exact location and whether you are recording indoors or out, but wherever you are you must always be ready to react to those problems.

Shooting indoors gives you most control over extraneous noise (although in a large building you may get unwanted door-slamming and telephone-ringing from some distance away), but the room can give trouble. If there are too many echoes – if all your subjects sound as if they are talking from inside an empty tin – the room is acoustically too 'live'. What you need is more absorbent surfaces, such as carpets, curtains and cushions, to cut down on the amount of sound reflection.

If, however, there are no echoes at all, and everyone sounds as if they are speaking through cotton wool, then your room is acoustically 'dead'. In this case, try to add to the number of reflective surfaces, for example by opening the curtains.

Only practice will tell you when the sound from any room is just right, but wearing headphones while you record will help you, or your assistant, to judge the strength of background noise in relation to the other sounds on the recording.

Outdoor recording does away with worries about acoustics, but you can be plagued by all manner of unwanted noise, from hammer drills to passing aircraft. Apart from choosing your site as carefully as possible, there are few simple remedies for these problems. You can wait for the offending noise to stop; you can move, or come back on another day. If none of these is practicable, put the microphone as near as possible to the subject.

Wind is another potential hazard in outdoor recording, and can produce roars and crackles on the sound-track. To reduce or eliminate this, fit a wind screen to your microphone, or improvise by taping a sheet of polystyrene foam or several layers of folded handkerchief on to it.

Microphone mountings are almost as important as microphones, so give them due care and consideration. Most important of all, remember that the time taken to rehearse at every stage, and to test equipment before you begin, will pay rich dividends.

Mixing sounds
Most domestic video recorders have only a single sound channel. This means that you are automatically restricted in the mixture of sounds you can record at any one time.

If you want to add a commentary or some background music to your tape, you will automatically wipe off the existing sound.

The answer is to build up your sound-track separately, by copying your video sound-track on an audio recorder. If you then rerecord it as you deliver the commentary, or as you play the background music, you can incorporate several sounds on a single track. If all this sound is on one audio tape it can be put back through the audio channel as a single recording.

The main problem with this technique is getting the sound synchronized with the pictures. As an alternative, you could consider buying a simple audio mixer. This device allows you to mix the inputs from different mikes, together with a pre-recorded track from an audio recorder, into a single track, which can then be fed into the VCR. The most sophisticated VCRs, such as the industrial VHS machines and U-matics, have two audio channels for more professional sound-recording.

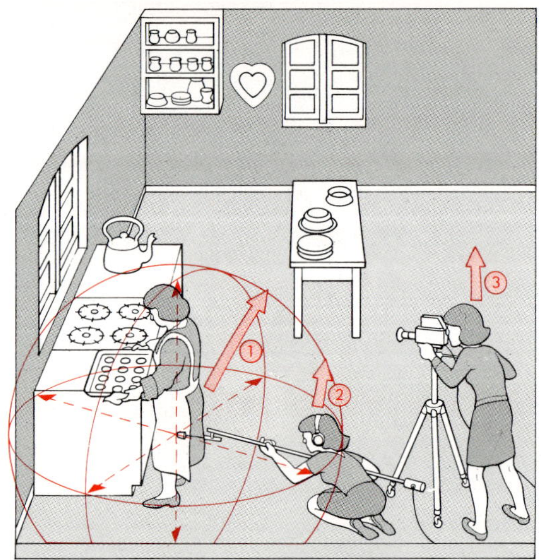

To shoot sound and vision of a sequence such as a cookery lesson, use an omnidirectional mike attached to a boom held by an assistant, **2**. Use an electret condenser mike to cut down wind noise as the mike is moved, **1**. Rehearse the scene carefully so that the assistant can kneel in comfort while keeping the mike close to the sound source, and so that the camera operator, **3**, can make the correct movements.

Outdoor recording
Making audio recordings outdoors can be problematical. To get a convincing soundtrack to accompany a recording of this canoe race, use the following guidelines:

1 If you are using a camera mike, film from the best, uninterrupted vantage point, in this case standing on a hill. Before shooting, check the sound through headphones to make sure it is loud enough.

2 If you are using a separate mike, plug the lead into the mike input socket in the camera, and take the mike as near as possible to the water. Again, test the sound level before recording.

3 Record water noises and shouts of onlookers separately on an audio recorder and dub this 'wild' track on later (see 6).

4 If you do not have an audio recorder, record 'wild' track by taking your portapack down to the water after shooting and dub the sound on the picture. Remember, however, that sound and vision will not be synchronized.

5 Another alternative might be to attach a lavalier mike to the canoeist, or let the canoe carry an audio recorder, but this is not likely to be practical.

6 To dub commentary on to your soundtrack, shoot the pictures and make a separate recording of sound effects on an audio cassette. When you get home, put the video tape into your VCR and spin it back to the start of the sequence, making a careful note of the footage. Plug a mike into the VCR and press the 'Audio dub' button. As you speak your commentary into the mike, play the audio cassette of sound effects in the background. Any sounds on the original tape will automatically be wiped off and replaced by the mix of commentary and sound effects.

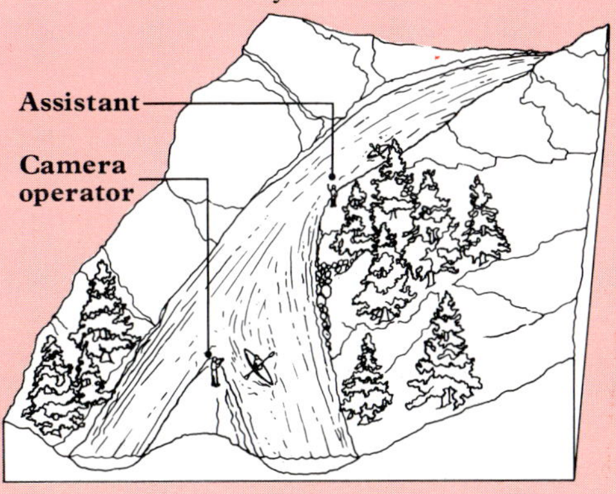

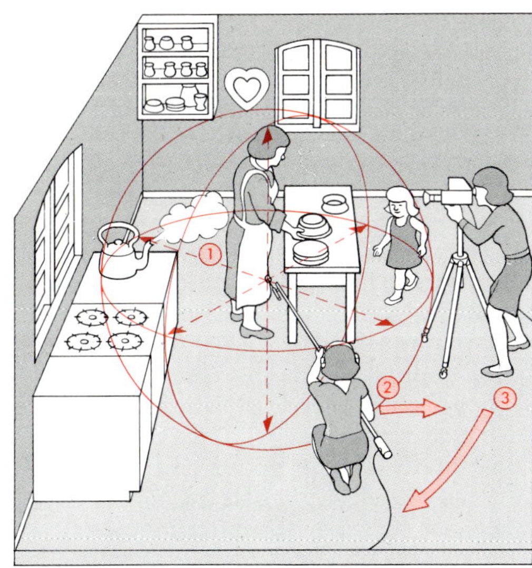

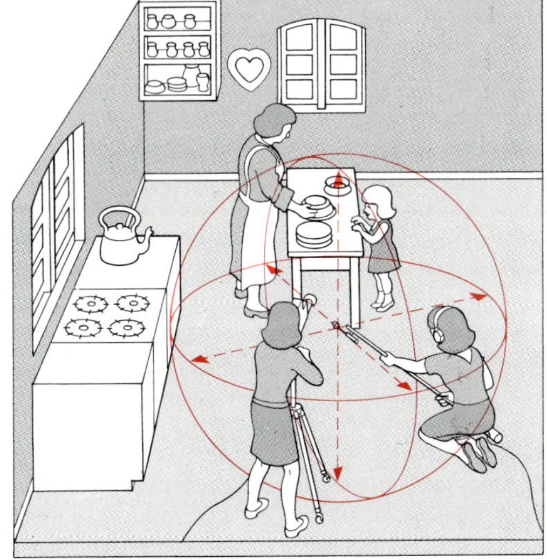

For realistic acoustics, aim to cut out all unwanted noise. If the child enters unexpectedly she will be out of shot, but you may record the sound of the door opening, thus making the sound on the finished film confusing. If an unwanted sound, such as the siren of an ambulance should intrude on your recording, repeat the sequence. The whistle of the kettle, however, can be incorporated into the recording.

As the camera operator moves to bring both child and demonstrator in shot, **3**, the mike operator must change places, **2**, without tripping over the wires. She must also ensure, by judging the sound in her headphones, that the audio level stays constant, relative to the background noise. To get the correct acoustics, particularly with noisy crockery, try using plastic plates.

Mixing sounds

The atmosphere of a location, together with any dialogue, will automatically be added to your recordings by the built-in microphone on your camera. Your major preoccupation when shooting will be with the pictures you are capturing on tape. Should some sound occur to complement your sound-track: should a seabird begin to call, for instance, just as you are shooting a yacht setting sail, it would be an unexpected bonus.

Yet sound can add all kinds of qualities to your pictures, and because it is under your control to a far greater extent when the shooting is finished, it offers a wealth of new creative possibilities. You can add atmosphere, blend in a commentary, improve the sound effects and add background music.

To appreciate the power of sound, try recording a completely neutral picture for a minute or so: the front of your house, perhaps, or a close-up of someone looking at the camera with a dead-pan expression. Then add sound. Play happy, up-beat music on your turntable; then try something gloomier with a hint of foreboding. Try a fast, urgent piece, then something nostalgic. In each case you will find the music imposes something of its own quality on the neutral picture.

Sound effects do the same thing. The choice is equally vast: sound effects libraries are stocked with sounds you could hardly imagine. A request for drums will yield drum beats from every corner of the world, drums in religious festivals, in war, celebrating weddings, in jazz combos and rock bands and solo, made by every possible kind of drum.

This overwhelming choice is whittled down to the needs of the amateur in sets of records and audio cassettes. They are released for home use by pre-recorded cassette producers, and by organizations such as the British Broadcasting Corporation. There are records of animal noises, cassettes of transport noises: steam trains; aircraft taking off and landing and even more specialized collections including tracks of people walking and horror sound effects which include screams, creaking doors and moaning wind.

Sound equipment is surprisingly cheap and simple. To record sound on location precisely enough to use afterwards, you need a microphone and an audio recorder. To mix different inputs, such as music, sound effects from a record or tape, a voice-over and a track of the background noise during location recording, you will need a record or cassette player to play back the library effects, and an audio mixer to bring together the various inputs. These can then be recorded on a single video-tape sound-track.

Alternatively, use two audio recorders and copy the sound between them, adding a new input each time. Provided the drop in quality is not too obvious after all the copying and recopying, you can build up complex sound-tracks. These can all be fed into a single sound input which, fed into the video tape's sound channel, becomes the final step in the production.

When you record sound as you shoot, the sound and video channels are automatically recorded in step. If you record sound for mixing and later transference to the video tape's sound channel, you must make sure you have a fixed reference point to help you synchronize sound and pictures.

Film crews do this by including a clapper-board shot in every take they shoot. At the beginning of the shot a crew member holds up to the camera a small blackboard with details of the shot chalked on it. Slamming closed a pair of wooden jaws on top of the board gives a clearly identifiable clap on the sound-track tape. This can afterwards be aligned with the frame where the jaws come together, giving precise synchronization between sound and pictures.

When there is no time to go through this procedure at the beginning of a shot, the clapper-board is included at the end instead. The board is held upside-down to show that it refers to the previous sequence.

If you want to edit your shots afterwards, use a clapper-board, and take it out of the sequence once you have synchronized sound and pictures. As an alternative, ask your assistant to tap the microphone. Provided you can see clearly when the tap is taking place, use it as a synchronization point.

If you are confined to shooting in sequence, and using the roll-back editing facility in the VCR as you are working, you will have to fall back on picking up any synchronization clues in the pictures. In most cases, this will only matter when you are shooting someone talking, either to camera or to an interviewer. Then you have to depend on your subjects opening their mouths clearly enough for you to synchronize on their first words.

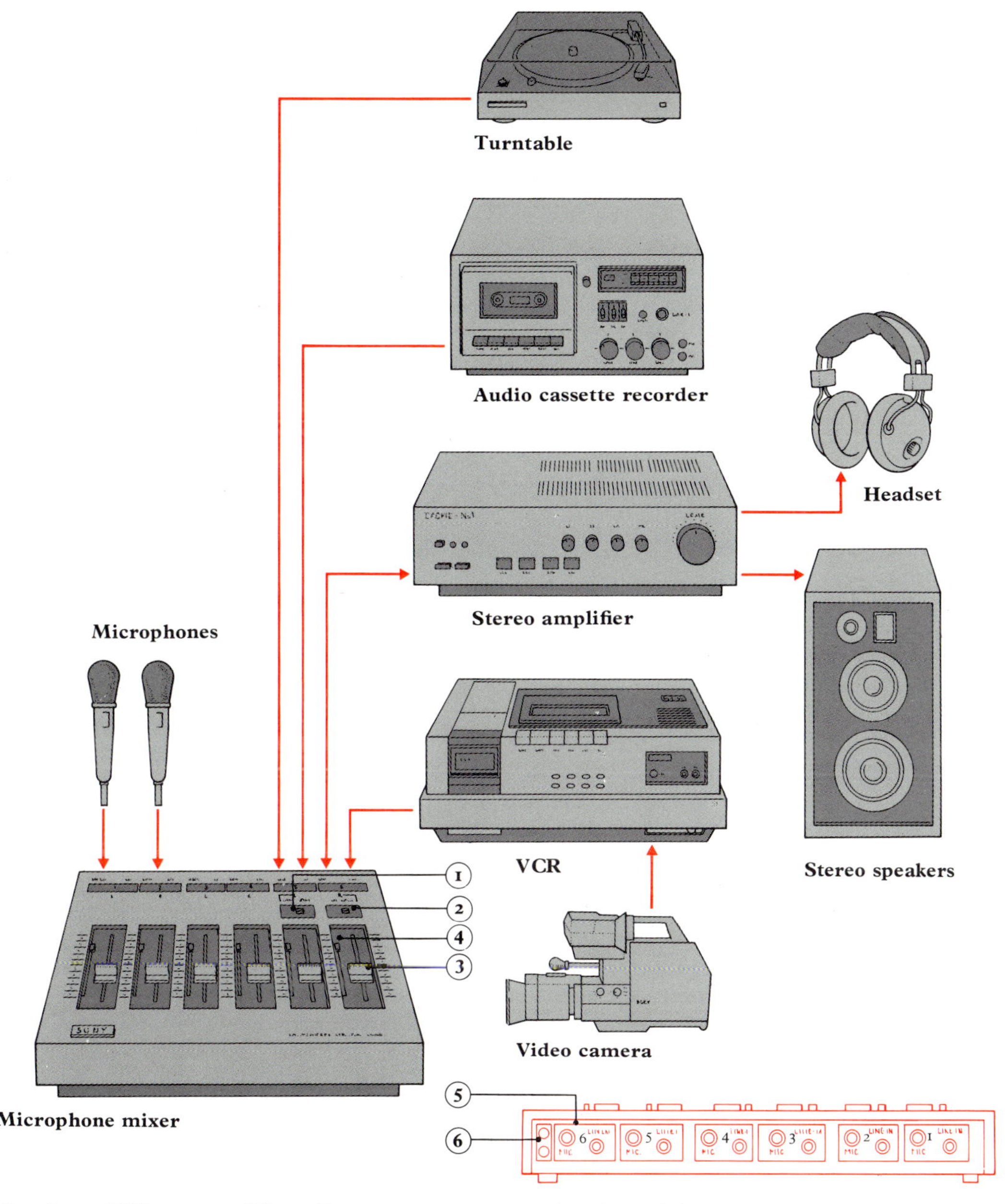

The Sony MX-7 mixer, *above left*, is designed for home video use. It can mix up to six channels of input and provide two stereo output channels.

The distribution switches, **1** and **2**, control channels 5 and 6, *above right*. They allow you to feed in sound to be put out through both arms of a stereo system. For mono sound they are set to the centre to ensure equal output from both speakers.

Each slider switch, **3**, is in direct control of the sound level coming through its channel. Beside each slide is a smaller version, **4**. This is used as a reference point, so that if you use the sliding switch to fade or amplify, you have a visual record of the level to which to return.

Each channel has its respective panel, **5**, on the back of the mixer, *above right*, with a microphone and a line-in socket. This takes input from a record player or audio cassette deck. The two sound output channel sockets, **6**, feed the mixed sound into the VCR.

Scripting

Some of the best video seems to evolve during production, so having a detailed script at too early a stage may inhibit you from recognizing ideas and opportunities while the action is taking place. On the other hand, working to no more than a sketchy storyboard can leave you floundering. Vague ideas which have not been thought through are difficult to shoot.

The job of a script is to relate the words (which may be a simple summary of what the production is trying to put over to the viewer, or a detailed script of the commentary and/or the presenter's words to camera) to the storyboard. A script for a home movie does not have to be as complex a document as a professional script, but it should give you an over-view of what you are trying to put across.

Your script should give some idea of shooting sequence. In most cases, where the tape is being edited in the VCR using the back-space facility, you will have to shoot the sequences in the order they are to appear in the final production, and a simple storyboard is all you need.

However, in a more professional production there may be all kinds of reasons why you would want to shoot in an entirely different sequence. You may, for example, want to show a particular location in several different scenes and it would obviously be more economical to make one visit and shoot all those sequences together, provided you can edit them into the tape afterwards. A shooting script lays out the whole production in the order in which the sequences are to be shot, giving numbers for each, and cross-references to help at the editing stage.

In every professional production, there will also be a sound script. This is divided up into columns, each one referring to a different sound input. One would be for the sound recorded on location along with the pictures, and this will vary from background effects to the voice of the presenter; another is for library sound effects from records or tape; a third for background music and a fourth for the voice-over commentary. The sound script shows where each particular sound is mixed into the sound-track, where it is faded down behind the commentary, and where different sound effects are put in.

As long as you have a detailed sound script it is surprisingly simple to build up a complex sound-track, even without the aid of a mixer. The sound recorded on location forms its basis. This can be transferred to an audio cassette, by connecting the 'Audio Out' socket of the VCR to the audio recorder input. Keep a close eye on the recording level throughout. The presenter's words may be coming through clearly but the background sounds may be too low, depending on your microphone position, and you can correct for this to a certain extent by adjusting the recording level control during the transfer.

Next, add the voice-over commentary. For this you will need another audio recorder and a microphone. Time your commentary first against the recording being played back on your television set and if necessary, cut and adapt it to fit. Play back the sound-track from the first recorder and talk over it, trying to judge that your commentary sound level matches that of the presenter, and that the background sound does not drown you out.

Once you are sure the words and pictures go together smoothly, set the second recorder to 'Record', set the first recorder (with the audio copy from the video sound-track) to 'Play' and deliver your commentary from the starting point. Pace yourself from the script so that your words mix well with the material already recorded.

You can now take the second cassette, which has a mixture of original sound-track and voice-over, and play that back alongside a record player (with your chosen piece of music or effects track providing the other input) on another tape on the first recorder. This will now pick up both inputs as a mixture. This time, you will have to build up the additional channels bit by bit – if you plan to ensure that music and background effects never occur together you will have to build up only one new channel. If they coincide you will have to build up two additional channels on the one sound-track.

This needs careful work, and takes a long time. You have to be sure that each new piece of sound-track comes in at the right spot in your already recorded track, fades up to the right level so as to mix well with the other sounds, and then fades out at the right point. This is where a good, clearly drawn up sound script is invaluable. It is complicated, but an imaginative sound-track can give your picture sequences an impact, and a professional gloss, they would not otherwise have.

In a professional production there will be several scripts, each with a different job to do. Apart from a shooting script and a sound script, there will be a camera script, presenter script for sequences to camera, and a dubbing script for the voice-over. The program script, *below*, brings them all together.

The sequence number identifies each paragraph in the script in terms of, say, pictures with voice-over. Some scripts split the sequences into shots.

'FX1' and '2' are sound effects tracks, and there could be half a dozen in a single script. In a battle sequence you would build up multiple-effects tracks from separate tracks of shouts, screams, shots, trumpet calls and so on.

PROGRAM SCRIPT. Page: 24
Program: THE MOGHUL EMPERORS

SEQUENCE:	TAKE:	PIX:	SOUND:	FX1:	FX2:	MUSIC:
45.		**Skyline –WALLS OF RED FORT IN L/S SILHOUETTE.**		**Cries and shouts.**	**Trumpet calls and gunfire.**	**Indian Music (track 7) fades out**
		MCU battlements – camera PANS RIGHT along battlements to towers at gate.		**F/Down**	**F/Down**	
		CU Flag flying from tower.	**V/OVER: But the world of the Moghuls, for all its wealth and luxury, was a dangerously unpredictable society.**	**F/Out.**	**F/Out.**	
		MLS Corner of battlements.				
		MS Gateway, camera pulls back to reveal PRESENTER.	**PRESENTER TO CAMERA: Anyone who seriously meant to make a bid for the supreme power,**			

A script evolves with the production. Any shot may need several takes; this column identifies the one used for the editor.

'Pix' are brief descriptions to identify the subject matter for the editor and camera operator, both of whom need on-the-spot information.

The 'Sound' column is divided into 'actuality' sound (recorded on the spot): here, the presenter to camera sequences, and the voice-over sequences which are dubbed on afterwards.

The music is taken separately from the rest of the sound-track and is usually identified by composer, piece and even track.

Instructions to camera operators and details of camera moves are abbreviated:

T/C = title card
COL = colour
RT = running time
2-S = two-shot (a shot including two subjects)
F/G = foreground
M/G = middle ground
B/G = background
L = left
C = centre
R = right
T.I. = track in
T.O. = track out
OB = outside broadcast
Repos = reposition (camera 1 to . . .)
X = exit

How to edit without an edit suite

So far, all the sequences discussed have been put together in the order in which the shots were recorded: in other words, they have been assembled rather than edited. While this may be all right for the majority of home video recordings, once you start thinking in terms of more ambitious productions, the ability to select shots and to combine them in a new sequences quite different from the order in which they were recorded, becomes essential.

The problem with video is that editing is much more complicated than it is with film. After you have shot your cine film and had it processed, you can edit the shots you want in whatever order you like by cutting out the selected shots and splicing them together with plastic cement or transparent tape. You build up a completely new version bit by bit. With video tape, this is not possible.

Video tape narrower than two inches in width must have the picture information recorded on it at an angle, in order to cram it all into the confines of the tape; cutting across the tape would leave some picture information on one side of the cut and the remainder on the other. Apart from that, each section of tape would have to be synchronized so that the picture remained stable across each cut.

So video tape editing cannot be done by cutting and splicing in the same way as audio tape and film editing. Instead, it becomes a copying process. You take a blank tape, and copy your chosen shots from the original tape (or tapes) in any sequence you wish so that you end up with a new video creation.

It sounds simple enough, and in principle it is, but there are technical problems to be solved before a smooth edit can be achieved. These involve the lining up of the synchronization signals in each part of the first tape with each another on the second tape, and there are various expensive and complex machines designed to do this.

However, there is a method of editing you can teach yourself with patience and a pair of home VCRs. This is crash-editing, and it involves a little trial-and-error and a good deal of practice. Even so, you will find that some problems with 'glitches' (picture break-up before a new sequence can establish itself) are inevitable with domestic VCRs, but it is useful practice for editing properly. With care you can turn out quite acceptable results.

Crash-editing check list

1. Choose as your master VCR a machine which gives the best freeze frame picture when in pause mode.
2. Connect the video and audio outputs of a second or 'slave' VCR to the video and audio inputs of the master machine.
3. Connect the master VCR's 'RF Out' terminal to the RF input of the TV.
4. Put your new, blank tape in the master.
5. Put a tape with your recorded material into the slave recorder.
6. Find a point on the recorded tape a few seconds before the point where you wish to start copying.
7. Set the master recorder to 'Record'.
8. As soon as the master is laced up, or within ten seconds, set the master to 'Pause'. Do not leave it too long, or you will have a blank area on the master tape.
9. Set the slave VCR to 'Play' and watch its progress on the monitor.
10. When the sequence you want to copy is beginning, release the 'Pause' on the master. The sequence will be copied on the new master tape.
11. When you reach the end of the extract, press 'Pause' again on the master. Then stop the slave machine. Do not use 'Stop' on the master, or you will get a bad noise burst.
12. Select another sequence from the recorded tape and repeat the process.

Crash-editing is so called because the signals from the two VCRs cannot be synchronized to one another, and the input signal from the slave VCR 'crashes' into the picture-scanning signals for the last sequence recorded on the master VCR, causing a glitch – or burst of video noise – on the picture.

It is impossible to remove this noise entirely when editing with home VCRs, but once you have had some practice with your simple edit suite of master VCR, slave VCR and monitor, you will learn how to time your edit entry and exit points for good results.

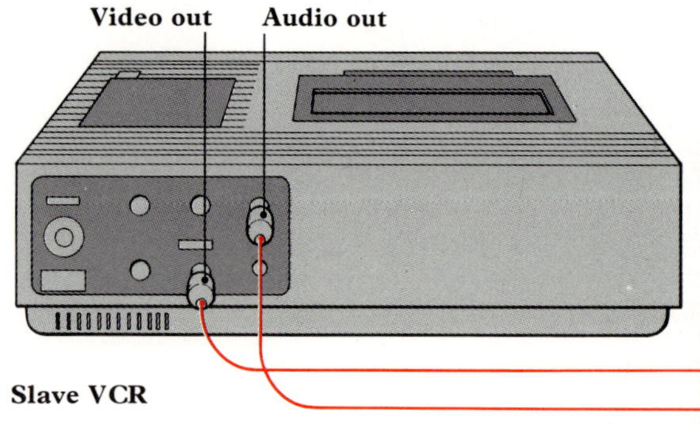

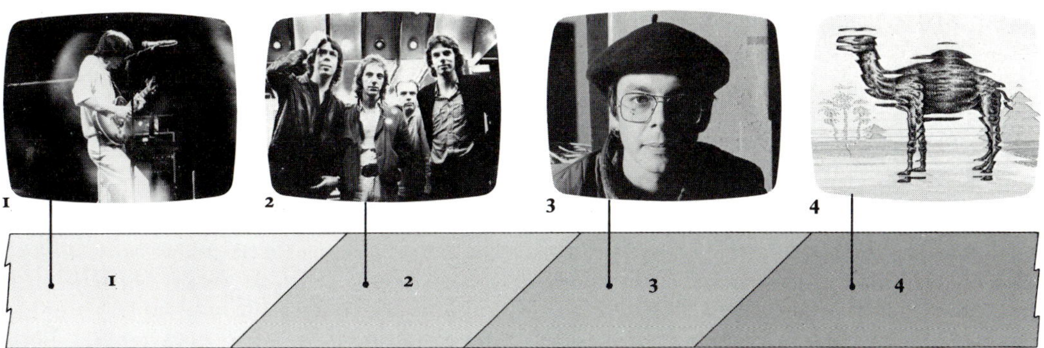

Use sequences recorded off air to enliven a location shooting which did not work so well. If, for instance, in a concert given by Camel, you could not move around to record from different angles, you might vary your sequences with an off-air recording from the local TV news.

A Camel poster, **4**, *above*, would make a good title card, **4**, *below*. You could open with a news shot of the band arriving at the stadium, **2**, then edit in a number from your live recording, **1**. Cut to part of a TV interview with the band, **3**, then edit in the next number from your live recording, and so on. End the tape with the band's best number.

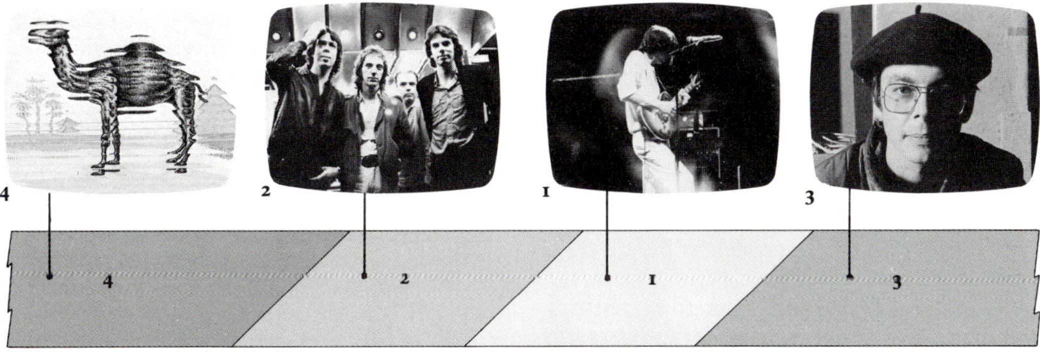

When you edit video tape you pass an electronic signal from one VCR to another, *below*.

Machines of any format may be used together. However, they must have video inputs and outputs, and not just RF connections. These are not suitable for editing. This eliminates some older machines, but older VCRs would not generally be able to provide a stable enough freeze frame.

If you intend to crash-edit on a Video 2000 VCR you need a special accessory (22AV5530) and a 22AV5002 cable to connect another deck.

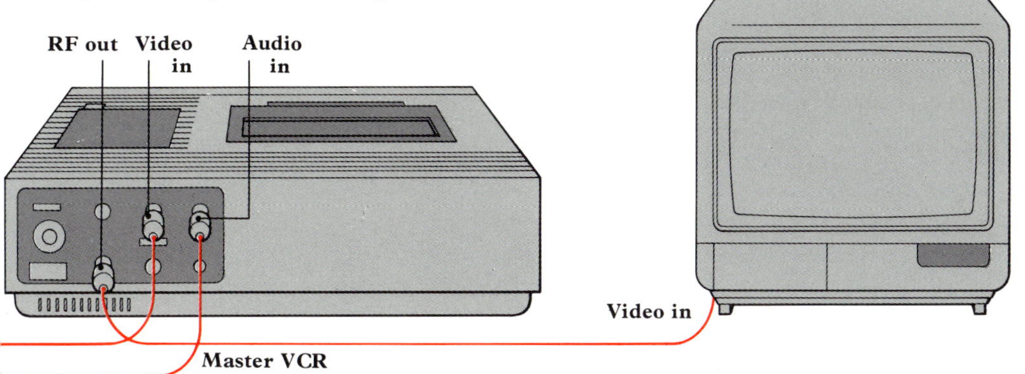

Professional editing

Although crash editing will allow you to copy sequences on a new tape, it has many drawbacks. However good the VCR and its operator, there will be some noise at edit points which will get worse if copies are made. Moreover, releasing the VCR's 'Pause' button is not accurate enough when you might need to be within a frame or two of a particular point. You sometimes need to be able to time the duration of shots more accurately than you can with a stopwatch or tape counter and you will often want to see whether a particular edit will work visually before committing it to tape.

All this and more can be done on a basic professional two-machine edit suite, linked by an edit controller. Sony's U-matic VCR editing system was one of the first and is still one of the most sturdy and simple to operate. It consists of two edit recorders (although it is not necessary for both machines to have exactly the same facilities), an edit controller and two monitors. One way of reducing the cost of the system is to use as a slave machine a cheaper, simpler player-only VCR, or even a simple player/recorder.

The process, which is simple by professional editing standards, is like this: you load the blank tape into the master VCR and then consult the notes you made as the shots were being recorded, together with their counter numbers, and select a cassette. As U-matic cassettes run for only one hour, there will probably be several sequences, shot in different places at different times, from which to choose. Use the picture-search capability, in U-matics and essential for an editor, to fastwind to the general area of the section you want to use, and then, by playing the tape in varying degrees of slow motion, locate the start of the sequence you want to use to the nearest frame. On newer U-matics, the digital time read-out (a sophisticated tape-counter) displays the position on the tape accurate to a single frame (a fraction of a second), reads the control track pulses on the tape and displays them as a digital time read-out in frames, seconds, minutes and hours.

If the master tape already has material compiled on it, you can find the point from which you want to start your next shot to the nearest frame.

Once you have both VCRs lined up, all you have to do is press the 'Assemble-edit' buttons, and the edit controller will wind both VCRs into reverse for five seconds. It then stops them, lines them up and starts them both running, correcting the run-up speed until they are exactly synchronized. When the edit point is reached, its memory tells it to start recording from the slave to the master tape, until you press the 'End' button.

If you want to try out an edit without copying it, there is a 'Preview' button. The control unit will go right through the motions without recording, while watching the monitor. The VCR then returns both recorders to their starting point, so that you can run the sequence by pressing the 'Edit' button, or change your ideas before trying another test run.

You can perform another kind of editing, called insert editing, using two machines. If you have a sequence which needs another shot inserted into the middle of it, you start copying the additional sequence using the 'Insert Edit In' control, but when you reach the point where the new shot is to end, you simply press the 'Out' button and the remainder of the original shot is left on the tape.

Two-machine editing

Editing between two machines using an edit controller does allow you to build up a production with a certain flexibility, but you still need to plan very carefully if any special effects are needed.

For example, all the editing system can normally do is cut from one pre-recorded sequence to another as a straight cut. If you want to fade out a particular shot, leaving a dark screen, you have to record it like that when you use the camera. The editor can copy it and insert it into the final scheme, but it cannot do fade-outs, nor dissolves, between two pre-recorded pictures at the editing stage.

Here again, if you want to mix for dissolve from, say, a shot of a rock band to the face of the lead singer, then back to a close-up of fingers on the guitar strings, you will have to record that as a single sequence in the studio, using the studio mixer and a special effects generator. Once it has been recorded on a single tape, you can edit it into the final recording on a two-machine set-up.

All professional editing machines can do insert edits as well as assemble edits – where a new shot is recorded over part of the duration of an existing shot. They can be used to vary visual interest, or to conceal some shooting error.

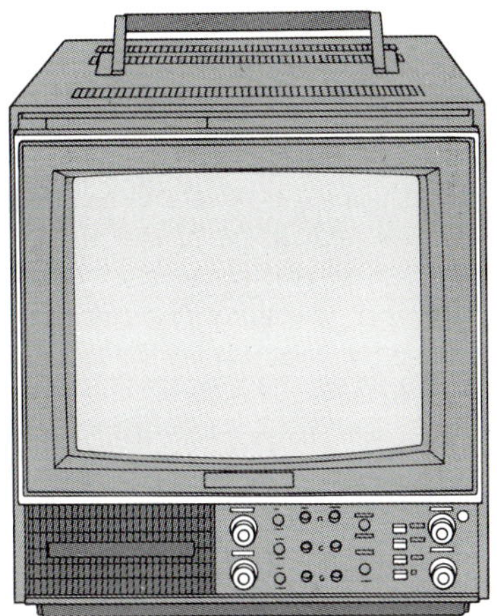

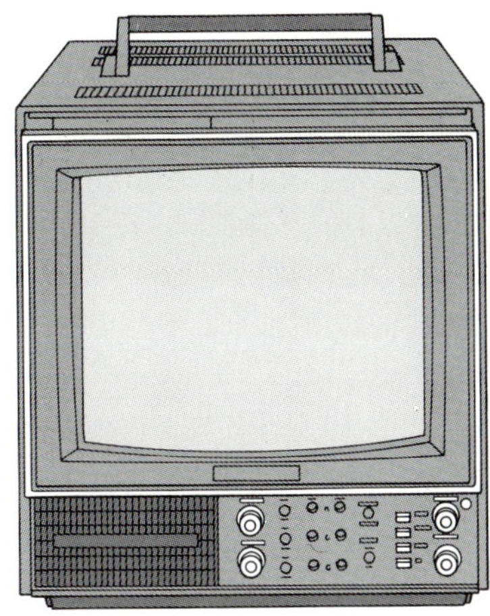

In a professional editing suite each VCR has its own monitor. At least one will be a high-quality colour set such as the Sony PVM-1300E, *above*. As the name implies, the monitor's role is to allow the editor to keep track of transmissions from a VCR or from a video camera.

The Sony RM-440 edit control unit, *bottom*, enables the editor to preview edits, program sequences into its memory for automatic editing, return the tape automatically to pre-set positions and keep a frame-accurate tally of its progress. It has a set of controls for both VCRs.

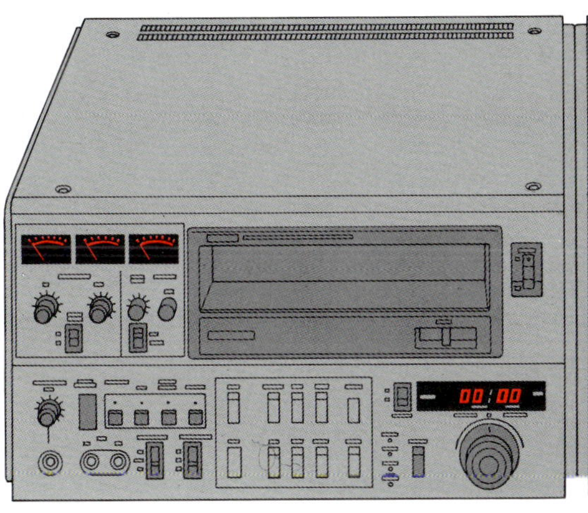

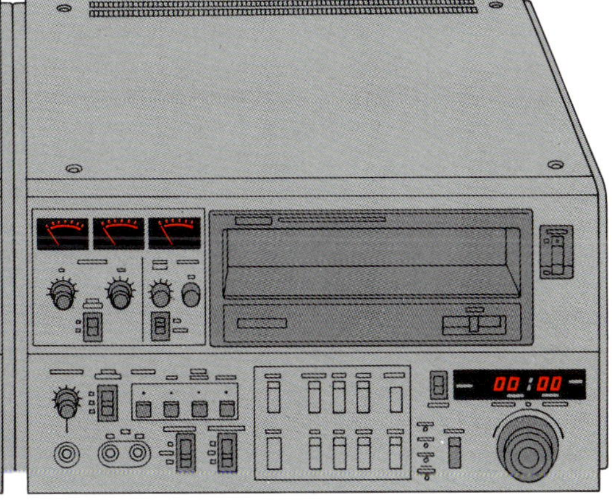

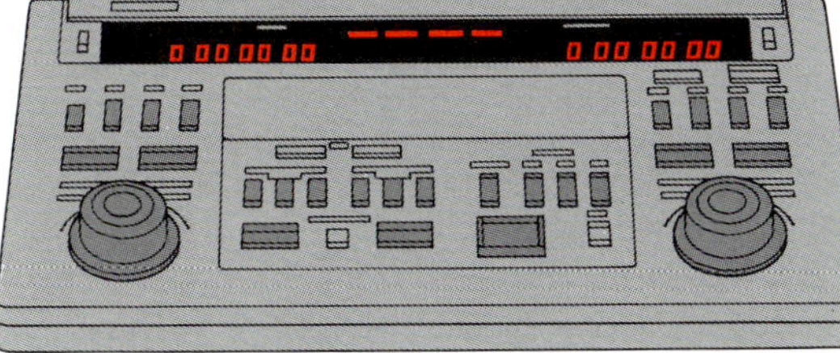

The densely packed control panel distinguishes editing VCRs from less complex models. Representative of the new front-loading generation, Sony's VO-5850, *above*, is so expensive that it is unlikely to be seen outside a professional studio.

The VCR, *above left*, is on 'Playback'. It contains your recorded tape from which you are copying precisely controlled segments, in the order you decide, on the VCR, *above right*. This is on 'Record'.

The operation is controlled by the edit controller, which decides the precise point at which each segment begins and ends. It is accurate in time to one frame, that is, 1/50 sec. (PAL) or 1/30 sec. (NTSC).

On the right-hand monitor you identify the end of the last segment to be copied on the master tape. Then you look on the left-hand monitor for the next segment to be added from the left-hand VCR.

The edit controller will rehearse the edit for you before recording, so you can see what you are looking for. It can determine for you the end point of the edit so that, when you press the 'Edit' button, it will automatically copy the segment on the master tape and stop where instructed.

The special effects generator

What happens if, when editing your tape, you want to mix between images shot with more than one camera, or at different locations on different days? Effects such as these have to be added with a full three-machine edit suite of two input VCRs and a master, with a professional special effects generator.

With a three-machine suite, not only can you compile sequences on your master tape but you can combine the pictures shown on the two input recorder tapes in any way the SEG will allow you; you can mix from one to the other; you can command split-screen effects; do a straight wipe; or even bring the second picture up through a shaped insert in the middle of the first, and transfer the resulting combination to your master.

Equipment as versatile and sophisticated as this is inevitably so expensive that even professionals rarely buy it. They rely on renting it by the day from production or facilities houses, and four-figure prices for a day's editing are not unusual. You can cut costs by first working out your editing sequence in detail on a simple editing suite, and then renting the full professional equipment to translate the sequence you have planned to the finished tape.

There are many places where time can be rented on professional machines, and unless you plan to do a great deal of editing it is hardly worth purchasing one. Should you find you need one eventually, second-hand bargains are constantly available as facilities and production houses update their systems.

However, with a two-machine suite there is an element of trial and error in the editing process: in the pace with which you cut from shot to shot, for instance, or the exact point at which you cut to a new picture. For this reason, an edited production has to evolve stage by stage, in a painstaking and often time-consuming process. Yet the costs incurred are so high that reducing editing time as far as possible is a priority.

One way of resolving this dilemma is to use yet another machine known as a Time Base Corrector (TBC). This is a computer which monitors all parts of a video signal. If a line is out of place the TBC can compensate. It is the only reliable source of accurate synchronization, and accounts for a high percentage of the cost of a full professional edit suite.

There are TBCs which can record, on a separate tape, the exact time and frame number on each input tape of the shots you use at each edit. When you finish assembling that production, the tape number, time and frame of each shot used is preserved as a time code.

This can later be used for building up additional copies of any part of the tape. The information has simply to be fed into the player machine as required, and the rest is done by the control unit, working on the time-coded tape's instructions. A new copy of the whole production can be made in a fraction of the time which would be needed to copy it by other means and, because it is a new original and not a copy of the first edited master, it saves an entire generation of copying; the finished quality is also much higher.

Production and facilities houses have more complex installations which can handle several input machines at once. If the tapes on which shots are recorded can be left permanently in an input machine during the whole editing process, the computer can be used to identify each shot by its time code, and to preview it to your instructions, until you are entirely satisfied that the length and timing are absolutely right.

If you prefer you can build up the whole production as a series of previews, telling the computer which tape the next shot is from, and at what points it begins and ends, so that the production is entirely built up as a set of coded instructions.

A further advantage of time-code editing is that if, for technical reasons, a producer has to shoot a production on expensive 1-inch (2.5-centimetre) or 2-inch (5-centimetre) open-reel video tape, copies can be run off on cheaper U-matic cassettes, which can then be used to produce a time-coded, edited tape. The time-code instructions can then be used to control the much more expensive open-reel editing unit to produce the same sequence, using the open-reel originals, for a fraction of what it would cost to do all the work on those machines.

Unfortunately, home decks are not yet made to be connected to a TBC – that is something only an electronics expert can do. Whether you buy your own equipment or rent someone else's, the most important ingredient of any video production is, and always will be, the originality of your ideas.

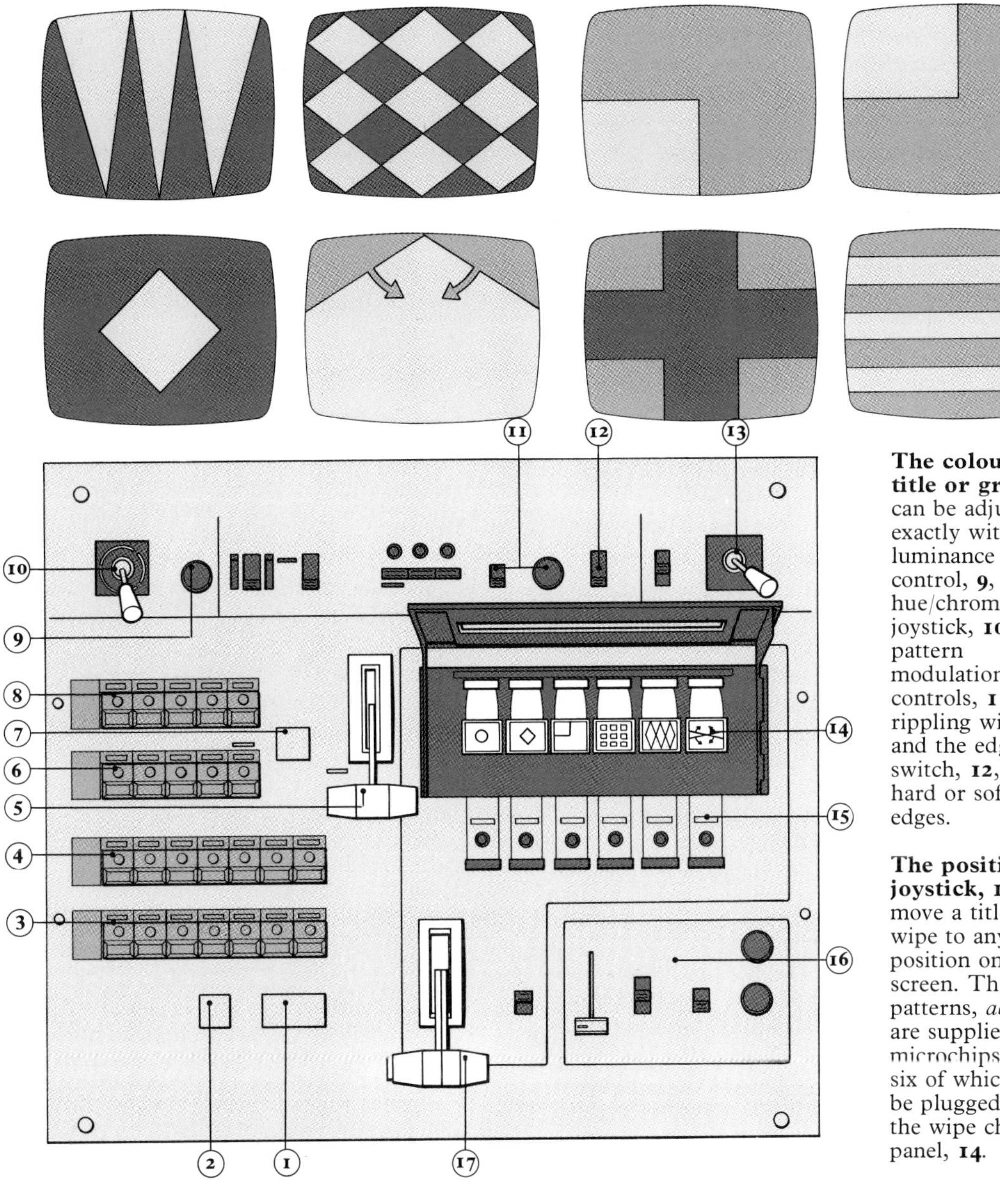

The colour of a title or graphic can be adjusted exactly with the luminance control, **9**, and a hue/chroma joystick, **10**. The pattern modulation controls, **11**, give rippling wipes, and the edge switch, **12**, adds hard or soft edges.

The position joystick, 13, can move a title or wipe to any position on the screen. The wipe patterns, *above*, are supplied by microchips, any six of which can be plugged into the wipe chip panel, **14**.

By adding a special effects generator to your edit suite you can mix between the input pictures from live cameras or recorded sources, or add graphics or decorative wipes for transmission. The 'Take' button, **1**, cuts from the output picture to the input picture which is to follow. 'Soft take', **2**, does this in dissolve.

The 'PVW' switches, 3, select an input picture from a camera or recorder to be displayed on the monitor. The 'PGM' switches, **4**, allow a picture which is being given out by the SEG to be monitored. The operator can run through a sequence of pictures, with effects added, on 'PVW' and then switch to 'PGM' for transmission.

A 'Mix' lever, 5, determines the rate at which the picture from one input mixes to that from another. Two rows of switches, **6** and **8**, can select any two inputs for the SEG to mix between at any one time. The 'AB' button, **7**, will swap one input with another in a wipe sequence, or restore the original order, if operated again.

A desired wipe is added to the input picture by using whichever of the wipe select buttons, **15**, is below the pattern.

On the downstream keyer panel, **16**, are controls which allow graphics to be added from a monochrome camera.

The wipe levers, **17**, position wipes accurately on any part of the screen.

Outside broadcast

Whether you are doing simple recording or an ambitious production, video is inseparable from television. Watching television with an aware and critical eye is one of the best forms of self-education for the video producer, but amateur enthusiasts tend, consciously or otherwise, to compare their work unfavourably with the output of trained specialists, equipped with expensive cameras and recorders, backed by every kind of advanced electronics, and on a budget infinitely higher than any amateur could ever contemplate.

There is a difference – it shows – in the ultimate quality of the pictures that all this makes possible, but it is heartening to realize how much broadcast television depends on ideas, originality and imagination.

Broadcast television comes closest to home video with location recording. The simplest method is ENG (electronic news gathering), which is one of the newest conquests of video over film. Reporters delivering material from distant trouble-spots used to shoot film, which had to be processed before it could be broadcast. Now, a camera operator and assistant, with a lightweight, high-quality video camera and recorder (costing five to ten times as much as a Betamax or VHS portapack) can record on video tape, which is then rushed, relayed by satellite, or microwaved straight to the studio for broadcast.

The same principles are used for EFP (electronic field production). Shooting an outdoor sequence for a drama production, the camera operator and sound technician are augmented by a crew of four to a dozen people, handling extra portable equipment: lighting, a colour monitor to check the material being shot, and a waveform monitor and vectorscope to give technically precise representations of the recorded signals.

The production techniques are more like those of film, or home video, than studio television. One camera is used, and the feature is built up, shot by shot, to be edited back at the studio. Unlike film, however, shots can be played back on location.

Finally, there is the full outside broadcast coverage of an event such as a sporting occasion. The technique used in shooting, cutting from camera to camera, is more akin to studio production, with the director and video switcher seated at control desks in their self-contained mobile control room.

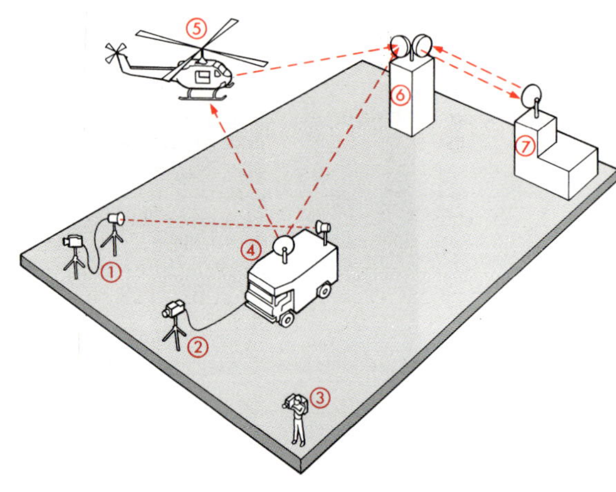

Video shot on location reaches the TV channels from a number of sources, *above*. A portable camera may have its own small transmitter, **1**, or be linked to a control truck by cable, **2**. A helicopter, **5**, may be equipped with portable cameras to shoot crowded or dangerous situations.

A mobile control truck, **4**, equipped with a microwave relay antenna, can transmit live material direct, or via a relay station, **6**. This passes the signal to the studio, **7**, for broadcasting.

An ENG camera operator, **3**, turns in news to a control van or studio.

The BCC-20 Digicam camera, *below*, can check and adjust its own circuits to give studio-quality pictures. By changing lens and viewfinder, this camera's versatility can be extended from hand-held portable use to full studio service.

The ACM-300 outside broadcast mobile, *right*, was fitted out for the Swiss Army Film Service, who use it both for teaching technique, and to make training tapes for other skills. Modern mobiles are versatile and compact.

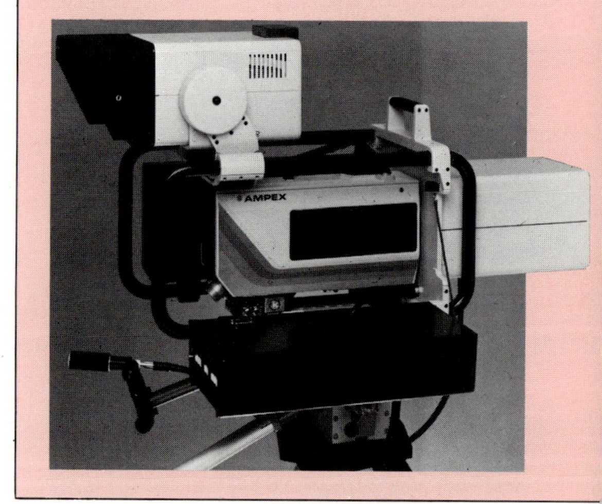

An outside broadcast vehicle, remote truck or mobile control room, *below*, holds enough machinery for full studio operation. There is an on-board petrol engine, **1**, to give an independent power supply, and the production desk, **2**, has full sound and video switching capability.

The colour monitor, **3**, is for colour control, and there are cheaper monochrome camera monitors. The VPR–2B VTRs, **4**, can also replay recorded material in forward and reverse slow motion. A waveform monitor and vectorscope, **5**, give an electronic read-out of the colour and brightness signals. Above the engineer's desk, **6**, is a full set of monitors, camera control units and video distribution amplifiers. Air-conditioning units, **7**, are essential with equipment running in a confined space.

A camera operator can shoot on the move from a hatch, **8**.

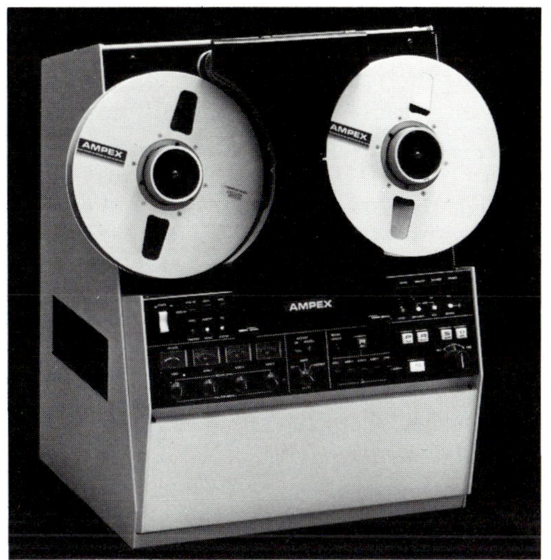

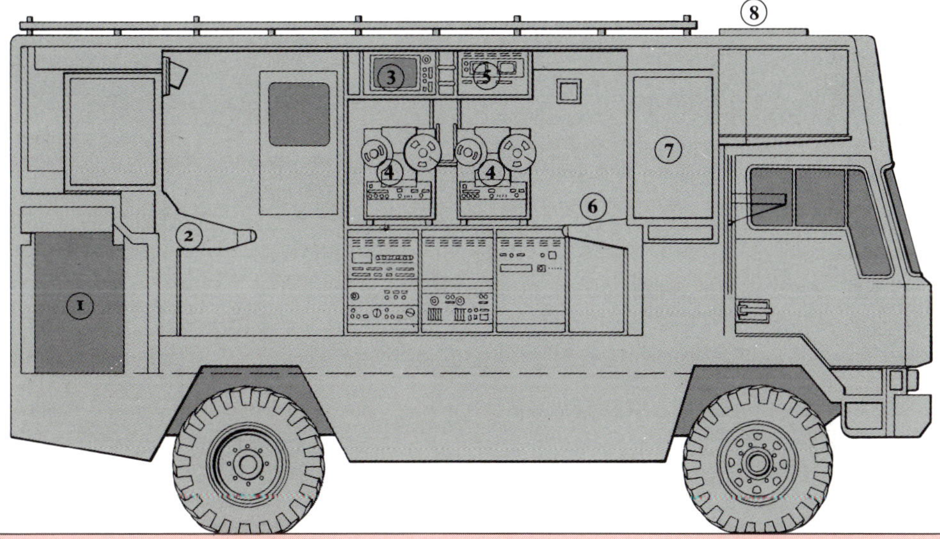

A camera operator working in the field is inured to dealing with the unusual. A helicopter, *right*, is often needed to make possible the shooting of dangerous events, such as large fires or civil riots.

To isolate the camera from the aircraft's jarring, special mounts support the camera in perfect balance, so that its movements are unaffected by the vibrations.

Precision technology cannot always help. One operator, required to record a singer while running down a woodland path, found it impossible to operate a camera while running backwards over fallen branches. The problem was solved only with the help of assistants. The camera operator had to submit to being half-towed, half-carried along the path, with the camera, pointing backwards at the subject, under one arm.

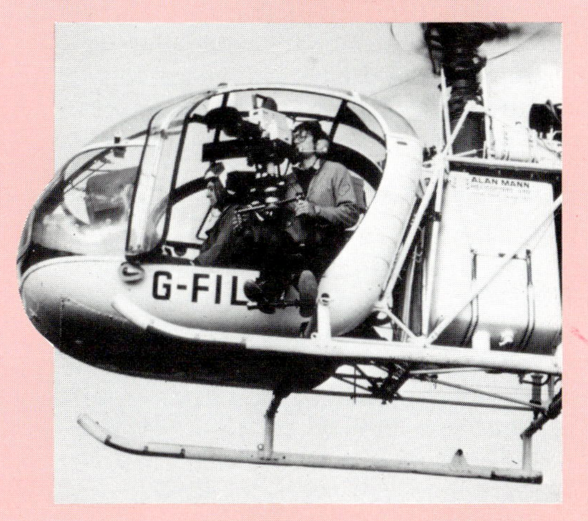

Directing the show

A television studio is a purpose-built environment for television production. Despite an ever-increasing emphasis on location shooting, studio technique is still geared to the assembly of a show in one continuous session, just as it used to be in the early years when most broadcasts were live, and equipment was bulky. The skill and forethought needed to effect an elegant production with a minimum of editing is still a yardstick of professional excellence for a television director.

In drama particularly, a mountain of preparation lies behind the smallest production. The set designer must excercise ingenuity in fitting all the sets into one studio, often making a scale model to test the set's efficiency. Movement of cast and cameras is minimized, and camera positions calculated so that the viewers never see a camera, microphone boom or other giveaway in their picture.

Studio lights are a world in themselves, suspended in their dozens from a gantry high above the action. While the cast rehearses elsewhere, the lighting director coordinates the lights with the required camera angles to achieve the correct light level and, where needed for atmosphere, shadows. During the final camera rehearsal, which simulates transmission as closely as possible, the director can check that the cast and equipment work together for the best effect. During shooting, the final camera script is followed by everyone, although small changes can be made if the director and camera crew agree on them.

The director orchestrates the whole operation from the control desk in the production room, watching the monitors and following a camera script prepared during rehearsals. It is the director's job to cue the vision mixer to switch between cameras, while warning the camera operators in turn what their next shots will be. While camera 1 is on air, camera 2 is picking up shots, which the vision mixer can choose from, and camera 3 may be moving around, setting up a new angle.

In the audio control room, audio engineers mix and adjust the sound from the studio microphones, and add sound effects. A different unrehearsed show, such as a talk show, is a challenge, demanding instant reactions from the whole crew, to pick up the best of what is happening as it happens.

It is the floor manager's job to convey instructions to the cast during shooting by gesturing silently. This sign language is a studio tradition.

On the nose. (We're on time)

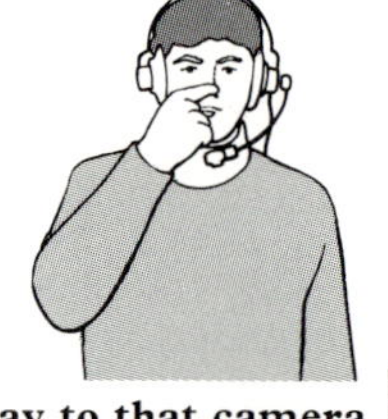

Play to that camera

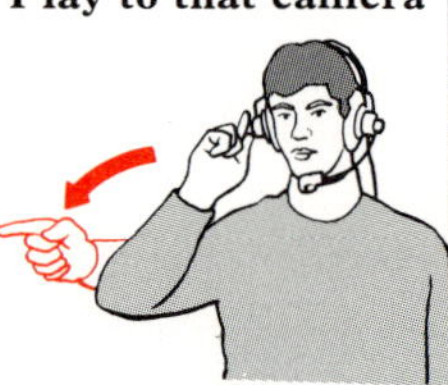

Slow down

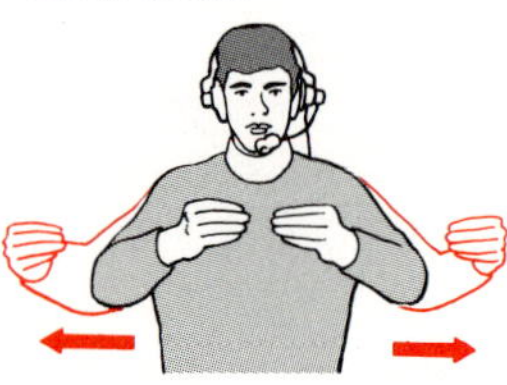

Stand by

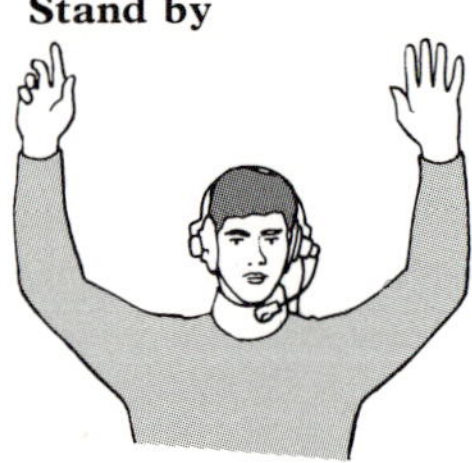

O.K It's all right

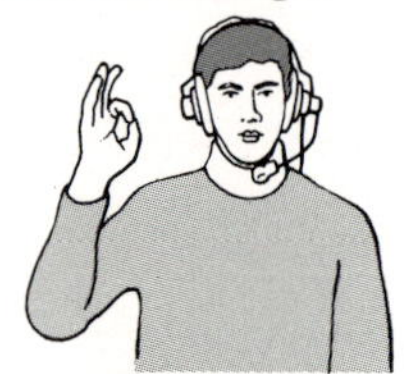

Cut; finish

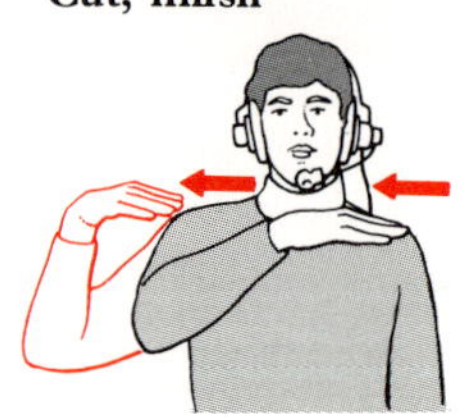

In a TV studio, *below right*, the sound and vision control rooms, and the production room, overlook the studio floor.

In the video control room, 1, engineers monitor the colour and contrast levels of the signals from the cameras.

The production control room, 2, houses the director and assistant director, the vision mixer and the technical director. A bank of television monitors includes a colour transmission monitor, a monochrome monitor for each camera, and others showing outside broadcast, telecine and recorded material.

In the sound control room, 3, are the audio controller and technicians who mix and switch live and recorded audio sources.

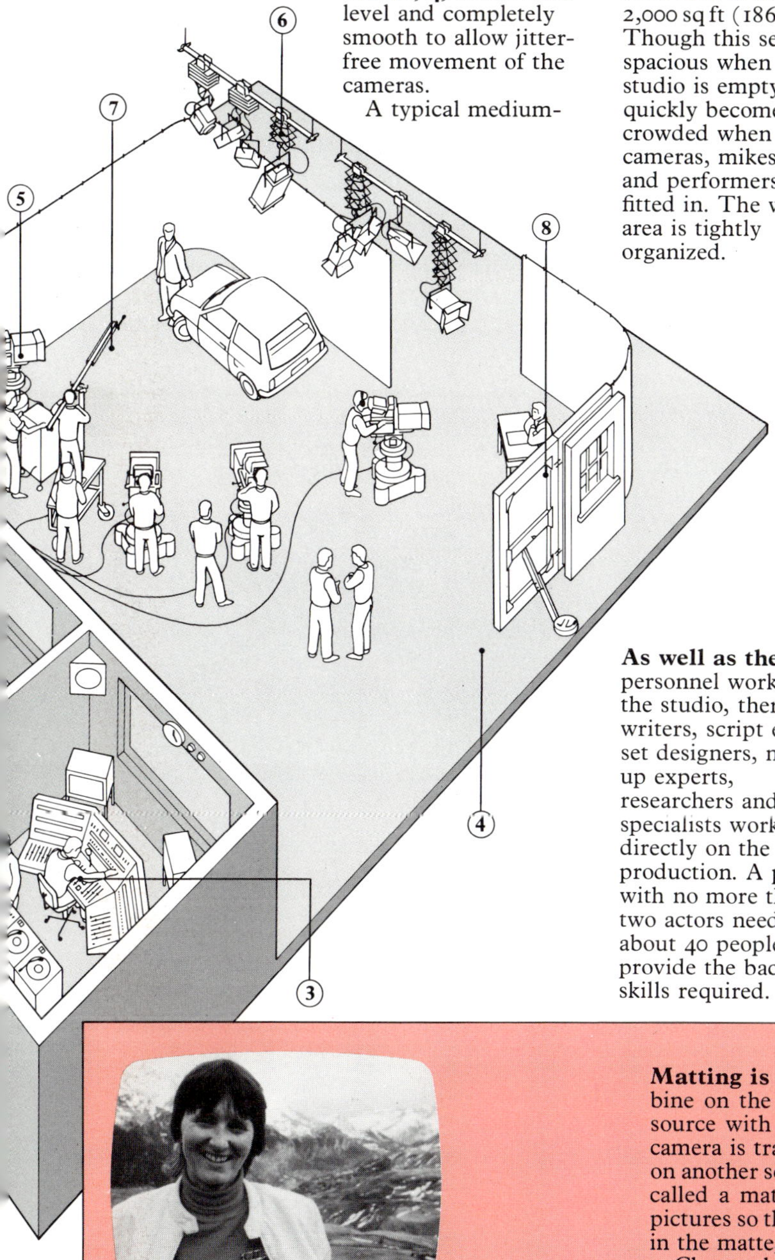

The floor of the studio, **4**, is laid dead level and completely smooth to allow jitter-free movement of the cameras.

A typical medium-sized studio has a floor area of about 2,000 sq ft (186 sq m). Though this seems spacious when the studio is empty, it quickly becomes crowded when cameras, mikes, props and performers are fitted in. The working area is tightly organized.

As well as the personnel working in the studio, there are writers, script editors, set designers, make-up experts, researchers and other specialists working directly on the production. A play with no more than two actors needs about 40 people to provide the back-up skills required.

A studio camera, **5**, is larger and heavier than a portable camera. It weighs about 112 lb (51 kg) and is carefully counterbalanced on a pedestal, whose hydraulics enable smooth movements to be made with finger-pressure. The pedestal is moved around the studio on dolly wheels.

Studio lighting is fixed on a permanent grid, **6**, well above camera level. Some grids are worked by hand, while others are automated. The changing light intensities can be computer controlled.

Mikes, **7**, are operated from the floor, mounted on booms or stands, or attached to the performers, bringing them close to the source of the sound.

Convincing scenery is based on 'flats', **8**, board constructions which can be made into complexes representing the interior of a whole building.

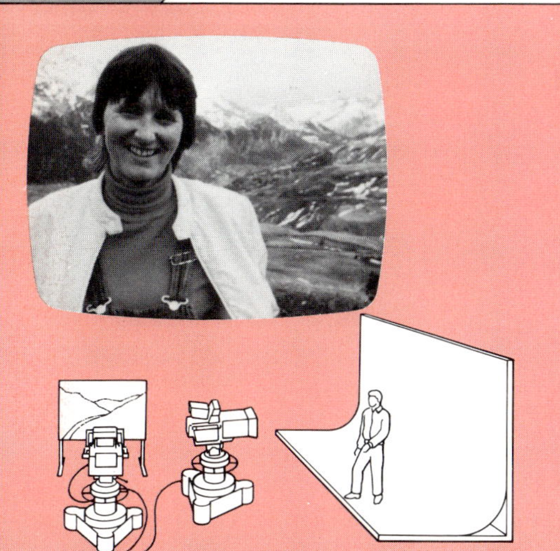

Matting is a technique which can combine on the screen a performer from one source with scenery from another. If one camera is trained on a presenter, a second on another scene and a third on a silhouette called a matte, an SEG can combine the pictures so that the presenter appears within the matte, surrounded by the scene.

Chroma key, *left*, is a more modern and widespread technique similar to matting. The presenter is shot against a pure colour (usually blue, the hue least dominant in human colouring). An SEG can then replace the blue area with a picture from another source, such as this alpine scene.

By means of these techniques a subject can be apparently transported to an entirely different setting – an airport, perhaps, or a tropical island beach.

Electronic wizardry

Drama may bring creative revolution; documentaries may break new barriers in communication, but many people say that the most adventurous television is to be found not in feature television at all, but in commercials. Their subject may be washing powder or after-shave, but their true aim is to pack into thirty or sixty seconds an impact that will remain after the rest of the night's viewing has been forgotten.

This is where money talks. Television audiences are vast; demand for air time is high, and so are costs. Customers are prepared to pay fabulous sums to have their message conveyed as persuasively as possible, and for their money they expect to receive not only flawless technical performance but originality, in a field where, it seems, everything has already been tried.

The shooting of commercials is the province of the production houses, companies which provide the production, editing and special effects equipment which only this rich trade can afford. Their clients may be hiring a two-man EFP team, or commissioning on a budget of millions.

Millions are also invested in machinery for recording and editing film, sound and video: mixers which can mix two camera inputs in dozens of complex patterns – ripples, lozenges, spiral wipes – or chroma key a model safely from a bare studio into a raging inferno. These are analog effects which modify the scanning sequences from the cameras by waveforms representing the desired patterns. Even more costly tricks are performed by computer-controlled frame-store devices. These encode the picture digitally and then stretch and squeeze it into distorted shapes or even rotate it. With the help of computerized edit suites, caption generators and colorizers, the permutations of images available seem limitless.

Perhaps the most precious commodity is the patience and practised judgement of the engineers, directors and editors who produce the commercials. Many essential processes from writing computer graphics programs, arranging props, testing camera angles, and drawing mattes, right down to the simple, but essential jobs such as cooking a fresh dish of baked beans for each new take, can only be done by hand. One professional estimated that he produced twelve seconds of finished work a day: twelve seconds of video perfection.

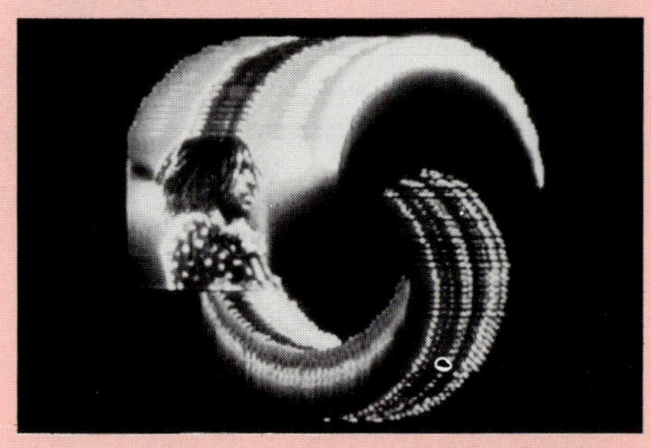

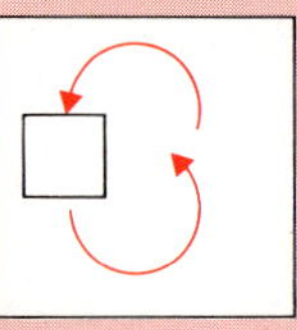

Using its frame store, a computerized SEG reduces an image to a portion of the screen and swirls it round, leaving a trail of duplicate images behind it.

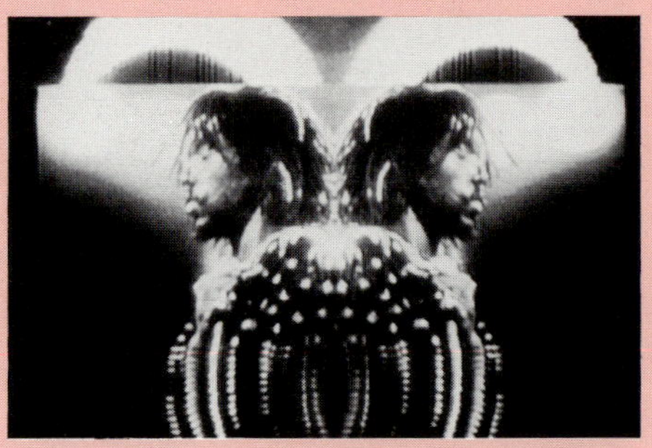

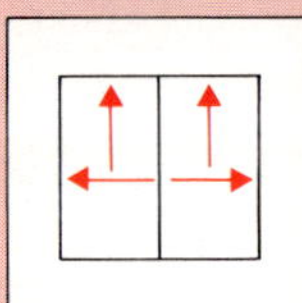

As it moves round, the image expands. The SEG's next instruction splits the screen and puts the image into one half, with a mirror-image in the other.

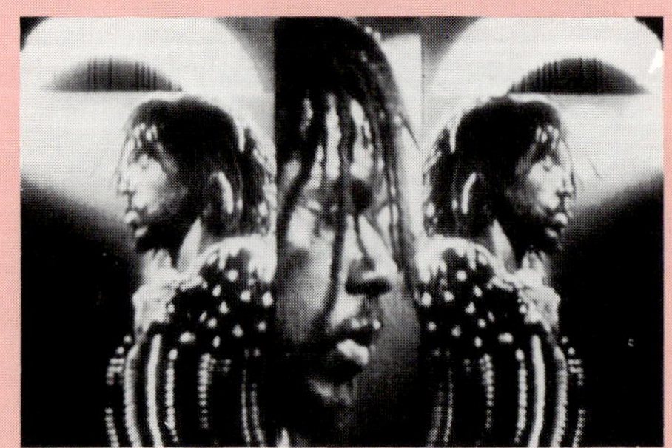

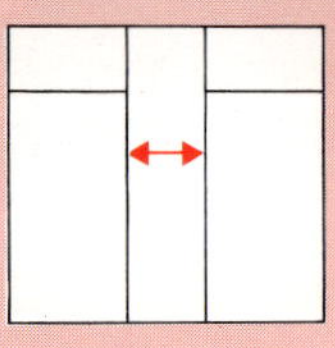

In a third stage, the machine opens another picture area. Taking about two seconds to run, this sequence is only a sample from among dozens of effects.

In this modern post-production suite, each piece of equipment has its own control panel, but central control is exercised from the computer keyboard to the left of the main desk.

From this computer, a stereo sound mixer and tape recorder, a vision mixer (video switcher), a digital harmonizer (a device for regulating sound pitch when the tape is running slow or fast), four digital effects devices, a 2-in (5-cm) and three 1-in (2.5-cm) floor-standing studio VTRs, and a dozen or so monitors are coordinated into a system that enables the creation of image sequences so complex and ambitious that to control them by hand might take years, instead of days.

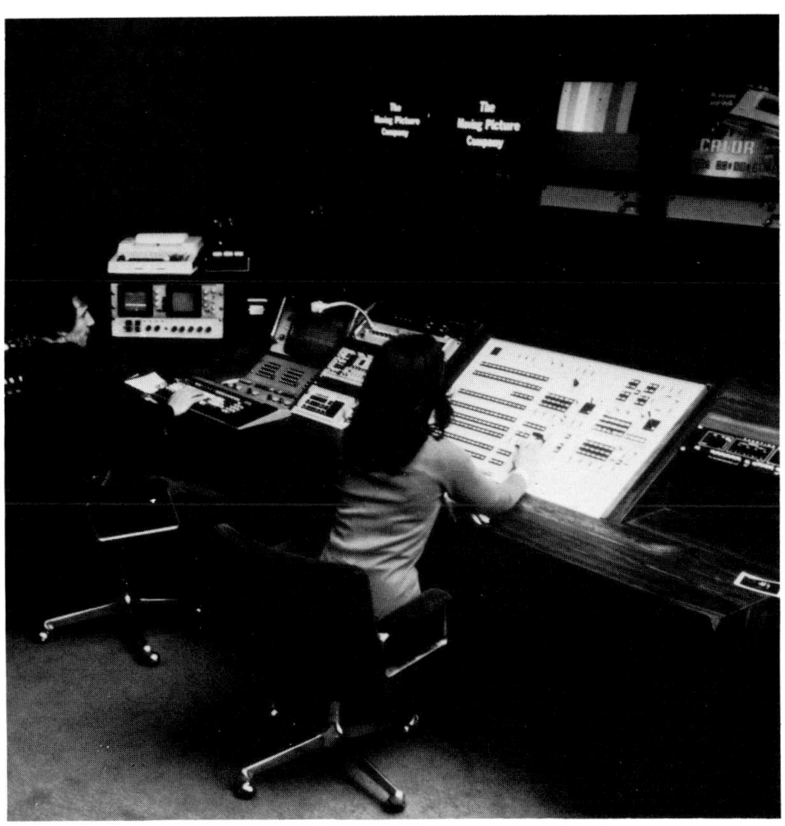

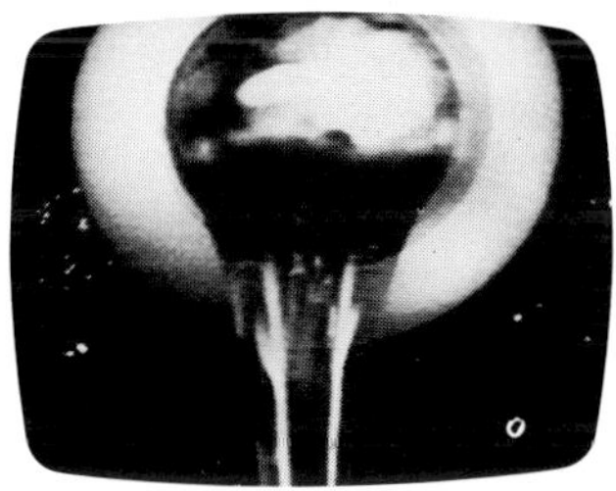

A commercial is built of images from many sources. Here motor oil is shown pouring from a bottle.

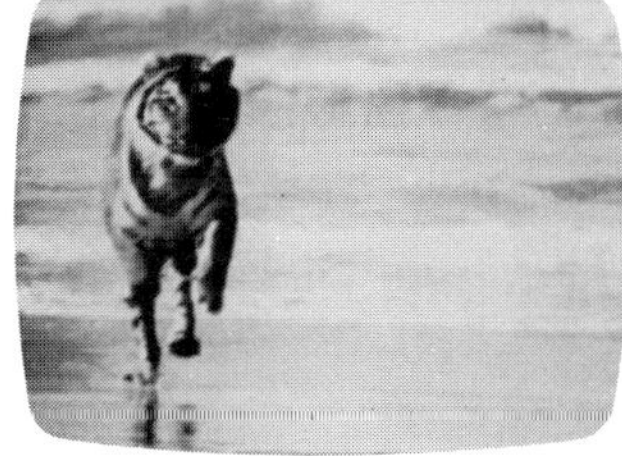

Film of a live tiger, shot from a distance to protect the film crew, has already been provided.

A matte is hand-drawn for each frame, to isolate the tiger from the unwanted beach scenery.

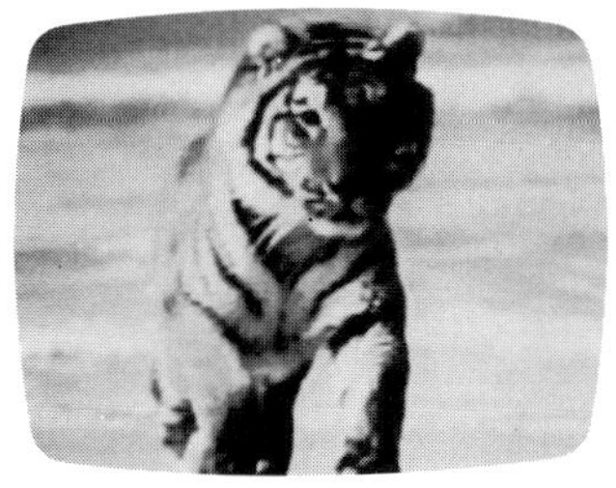

With a digital device the editor can enlarge the image of the tiger as if a camera were zooming in.

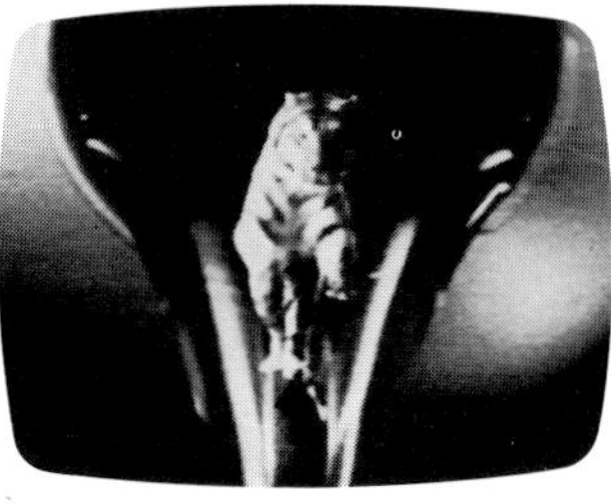

The whole tiger sequence, from long shot to close-up, is matted on to the background of pouring oil.

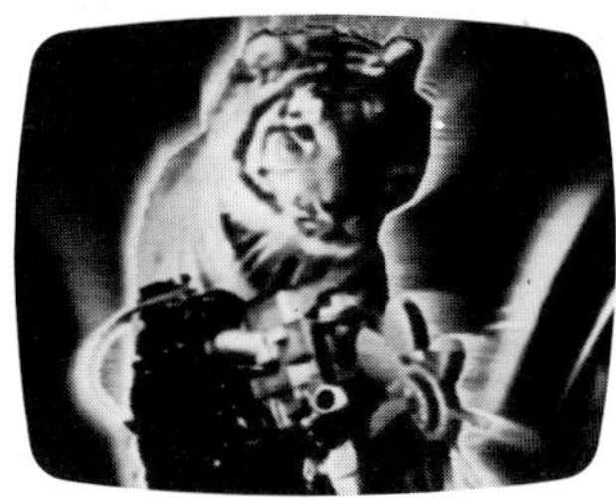

In the final stage, the tiger, now standing by a motor engine, is surrounded by an electronic halo.

Making video work for you

Video makes a splendid hobby. Its endless fascination and its limitless challenges to the imagination make the prospect of becoming bored with it seem remote.

Enjoyment is not the only benefit video production can offer. It can be used to help make a living – either indirectly, by contributing to the success of your business or profession, or by becoming your business. Video is already an accepted and increasingly important aid in a variety of fields, from education to salesmanship. Anyone with the skills and equipment to supply polished video recordings to paying customers could easily find their hobby expanding into a profession.

People in commerce and industry are discovering a need for video recordings to sell new products, train workers, communicate with far-flung divisions and branches, reassure shareholders and drum up new customers. So far, the demand for small outfits that can handle video with flair, with imagination and with dedicated professionalism, far exceeds the supply.

Even in the non-commercial world there is immense scope for making many of the tapes, described in the previous section of this handbook, on a profitable basis. People are becoming aware of video's potential for providing an imperishable record of unique family events: weddings; bar mitzvahs, christenings or anniversary celebrations. However, lacking equipment and expertise they need experienced producers to take the whole process off their hands, and to give them the results they want at prices they feel they can afford.

In between these two extremes fall a host of other opportunities ripe for exploitation. Travel agents are beginning to use video to show off hotels or resorts to customers, more vividly than is possible in a brochure. Colleges are using video as a teaching aid wherever repeated demonstrations would be difficult to set up, or expensive. Tapes cover subjects ranging from complex surgical operations for medical students to *haute cuisine* for trainee chefs, and from particle physics to drama. In education and training there are as many possibilities as courses.

At this stage, however, a word of warning. Wide as the market is, it demands a new attitude to video production. From the moment someone commissions you to make a video recording, you are a professional, and you need to bring to your work the attitudes and priorities of a professional.

You cannot afford, for example, to practise at your customer's expense, for video is an expensive item for both individuals and professional organizations to fund. They want competent results, not inspired ideas that failed to work. Stick to methods you know and techniques you have practised.

Eliminate problems as far as you can by pre-planning. Always check your equipment before and after each assignment to make sure everything is working properly. Make check lists. If you can run down a list of numbered points before setting out on location you can be sure you have everything you need. Err on the side of caution. It is easy to forget a vital item when relying on memory; we once drove 200 miles (around 300 kilometres) only to find a tripod was missing because it had been put away in a slightly different place the night before. When you are professional, mistakes like this become expensive.

Begin from the beginning to keep account of the smallest expenditure. Cost each job properly, so that you can be sure you are making a reasonable profit for your skill and effort. It is easy to ignore items such as telephone calls or car mileage which, once you become involved in a busy working schedule, mount up only too quickly. An accountant will advise you about setting operating costs against income tax.

When you start to make money, think about investing some of it in better equipment. A more expensive camera will produce better pictures and may also save you some effort: it may be more simple to operate than a cheaper one. More lights will give a more polished finish to indoor productions. A U-matic machine may be an asset. Even if you know that most of your customers will want cassettes in VHS or Betamax, it will enable you to make higher-quality original copies.

There is no need to buy a lot of equipment to begin with. Rental companies will lend you almost any combination of equipment to suit any assignment, and this gives you an excellent opportunity to try out different formats, recorders, cameras and other equipment. You can also rent the use of an editing suite to liberate yourself from the tyranny of having to edit as you shoot. If the production calls

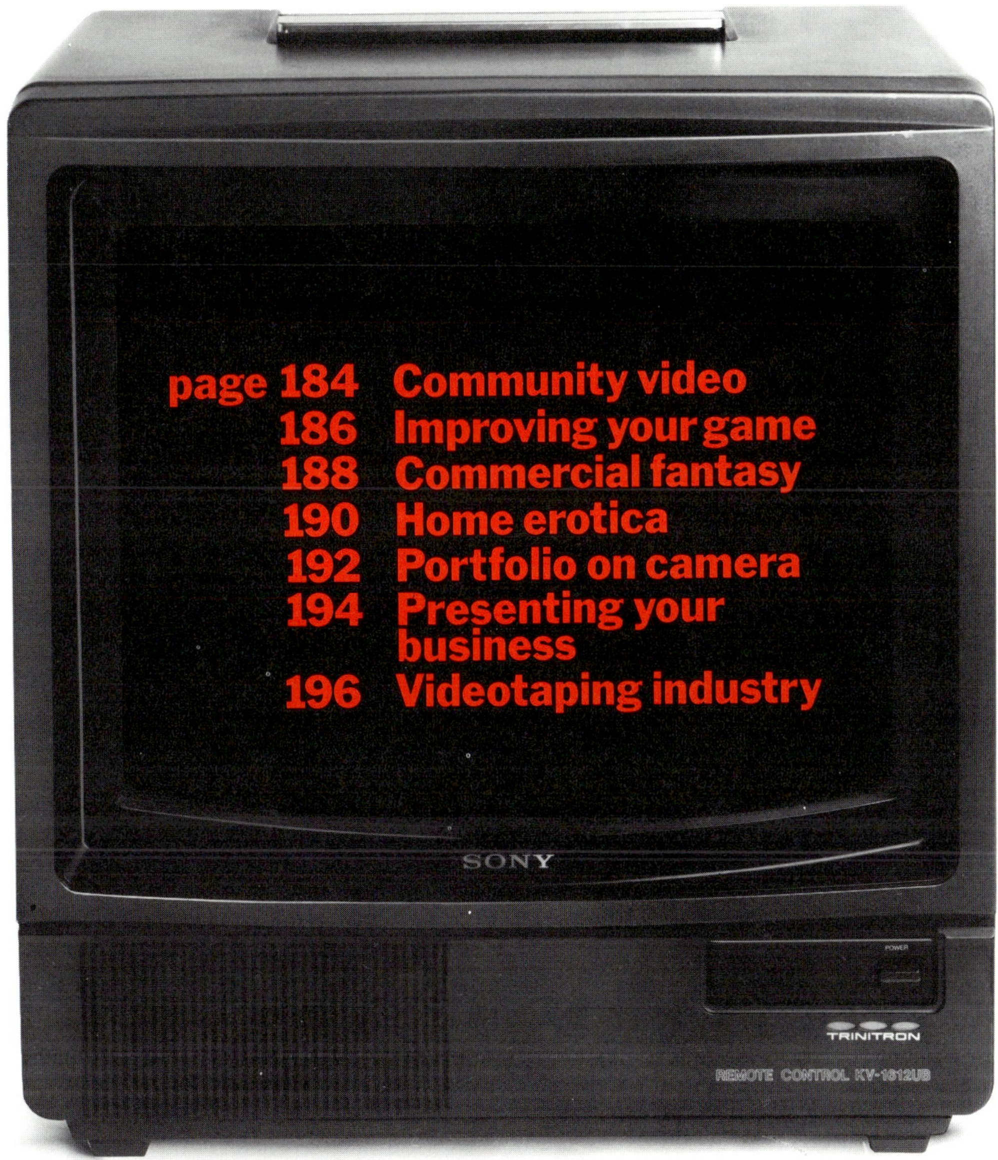

for it, you can hire elaborate facilities which will allow you to include fully professional special effects – perhaps in a demonstration tape of a rock music band.

Buy new equipment only when you find you need to use it regularly. Putting rental charges on to a bill is a useful reminder of the real cost of attractive extras, and the need to justify the extra expenditure is good practice for salesmanship and self-discipline.

However good your equipment, you will find you need helpers. At first you will need help in carrying equipment on location, and with setting up difficult shots. For this you can call on people on a part-time basis, extending their employment to a more permanent arrangement later, when the volume of work justifies it.

Later, when you have gathered a small team together, make sure it is well looked after. Remember to take enough food on location. Make sure that everyone wears, or takes along, warm clothing: location work is invariably colder than you expect.

In the following pages we explore some of the ways of making video your profession. The field is wide open; there has never been a better time to make video work for you.

Community video

Perhaps the biggest opportunity to make money from video is in recording weddings, birthday parties, christenings, bar-mitzvahs, and other once-in-a-lifetime occasions where a family is quite happy to pay the going rate for an imperishable record of an important event. It is a good place for the enthusiast to start being professional, since the sequence of events is usually preordained.

Do not, however, fall into the yawning trap of thinking that celebratory occasions are easy. There are all kinds of problems and pitfalls for the unwary, and even the experienced can be caught out, because no two locations are ever the same. If you can, try out your ideas on a suitable family occasion before going out to look for commissions.

First, try breaking the event down into a set of sequences. Talk over with your clients the kind of scenes they want to include – for instance, guests arriving, significant highlights of the ceremony and celebrations, activities to be performed by the principal participants, and so on. Then try to enlist their aid in making sure that you can deliver what they want. For example, if you are asked to record guests leaving one location and arriving at another, will you be given enough time to set up the camera at the required place?

Find out in advance whether you will be allowed into the building where the ceremony is to take place. Non-believers are not allowed inside, or are resticted to certain parts of, most places of worship. In others, religious officials may be reluctant to permit lights to be brought inside the building, or to have a camera operator near enough to record the ceremony.

If a camera is not allowed, it may be possible for an audio recorder to be placed where the ceremony is to take place so that at least there can be a clear sound coverage. This can be dubbed on to the video tape later. If you are permitted to record the ceremony, your aim will be to find unusual, interesting and well-composed shots without intruding.

Recording the celebratory meal afterwards poses different problems. Since the whole proceedings usually take far too long to record on one tape, try and capture the highlights, and the most entertaining and amusing impromptu happenings, as they occur. Never relax. Remember you are working and cannot afford to lose concentration.

Planning shots and angles

You need a solid establishing shot on which to fade up and begin the recording. For instance, you might start by framing the building in which the event is to take place and then zoom in for a close-up of an announcement, or banns, or an invitation.

At the celebration, plan to use shots and angles which will suit the atmosphere and mood of the event.

Make the recording of a ceremony more interesting by having close-up shots of individuals interspersed amongst compositions of the whole group. Take into account any existing sources of bright light, such as windows or candles.

If there is to be a change in location, make some link between the two places; for example, fading down on two newly-weds, or parents and child, leaving the church, and up again on the happy couple, or family relaxing at the reception.

During speeches you could fade down the sound and cut to shots of listening faces before coming back to record the final words or toast.

End the film on an appropriate subject which sums up the happy event. Zoom in to close-up, and fade to black.

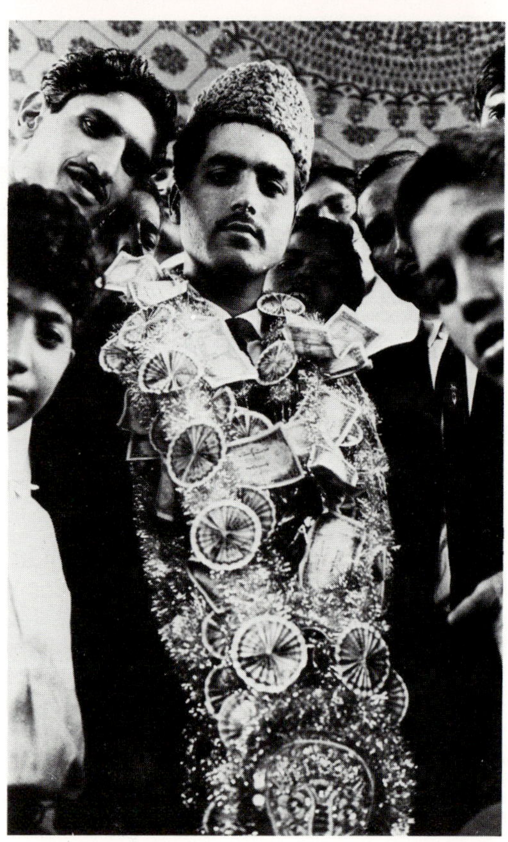

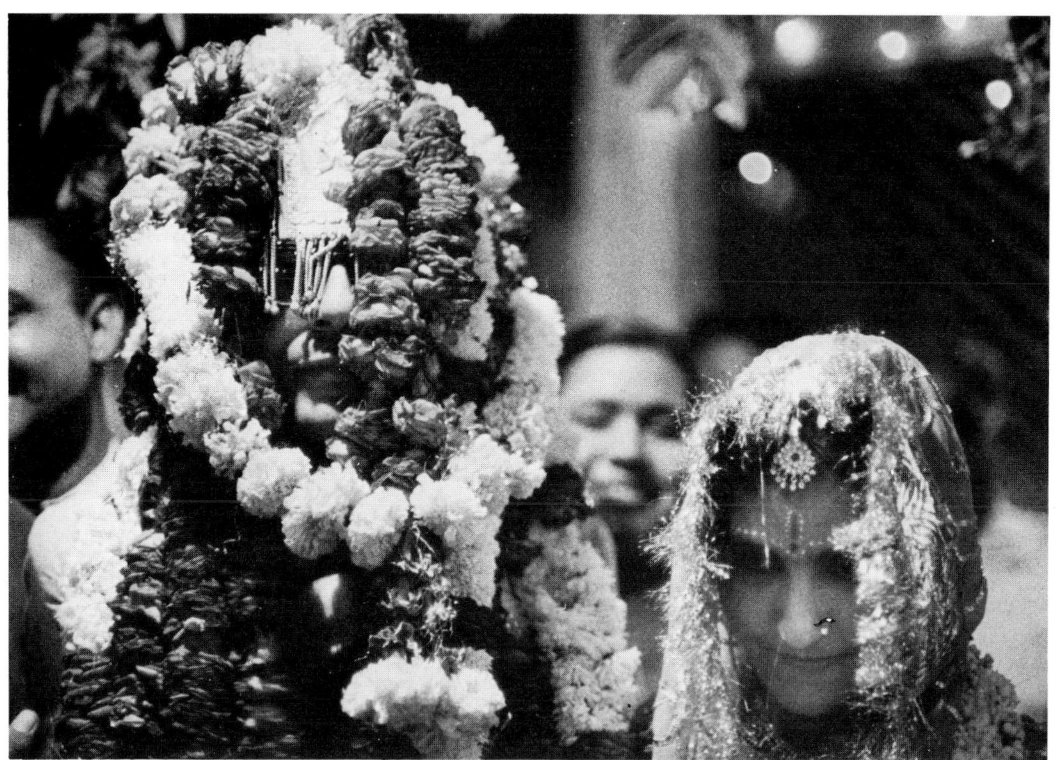

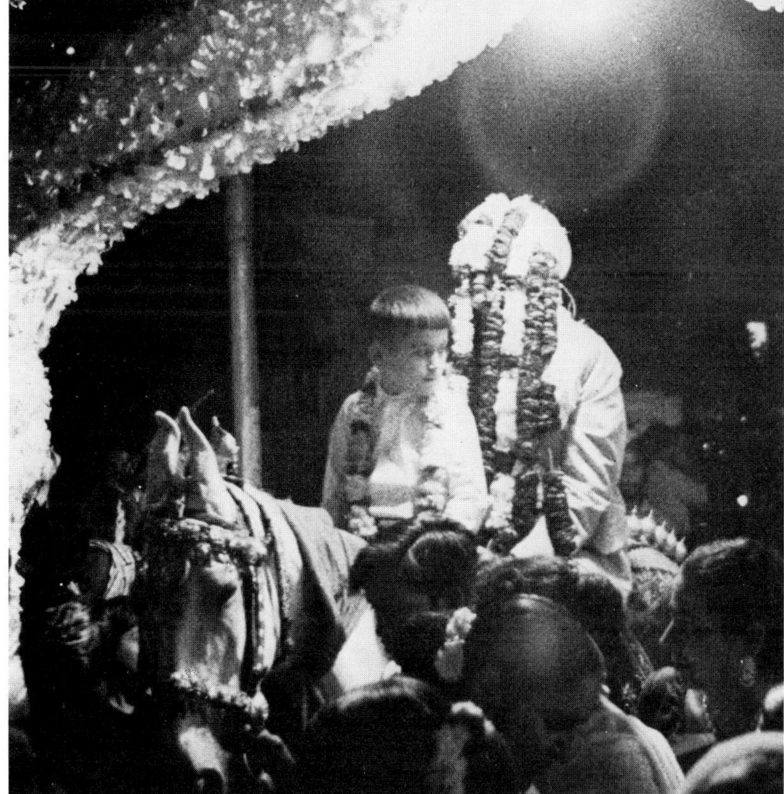

Pick a suitable angle, either directly in front, *above*, or sideways on, for recording the highlights of a religious ceremony, and slowly zoom in, where appropriate, for close-ups.

Make the most of any planned dramatic moments, *left*, by focusing on the expectant faces of the guests before panning out to the action.

The protagonists: child, parents, bride, bridesmaids, groom, *far left*, should star in the recording, but lack of time may force you to choose between two different events: the groom, or the bride, arriving; the couple leaving the place of worship, or arriving at the reception.

Improving your game

Most professional productions involve the making of edited, planned tapes. There are times, however, when you may want to make use of raw video, unedited and unscripted, which can be played back as soon as it has been recorded. If people can see themselves on tape as others see them, they can get a more objective viewpoint of what they are trying to do, and a clearer idea of how they could improve. As well as being useful for actors, public speakers, teachers, interviewers and interviewees, this approach can be of particular value to both professional and amateur sports enthusiasts.

All the equipment you need is a video camera and VCR, and a monitor on which to play back the results. The extra refinements of slow motion, frame-by-frame advance and a still frame hold on the VCR will obviously be advantageous to anyone trying to improve their sporting abilities, since they make possible an analysis of the various bodily movements involved.

Perhaps your local tennis club might be interested in renting your services for the use of its members. Having ascertained the particular action a tennis player wants to study, set up the camera at an appropriate angle so as to capture the whole movement clearly. Once you have recorded a few demonstration shots, wind the tape back and play it over again so that your subject can see where the fault lies.

If the club coach is there, you can shoot before-and-after shots of the subject's service or backhand, for example, at each stage of a fault-correcting talk, so that there is some visual evidence of your client's improvement. As a refinement, you could edit in an excerpt from a recording of a professional's service or backhand alongside that of your subject, for further comparison.

Having recorded a player's service, forehand and backhand, you could then suggest following through part of a match, so that the club coach can highlight more subtle points of tactics or positioning on the court. It may be worth taking a look at one of the pre-recorded tennis cassettes for further ideas.

The list of sports activities that can benefit from visual analysis is almost endless. Golfers, karate or judo students, divers, gymnasts and trampolinists can all be given a view of their performance they would otherwise never be able to see.

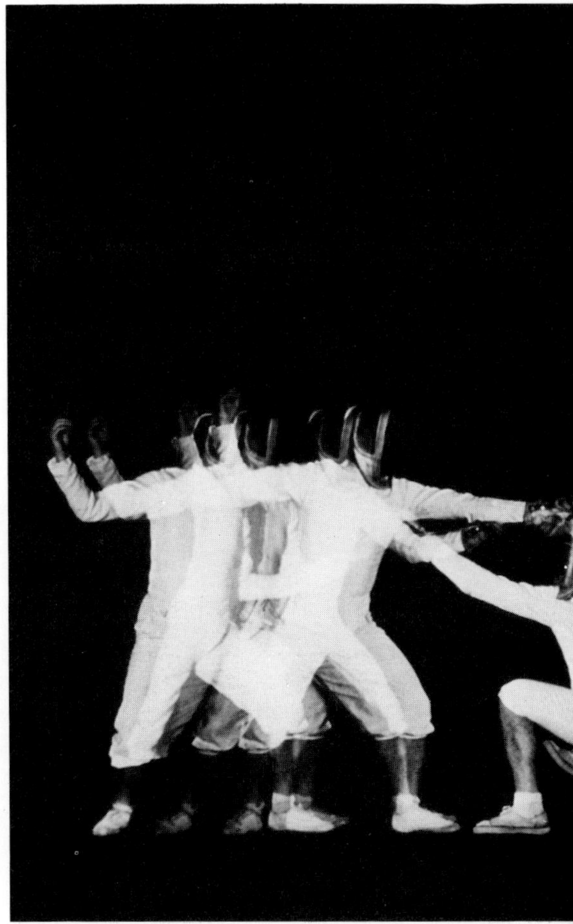

Viewing themselves on the playback tape, fencing partners get a chance to examine their interacting movements as well as each other's foot and blade work.

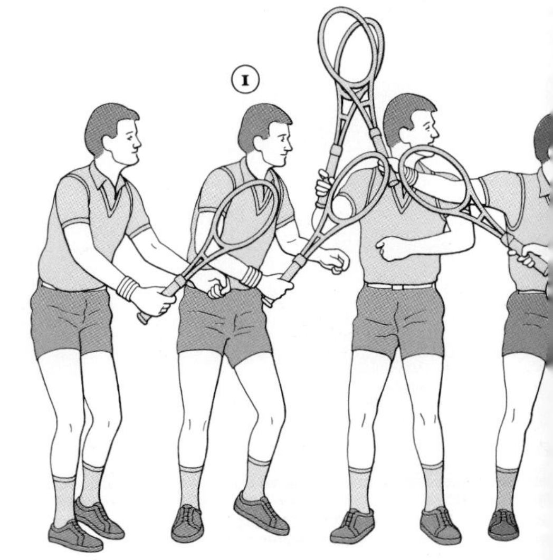

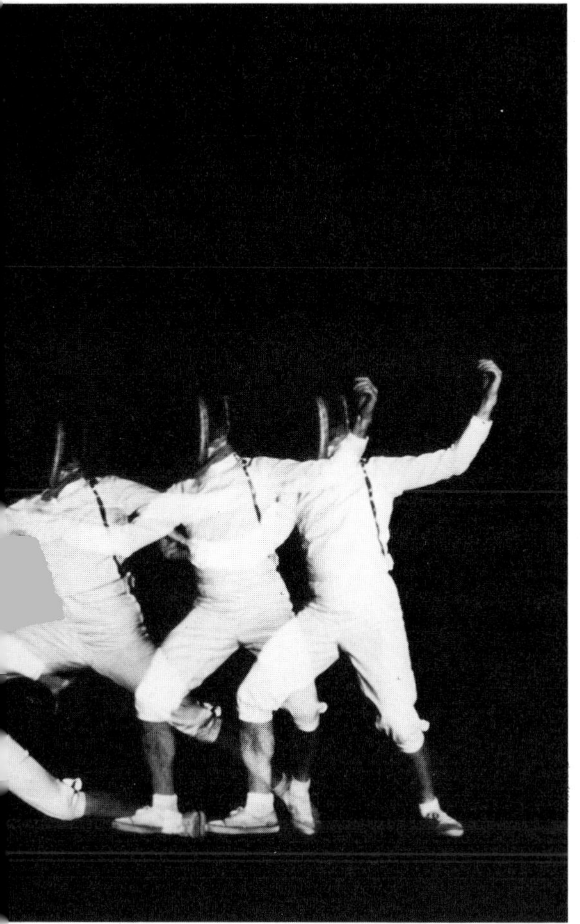

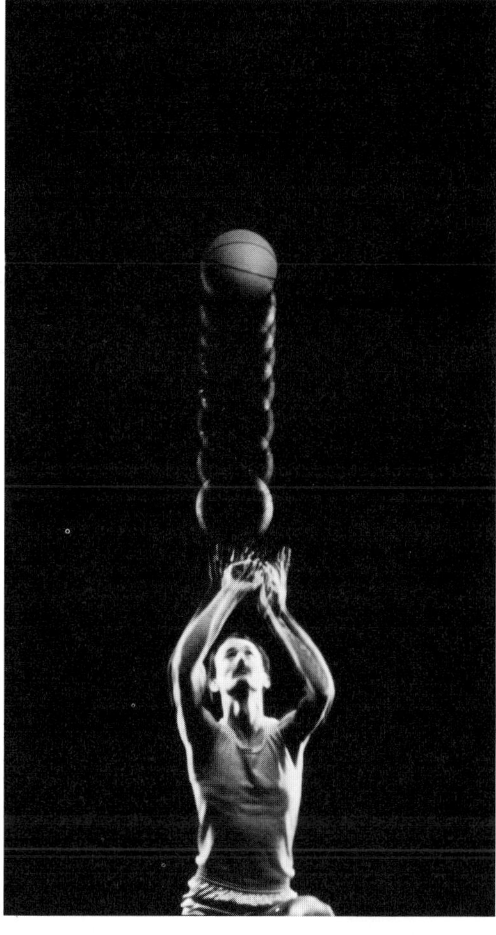

Correct positioning of the arms and hands is vitally important in basketball. Studying movements frame by frame can give a player greater insight into his or her own game as well as into team strategies.

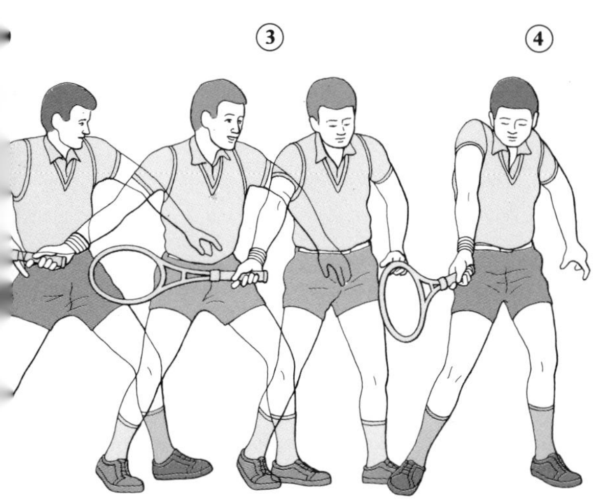

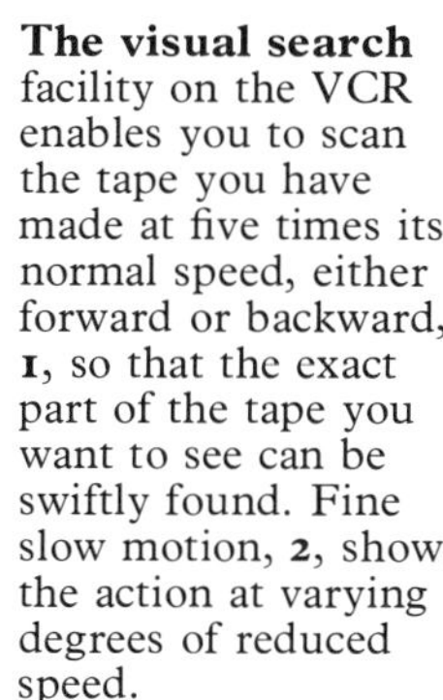

The visual search facility on the VCR enables you to scan the tape you have made at five times its normal speed, either forward or backward, **1**, so that the exact part of the tape you want to see can be swiftly found. Fine slow motion, **2**, shows the action at varying degrees of reduced speed.

In addition, the tape can be made to advance frame by frame, **3**, so that every angle of the action can be analyzed in greater detail. For even closer examination, one single frame, **4**, can be held completely still to be studied.

Using a tripod will help ensure clearly defined pictures, as it keeps the camera steady. Avoid having to zoom out to capture the movement, and do not film against a bright sky or sun: a silhouetted figure will lose some of the important details needed for accurate analysis.

Commercial fantasy

Television advertising is one of the most persuasive weapons a salesperson can have. Although the world of the broadcast television commercial, with its six-figure budgets and trail-blazing ideas, may seem light years away from anything you could do, the difference is one of degree rather than principle.

Television commercials command such huge audiences that the most expensive productions become worth while on a cost-per-client basis. However, many advertisers want to reach a specialized, or local, audience. The owner of a small vacation hotel, for example, may want a tape to show travel agents the attractions of the locale and the accommodation, but will not want to pay television advertising costs. Aim at this market.

Outlets for video advertising are opening up fast. Already, department stores are investing in video playback equipment to advertise new sales lines and promotional events. Manufacturers are beginning to use video cassettes as a form of direct-mail advertising to outlets such as stores. Soon, video cinemas may turn to running advertisements between features, and these are likely to attract local industry and services.

A local hairdresser, or an automobile dealer, would need the simplest of advertisements: perhaps an assembly of slides. These might show the premises; three or four hair styles, or the latest and most expensive automobiles in the showroom; and title cards giving all the relevant information.

This may seem unimaginative, but if it is what your client wants, remember that slides are a useful discipline for thinking about the essentials of a subject. Make sure the slides meet the highest aesthetic and technical standards, and use your imagination in presenting them. Record the advertisement using two projectors and a dissolve unit, so that you can cut and dissolve from one slide to the next. A series of slides of a slowly moving object can be made to appear to move; or, using a battery of projectors, you can work on a complex sequence to make a display of constantly changing segments within a main picture.

If you are asked to shoot a moving sequence, remember that advertisements are always short. Think in terms of each shot taking only a second or two. Look even more carefully for the unusual or unexpected shot to add into the essential information.

Budgeting check list

1. Before you bid for or agree to an assignment, work out your costs as carefully as you can, and include some provision for unexpected expenses.
2. Work out your charges in advance, aiming at a fair return for your time and expertise, and including an element of insurance against depreciation of equipment.
3. If at the beginning it seems necessary to cut profits in order to charge an attractive price, do so, but not too often.
4. Plan out your assignment, and make sure your client agrees to the content, the treatment and the script.
5. If, as you progress, your client wants to make changes, cost them out and quote for the extra work; then let the client decide whether the changes are worth the cost.
6. Remember that the most expensive production may be no better than the cheapest, and that hiring sophisticated equipment adds to the bill, while profiting the hiring company. A cut-price production which pleases your client will give both of you a bigger return.

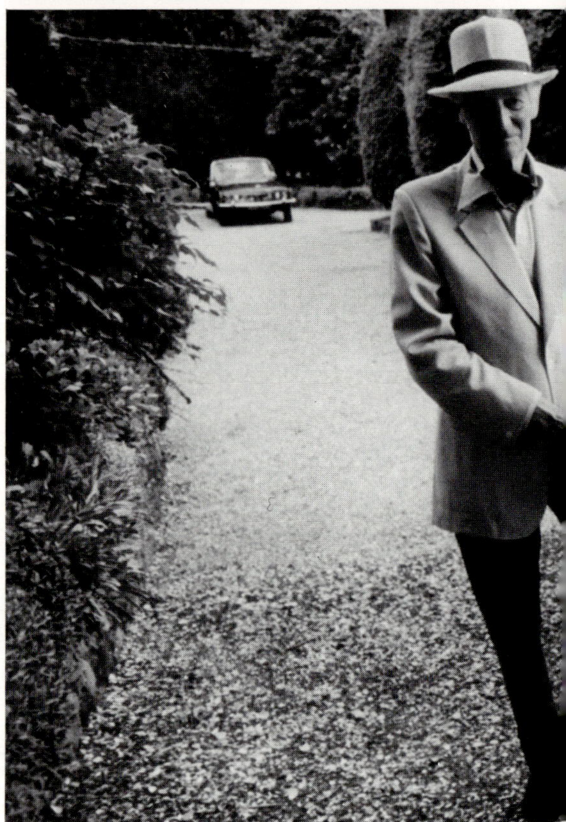

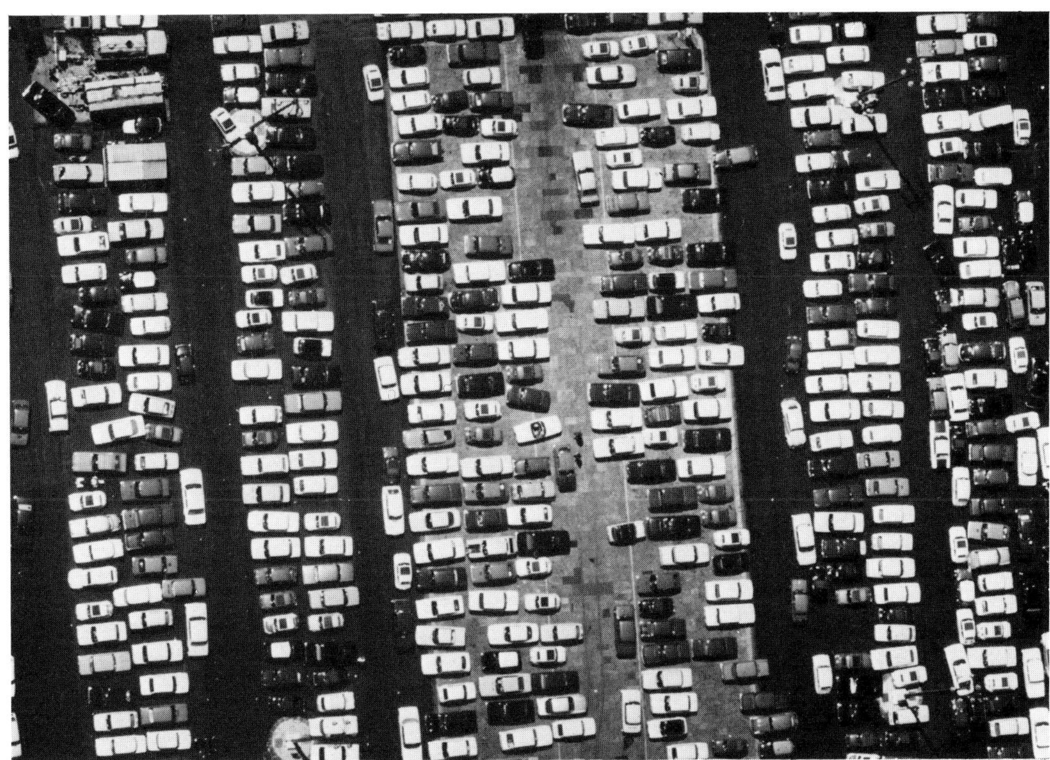

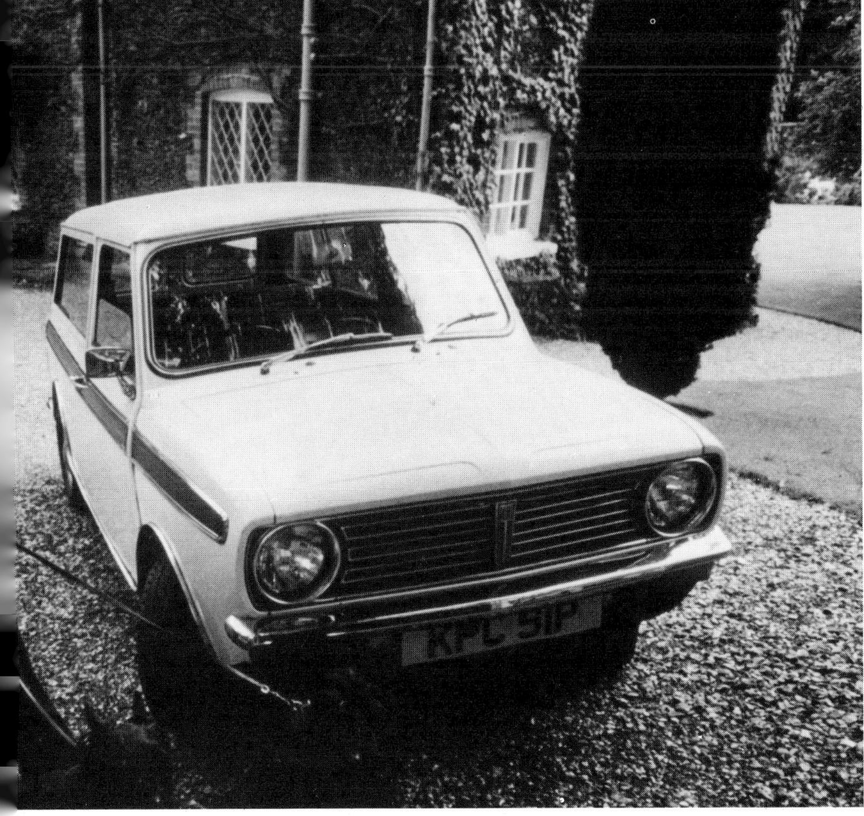

Try to think of ways of enlivening your production at minimum cost. Advertising agencies and manufacturers' publicity departments often release slides, film and video tapes to anyone promoting their products. The aerial view of an automobile dealer's stock yard, *above*, might be incorporated into an advertisement gratis.

Local celebrities, *left*, are often willing, for a small fee, to appear in an advertisement endorsing a product, especially if it is produced locally. The problem of inappropriate weather on the day when the recording is to be made is an expense you will have to provide for when drawing up your budget.

Home erotica

Video, for many people, means one type of production above all others: that of the so-called adult-only movie. Not only are video magazines packed with advertisements for professional pre-recorded erotica, but the making of erotic home movies by amateurs is now a popular hobby and art form. Some magazines offer prizes for the best and most imaginative recording, which may then be placed on a retailer's list.

Making erotic movies needs more trial-and-error experimentation than perhaps any other video production. It is essential that the person or people being filmed feel relaxed if they are to give a convincing performance, and you will need to give all the help you can from behind the camera, but without acting as an inhibiting influence.

Good results often need more light than seems romantic, however you can always experiment with shooting in low light to start with.

Getting your subjects to over-exaggerate their actions and play for laughs will also help them to throw off inhibition. For example, a girl could try a mimic of a striptease artist's routine. As she warms to the role, the chances are that natural sensuousness will start taking over.

From a technical point of view, you will be using many of the same camera angles as you would for any other kind of video portrait. You want to make the most of your subject; you want to make him or her look as good as possible. Use changes of clothes, locations, or situations to suit several different moods.

Many apparently explicit films, when analyzed, can be found to show nothing that could not be seen on the local beach. So much depends on the context, atmosphere, suggestion and the things the camera does not quite manage to show you. The mere sound of a zip being parted can be extremely evocative, given the right setting.

The erotic home movie is one in which amateurs can almost always turn in better results than the so-called professional movie-makers. Good results depend on originality, a fresh approach, and involvement, and you are more likely to get this when experimenting at your own pace, than turning out a commercial product on an assembly-line production schedule, with models to whom the experience is just an everyday occupation.

Many manufacturers offer a remote control unit as an optional extra which enables you to operate the VCR from a distance. It carries all functions, including tape stop/start, pause, rewind and fast forward, so you can use it to operate a camera positioned to face you. Sony manufacture cameras with detachable viewfinders, and offer a viewfinder extension cable and a remote control unit with a zoom control switch, *below*, so that you can operate the camera at a distance, and still see into the viewfinder.

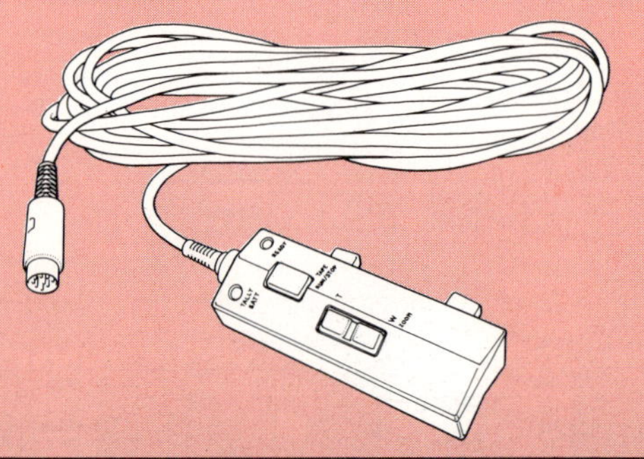

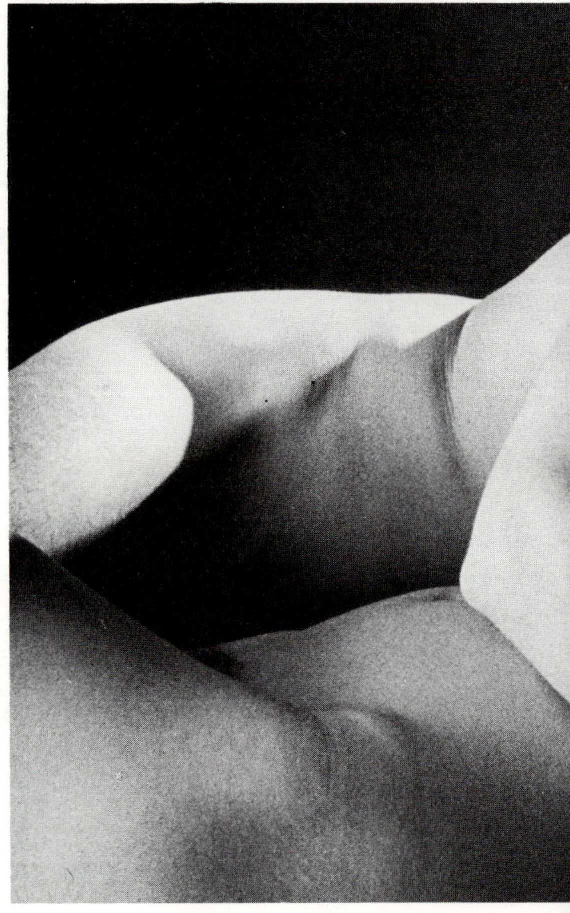

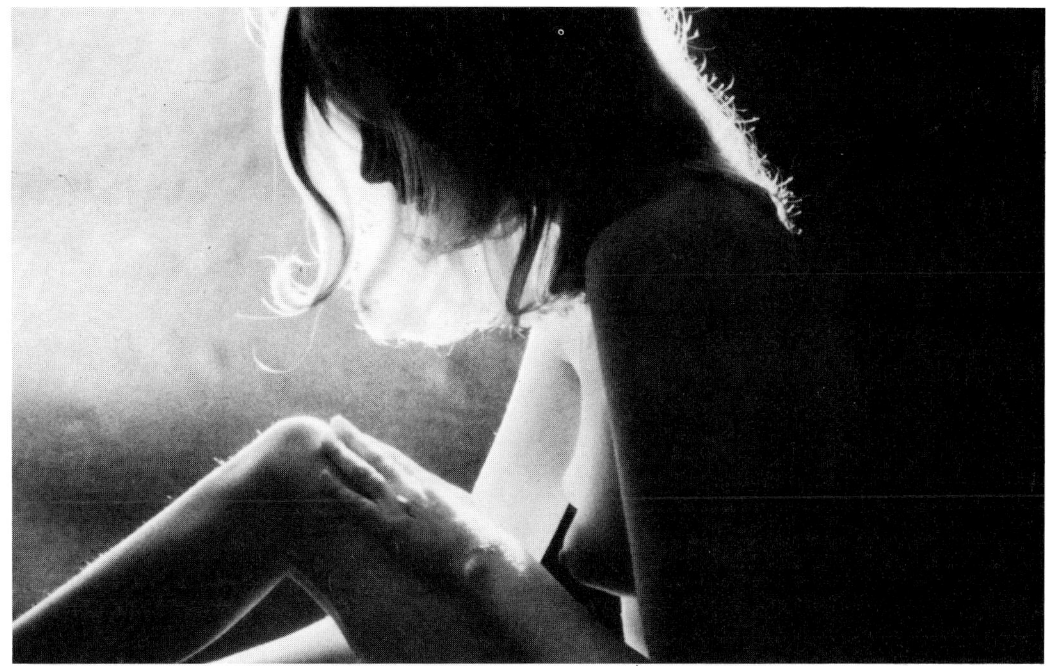

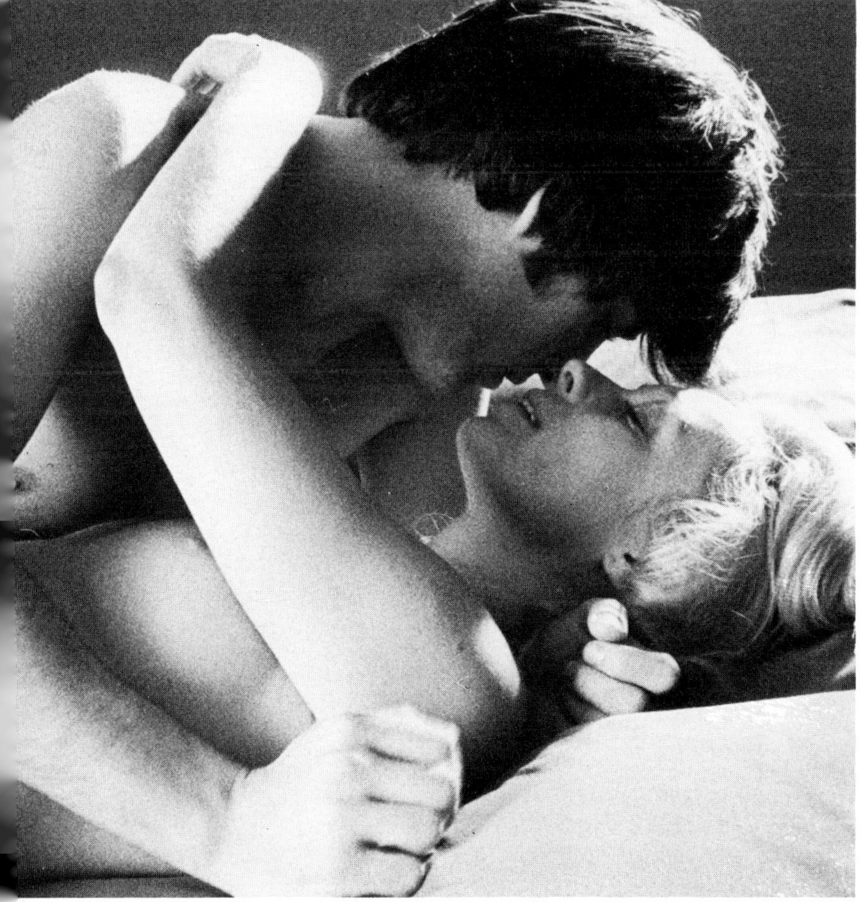

Lighting, background colour and camera position should combine to create the skin tones you are looking for, *above*. Increasing the red response slightly can make your pictures look warm and romantic. A bluer cast projects a more clinical, cold picture. To prevent unsightly close-ups of gooseflesh, heat the room first.

A story line is not essential. You can simply aim for a succession of shots, blending movements and gestures with a suitable sound-track in a way that you find attractive or arousing, *left*. Ask for any changes or improvements that you can think of, and be prepared to throw away, or record over, any unsuccessful first efforts.

Portfolio on camera

Anyone who has to put on a performance as part of their professional work – singers, models, dancers, actors – could benefit from using video to sell themselves. A real performance might be immeasurably more persuasive than an audition held under conditions which inhibit concentration.

Professionals will be used to turning in a polished public performance, but they may not be used to working for the camera, so do not try to persuade them to change a well-rehearsed routine without good reason. Let them run through their performance while you work out how to translate it into video as practically as possible. How will you light it? Where should you shoot it: on location where the immediacy of a real performance may make it more convincing; or in a studio where there are none of the restrictions of fitting in with someone else's show? What background does it need? Where do you need close-ups and where would a long shot be better? How can you maintain variety and interest.

There will be room, even in a five-minute video tape, to demonstrate your subject's versatility. A model will need to be recorded in as many different styles as possible; an actor in a variety of roles. You will also want to vary the pace by lingering on some aspect of the performance of which the subject is particularly proud, to give time for the quality of the action to become apparent.

Although professionals are used to interrupting their performance in rehearsals, try and keep recording sessions relaxed. If you have the short shot you wanted, but the subject is deeply involved, let the performance carry on. However, if a take is wrong, say so and ask for a repeat.

You are unlikely to need a script: most performers will want to speak for themselves, but a model demonstrating lounge suits or teenage gear may be enhanced by an appropriate musical background. There may be a case for including relevant personal information as a voice-over and the subject will probably want to be the one to speak.

Always remember that this is one of the most difficult of sales markets to cater for. The professionals who use this kind of portfolio usually work in a buyer's market so the recording should be as polished as possible. On the other hand, avoid overselling your subject; try to let talent speak for itself through the sequences you record.

Video portability

Before video came along, models were restricted to assembling portfolios of still photographs to give advertisers a taste of how they could look in a magazine or on a poster. Almost nothing of the quality of their performances – of how well they could work to demonstrate a product, or how convincing they sound, comes through a photograph. A video tape can be a walking, talking portfolio.

A cassette carefully edited to project a subject just as he or she wants to appear, means that the client is able to turn out a perfect set of performances to order. The cassette may be kept at home to be played back on the client's own machine, or copied and taken to interviews and auditions, even sent to agents or prospective employers, to be played back on their machines and analyzed at leisure. Turning out copies on each of the major formats will not be prohibitively expensive and tapes are not heavy to carry around.

Video is a conveniently portable medium. There are even miniature sales executives' brief-cases equipped with playback machines and tiny monitors so that the whole operation can be self-sufficient from start to finish.

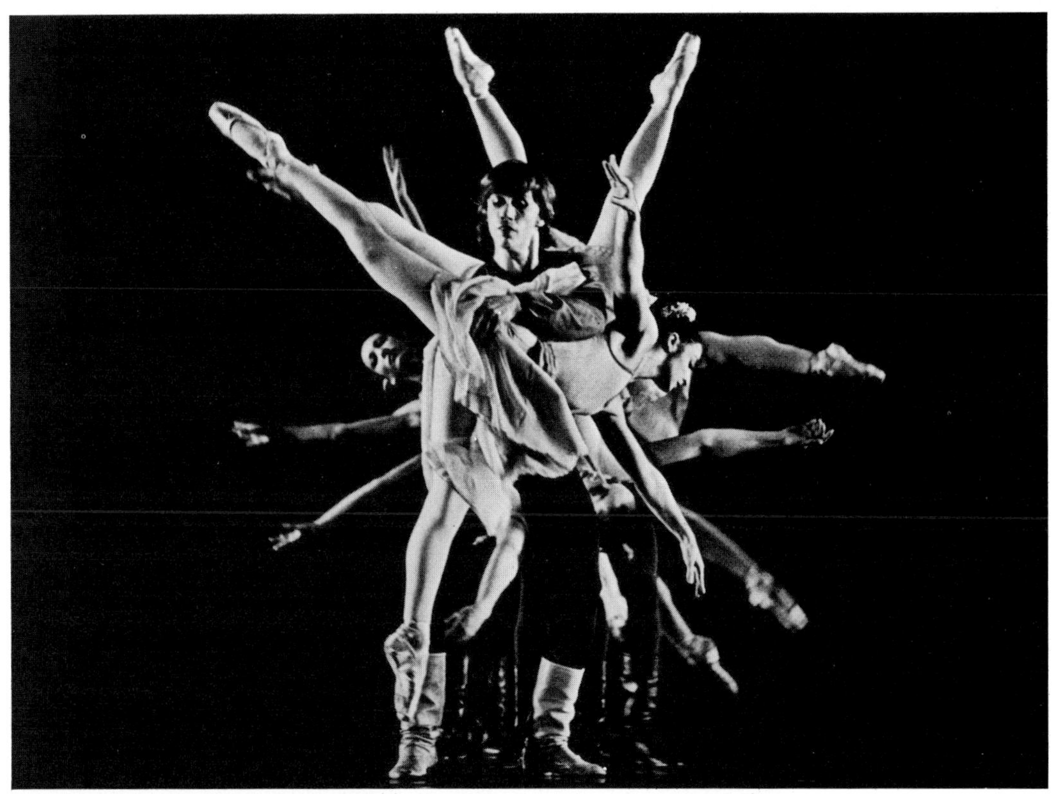

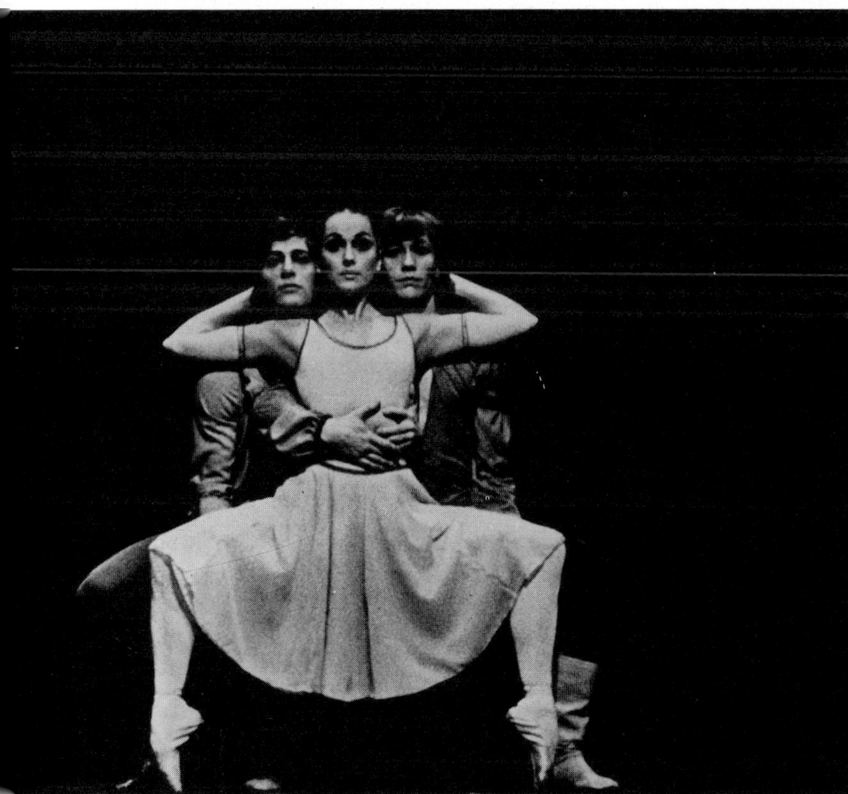

The Nederlands Dans Teater have long been using video as a training aid. Instead of learning roles only in rehearsal, dancers practise alone, using a video recording of a rehearsal.

Recording dancers on stage gives little scope to the camera operator. These shots from 'Miniatures', choreographed by Nils Christe, were taken head-on by a camera at the back of the theatre. They are what the dancers need. If it is to be used as a teaching aid a video recording must be a true, functional record, not an arty rethink of a performance.

Shots like these put over the skills of the subject, rather than of the camera operator.

Presenting your business

A video cassette can be a useful selling tool in many different ways. It can be used to give brief glimpses of holiday resorts for a travel agent to show clients; a real estate agent might consider using video cassettes to interest prospective buyers in certain properties; or an architect, a home builder or an interior designer might want to build up a library of video tapes displaying and analyzing their work.

This kind of production has unusual objectives. It might best be approached as a series of sequences of standard length and format so that each resort, house or room is given fair viewing. Your client will have an idea of the number of homes, say, he or she will want to include on each cassette, so effectively this will decide the length of each entry.

However you, as production expert, should assess the wisdom of your client's views. Too brief a coverage may not do justice to the subject's merits. Moreover, the cost of the recording will depend more on the number of properties your client wants to include than on the time devoted to each.

To cost the job realistically, you must work out how long it will take you to visit each location, set up the equipment and record the material. Include time required for planning shots and packing up, as well as fuel expenses to and from the location. Clearly, it is in your interests to work quickly and efficiently.

Work out before you start shooting all the shots you want and how long each should take. You will be less likely to miss worthwhile pictures, or to end up with too long a recording. Present all the essential details the client's customer will need to see in as objective a way as possible, while making the most of any positive features the property has to offer.

Keep as close as you can to the timings you have estimated for each shot. If you can choose when to shoot the material, make sure the weather is good: holiday resorts and housing look depressing on a drab afternoon. Remember that your client may be the best person to act as presenter or as a voice-over.

Present the edited recording to your client with a list of the subjects and timings of the sequences. If you then charge your client for each property entered, you will find it easy to quote for additional work.

Selling a home
Begin the recording with a bold and simple title card, followed by an establishing shot of the house taken from the road, with perhaps a pan or a zoom up to the entrance.

As the camera pans around the hall, the commentary could establish the numbers and types of downstairs rooms. Extra lighting should be used if needed.

Going from room to room, pick up any attractive features: for example, a long shot down the length of the lounge to emphasise its size; close-ups of bookshelves in an alcove; a shot of central heating radiators or a feature fireplace; the serving arrangement between dining room and kitchen; the shower fittings and bidet in the bathroom.

Try a shoulder-held shot to take you up to the first floor, and include any attractive views through windows, but be sure that the contrast does not make the interior look dark and gloomy. Having filmed the best features of the garden, sum up by explaining in a combination of pictures and words, the proximity of the house to shops, schools, means of transport and other local amenities.

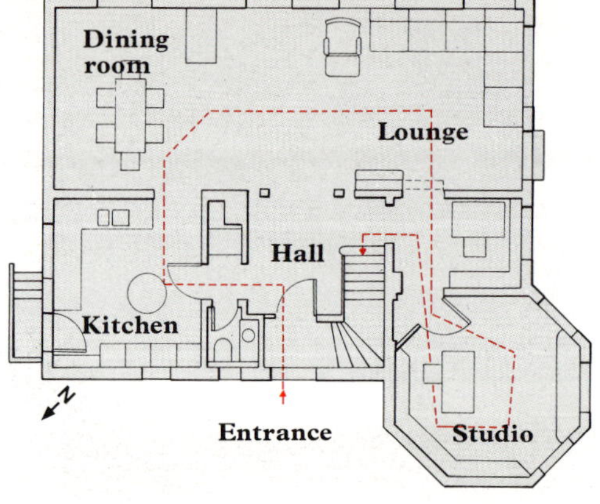

Try to show the viewer how a house fits together as a concept. Record each room in sequence, with the camera shoulder-mounted, so that the viewer sees the dining room through a serving hatch from the kitchen, and the lounge through a doorway.

Set the stairs in context with an establishing shot before recording on an upper floor. In sunny rooms, shoot at an angle to the windows.

A periscope lens can see objects as if through an insect's eye. A detailed architect's model, for example, looks on the screen like a real building, when shot through a periscope lens.

To make one, cut out a piece of smooth-sided corrugated cardboard exactly as in **1**, *below*. Run a finger-nail along the dotted lines in preparation for folding. Cut out a hole large enough to fit over the camera lens; then cut a thin metal sheet into a T shape and wrap the top of the T around the edge of the plano-concave lens (PCV) with masking tape, to form a collar. Mount the lens over the PCV hole with the curved side facing toward the inside of the periscope. The stem of the T can be attached with masking tape above the PCV hole, **2**. With glue, stick a mirror to the inside of each end flap. Close up the periscope body along the folds, and secure with masking tape. To mount the periscope on the camera, bend the aluminium strip into an L shape; then drill a $\frac{1}{4}$-in (6-mm) hole in one end to accommodate a camera retaining screw. Attach the other end to the periscope body with masking tape.

For full-frame magnification you may also need a close-up lens adaptor.

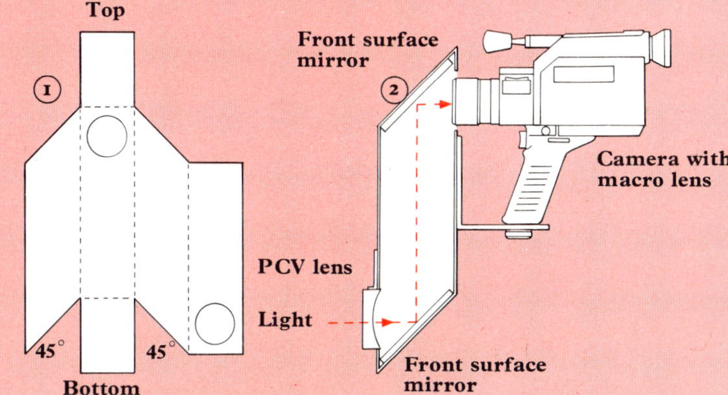

Equipment
Cardboard; aluminium strip, 1 in × 8 in (2.5 cm × 20 cm); thin sheet metal 4 in × 5 in (10 cm × 12.5 cm); masking tape; strong glue; front surface mirror $3\frac{3}{16}$ in × $3\frac{15}{16}$ in (8 × 10 cm) cut in two; PCV lens min 2 in (5 cm) dia.; 56 mm focal length camera retaining screw.

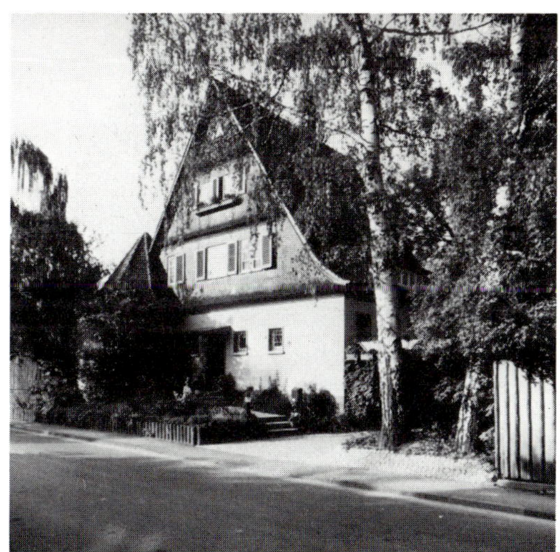

Open your recording by showing the house in its surroundings. Shots of façades are best taken with the light behind you. Try and work out the orientation and plan your visit to coincide with the best light; the sun will bring out details.

If the house has a view, zoom through a window to capture the sweep of the garden, or the landscape. Take various shots to show how the house fits into the neighbourhood.

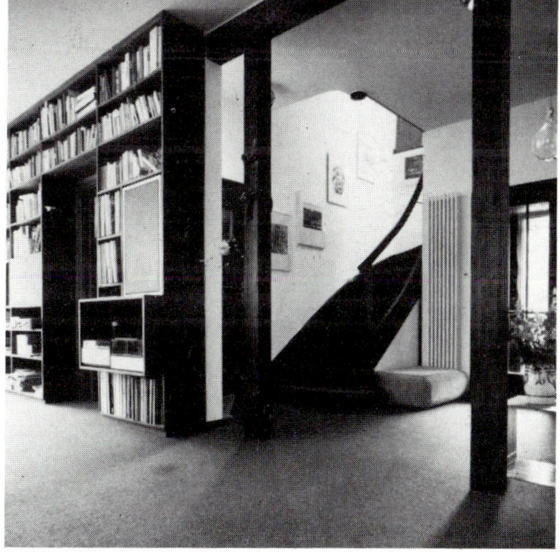

An interior designer would have you linger on a tablescape, or in an alcove, for subtleties of décor to make their full impact. To avoid extensive post-production editing, shot in 35 mm. It will give you more control over lighting and composition and, when you have worked out an attractive sequence, you can transfer the pictures to video tape.

Alternate stills with shots of plans and preliminary sketches.

Video taping industry

Video can work for commerce in all kinds of ways. It can be used by companies to put across information about their organization, or its product, to potential customers, employees, suppliers and shareholders, and it can be used as a publicity device.

The chief problem in deciding the shape and content of the recording is to convince the customer, who probably has preconceived ideas as to content and presentation, to listen to your suggestions. Planning out the sequence of shots on location, and drawing up a convincing storyboard, may help you to overcome this obstacle.

Whether you are announcing a new product line, or presenting a major extension to the production shops, your main priority at the beginning of the production is to capture the attention of your viewers. The opening shot should be something attractive, or something intriguing. For example, if your subject is the launch of a new automobile, you could begin with a really tight close-up of part of the body emerging from the paint-dipping bath, with the light sparkling on the wet paint. As the camera pulls back in a slow zoom, the body shape is gradually revealed.

Picture possibilities can be found in the most unlikely surroundings. One production made in a television factory produced an almost balletic sequence of sets swaying and gliding past one another on overhead conveyors, which, shot from an unusual angle, had all the grace and formality of a dance.

Another sequence in the same production showed a printed circuit board being conveyed through a bath of molten solder to have all the connections made. Looking along the conveyor, one brilliantly lit area was held in sharp focus, so that the resulting shot showed a dark, unidentifiable shape coming into brilliant illumination on the instant it came into focus. Together, the colours of the components and the bright sheen of the molten solder created a picture with a rare abstract beauty.

Lighting can be a major problem in an industrial location; so survey the premises in advance of the shooting day, to check what you will require in the way of lights and stands. You will have to carry all your equipment, probably along miles of corridors and workshop walkways. Bring a folding cart, or borrow a cart small enough to negotiate through doorways and steps.

Making a training tape
Clarity should be your major preoccupation when making a recording that is to be used for training personnel. Your object is to put over to the viewer clearly and succinctly exactly what he or she needs to know in order to be able to carry out a specific operation.

The information being taught may be complicated and highly technical, and the success of your recording will depend on your assessing such factors as the pace at which you impart the information; how much you can assume the viewer to know already, or whether and where you should pause for a demonstrator to play a sequence over before proceeding to the next stage of the operation.

Ways of tackling problems such as these should be worked out in close collaboration with your client, the client's training officer, or some other person with the specialized knowledge you need. Plan the operation out together, shot by shot, looking out for and solving problems as they arise. This way, the finished product should evolve as an integrated whole. If you can also use your creative abilities to help make the training tape interesting and attractive to watch, you will contribute still more to the success of the project.

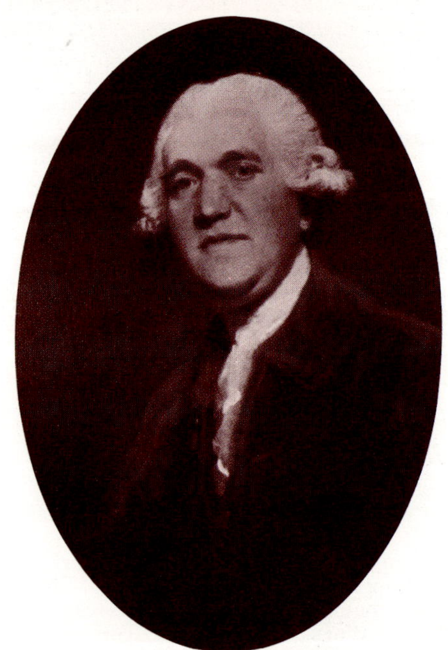

Discourage the company directors' likely insistence on the well-worn opening shot of the factory, or office, façade. Corporate pride could be satisfactorily stated by other visual means – perhaps a portrait of a celebrated founder, *above*.

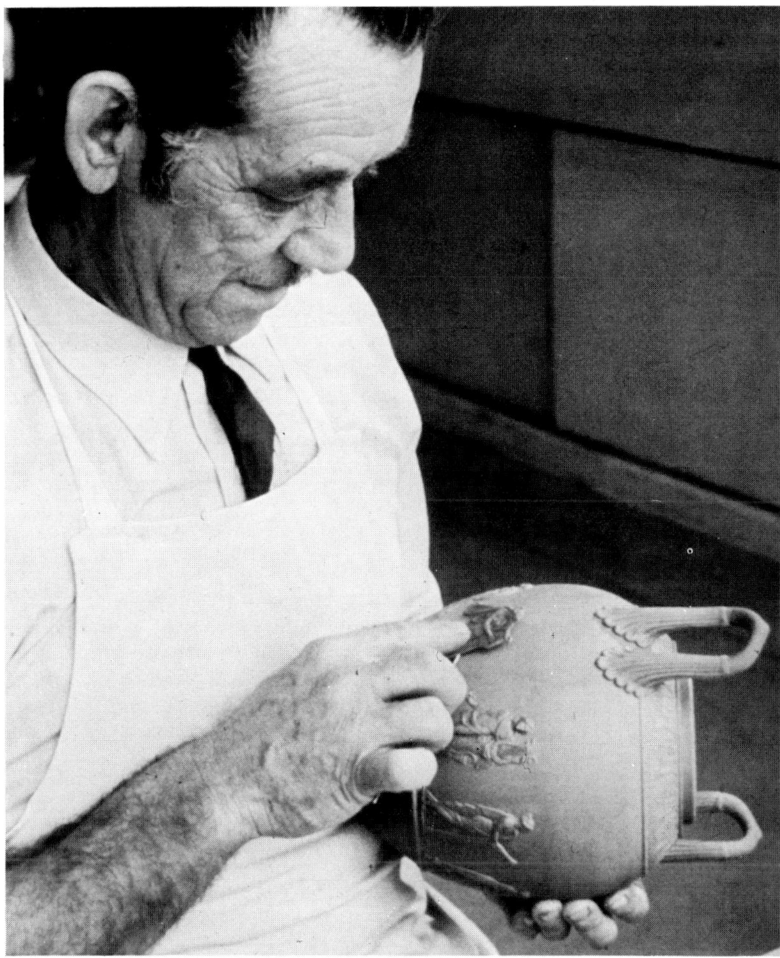

People always make good subjects, especially when engaged in tasks involving skill and concentration. Shoot their faces as cutaways, then shoot their hands, *left*.

If your recording is to be used for training, the picture must be sharp. Work out in advance how close you can position the camera without interfering with the operation you are going to record, and which angles will give you the best view.

Factory and workshop ceilings are usually too high, and the walls may be too distant, to give much reflection, so work out beforehand how much extra lighting you will need to eliminate shadows and boost the light level, and set up well in advance.

Plan any commentary as carefully as the camera angles so that it matches the pictures exactly.

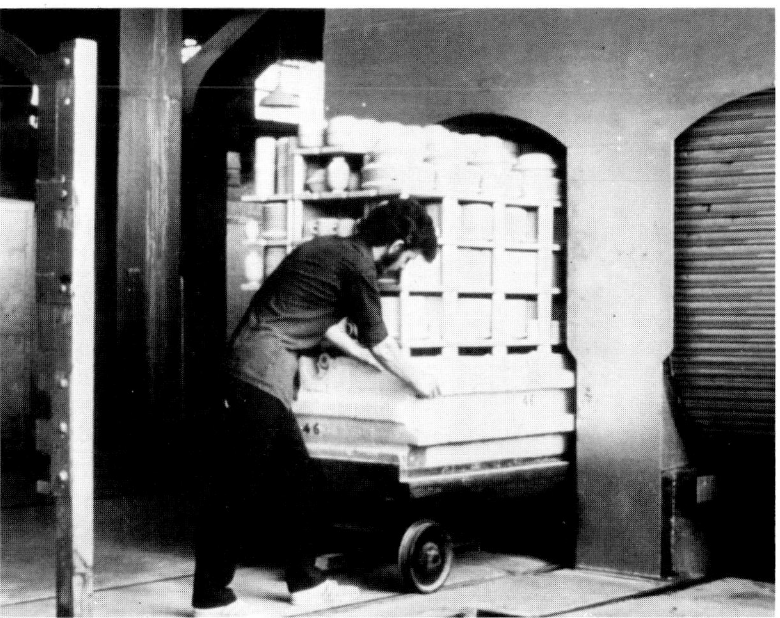

A pottery kiln being loaded could be a spectacular shot, but beware of the danger to the camera. Heat, paint spray, vapour from chemicals and thick dust can damage your camera.

If necessary, film from a distance or from behind a glass screen, searching for long, wide or angled shots to substitute for close-ups. A long shot of a row of objects or people is much more interesting than an unchanging view, especially if you focus from time to time on some specific part of the process.

Optional extras

So rich is the wealth of video, and so wide its potential scope, that some of its off-shoots have already become separated from the main trunk of the video business, and have become rooted as growing entities in their own right. These are the 'optional extras' of any video system, and they range from lighting and security systems, through the arcade of video games to the small computer, the child of the silicon chip, which promises to become as common in the home as the television screen on which its store of information can be displayed.

In the world of rock music, video tape linking the action with the sound has become a must for every group or solo performer with an eye on success. Video juke boxes made their début early in the eighties, and video has yet more to contribute to the disco age: those pulsating strobe and pulsed coloured lights, which set the atmosphere in a disco, can now be reproduced to the same quality in the home, not only with a sound-to-light converter but also on a screen.

Connected to a regular television screen, a video synthesizer (an item of video equipment hitherto found only in a professional studio) produces signals which create constantly changing, multi-coloured, abstract patterns in a random sequence having more than 2,000 million possible permutations. By means of a timing button, the pattern on the screen can be made to change as slowly as once every ten minutes, or as fast as twice a second. A synthesizer, looking like a simple audio amplifier, and about the same size, is most effective connected to a large screen.

Apart from the random generation of colour patterns, such devices have much wider possibilities. By connecting a video synthesizer to an external video source, such as a video camera, it can produce identifiable images, but in ever-changing hues. Moreover, by linking it to a hi-fi system or an electric guitar, the colours on the screen can be made to throb in time to the music.

A similar link-up of sound and vision promises to forge another art form. Using a video synthesizer with an audio input produces dazzling but formless patterns. The next logical step forward from this is video art: abstract but at the same time purposeful patterns created on a video screen by the deliberate action of a video artist. This new dimension is given to music by using a computer to program a video disk with a visual pattern, designed to match a specific musical accompaniment. Another possibility for the future could be to program a video synthesizer to make specific visual responses to audio input.

As a kind of hybrid between the fantasy oil projection systems of the sixties and the flashy disco lighting systems of the seventies, the video synthesizer has already succeeded in supplying a vehicle for transporting the public medium into the home. If its price (between £200 and £500; $100 and $500) seems beyond the scope of a personal budget, it may be an attractive item of hardware to hire for a disco party, since it can be linked to a screen to give a wall-to-wall, ceiling-to-carpet display.

Even more than video synthesizers, video games, once the sole province of the amusement arcade, have now begun to find a permanent place in more and more homes. The drug-substitute for kids and adults alike, arcade video games are so compulsive that it was inevitable that they should move indoors to the comfort of the homestead.

Video games are big business, and attract cut-throat competition, with new games (backed by millions of dollars of investment) being pirated by special copying machines virtually overnight. As video games become more and more highly evolved, so their appeal widens and the competition hots up. With tests of logic, numbers and spelling, they are proving their worth medically as aids in the rehabilitation of the disabled.

In an increasingly violent world, amply reflected in the intergalactic wars of video games, guarding the security of persons and possessions is becoming an ever-increasing challenge. Video has provided the answer to security problems everywhere, from the high-street bank to the private house. A better eye witness than any human observer, a video screen connected to a security system can record every last detail of criminal and deed and, simultaneously, note the exact time at which the crime was committed.

Fed information by cameras seen and unseen, the video screen of a security system can not only project a criminal's image but can also be used to check out the innocent by their faces or even their voices. Linked to an alarm, such a security device has already proved its worth in factory and home.

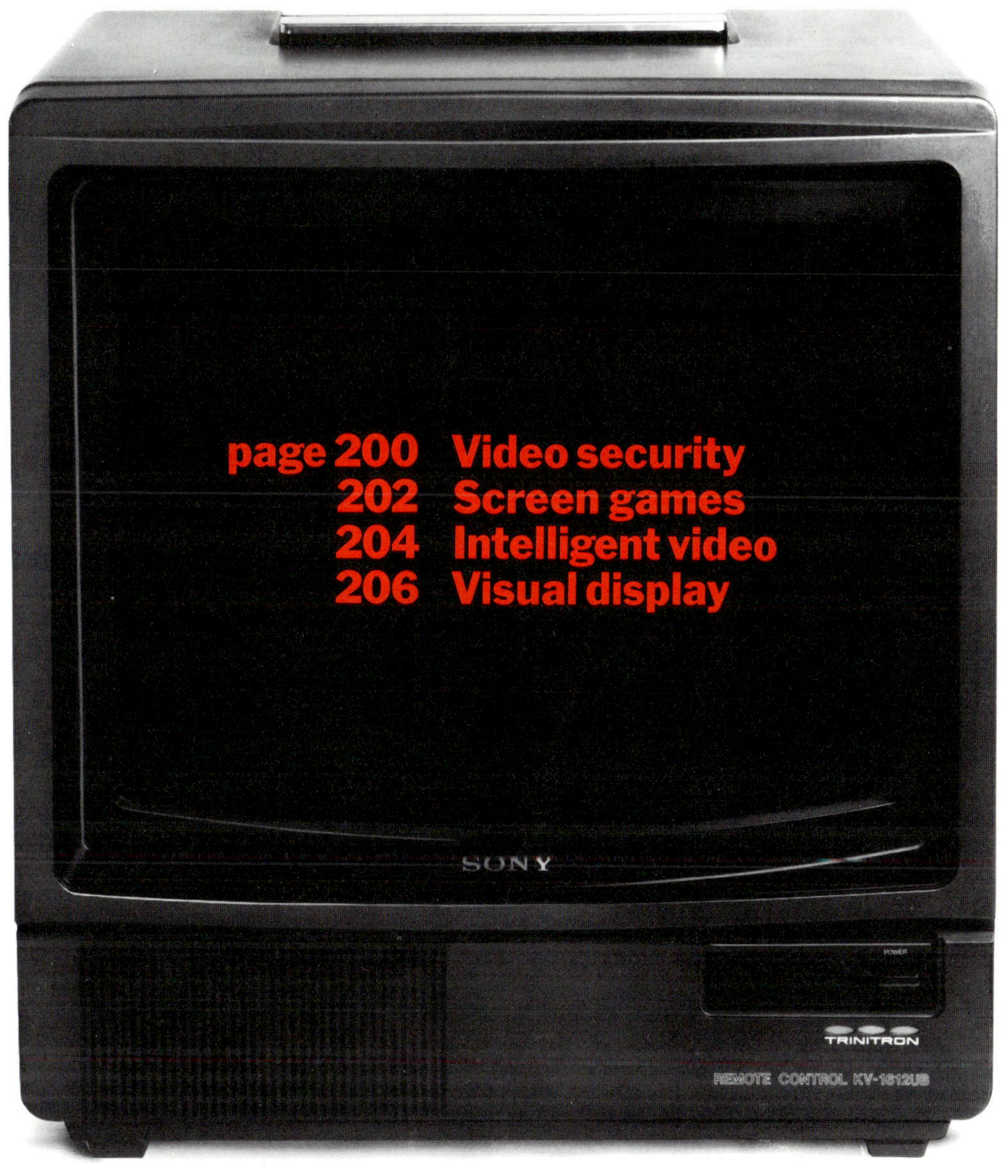

Of all the optional extras that can be attached to a video, the use of the video screen as the essential component of any computer system is undoubtedly the one that seems to have the most potential. When connected to even the smallest of home computers, the video screen, fed information from a floppy disk or data bank, can put before your eyes, at the press of a button, a lesson in music theory or a foreign language, the latest situation on the stock market, your bank balance or a list of reference books. Or the computer can be used as a word processor, displaying words on the screen for editing line by line, without the bother of retyping. It can then transfer the edited copy to paper by printing it out.

The possibilities for the computer-video partnership are almost endless. Its only limits are those of inventiveness and imagination. The system is so flexible that it can be tailored to your own personal needs: to supply a recipe idea for tonight's dinner, to help teach your child to do mental arithmetic or to spell, to turn on your television in time for your favourite show, or to list your video recordings. Whatever the future holds, it can only improve the scope of the partnership.

Video security

To set a machine to catch a thief: this is the challenge facing security science as criminals update their techniques to avoid detection. As a visual medium, video can do what other inanimate detectives cannot: a thief recorded on video is not only identified, but actually seen in the act. Security recordings must include a time and date on every frame. It is this which connects the suspect conclusively to the time and place of a crime. Video recordings are now accepted as evidence in courts of law in the USA and the UK.

Security video usually takes the form of CCTV (closed circuit television), where the cameras are connected to the monitors and recorders by cable. Connection by radio is less common. In large stores, overtly displayed cameras and even monitors are a common sight. Elsewhere, a row of monitors, fed by several cameras, will be under surveillance. Many of the visible cameras are dummies; while the live ones keep watch, the dummy cameras act as an effective deterrent. Often staff, as well as the public, are monitored. Cameras with lenses which can film through a hole the size of a pinhole are installed in offices and near checkpoints to combat pilfering; more conspicuous cameras are used at delivery points.

For private homes, one of the most useful devices is the video door telephone. A camera set into the wall by a door is activated by the bellpush; a small screen indoors displays the caller's face. Occupier and visitor can speak through an intercom.

Another form of home video security is a network of concealed pinhole cameras inside the house, which can be turned on from a distance by a radio transmitter (such as a Citizen's Band radio) and the whole house surveyed with a portable monitor. If the network is connected to a telephone, the user can phone a radio signal across the world, and receive pictures across the air, without anyone having to lift the telephone receiver in the house.

Complexes such as oil refineries may use cameras to aid their security staff. If an alarm signal sounds, the guard checks the monitors to make sure the alarm is not false. Video is rarely used to trigger alarms; its movement sensors are indiscriminate, and false alarms result. However, in places where movement is unusual, such as bank vaults, it can be used in this way.

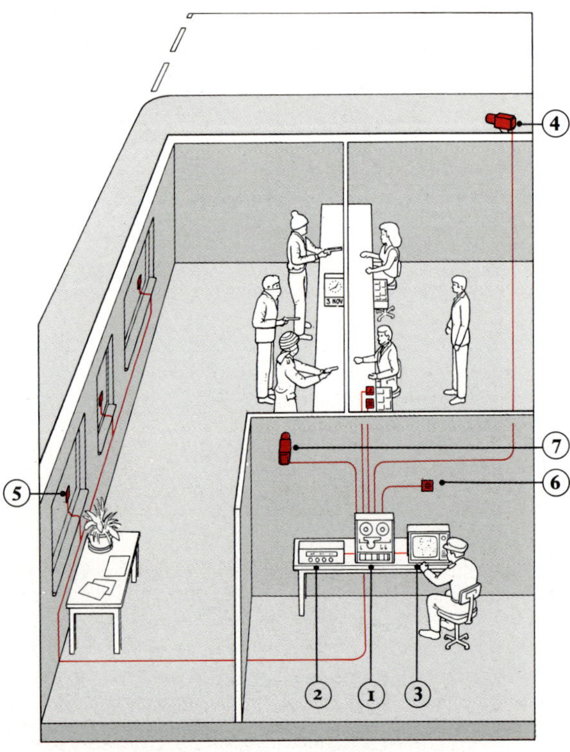

Caught in the act, a team of bank raiders in London are captured on video. A concealed camera was set in operation by a bank employee. Video cameras were not common in banks in the UK in the early eighties, when this raid took place, and the raider was not cautious about keeping his face concealed. The vital time and date appeared on a clock in the background.

This kind of video security must be switched on manually, as the camera cannot tell suspicious movements from everyday ones. A foot switch under the counter, well hidden from view, is sometimes used.

Real-time recording is helpful in this kind of incident, but a recorder left to run by itself would be a time-lapse machine, taking a few frames a second, or minute, allowing it to record for up to 100 hours.

The diagram, *above*, shows how a security system might be arranged. In the room are an open-reel VTR, **1**, a time-encoder, **2**, and a monitor, **3**.

Inside the bank, a small camera, **7**, is concealed. Security cameras are usually monochrome, making them cheaper than colour movie cameras. By the back entrance, a larger camera, **4**, in a weatherproof case, is a visible deterrent.

Alarm sensors, **5**, on doors and windows will set off an alarm, **6**, if there is a break-in while the bank is empty, and can be set to start the recorder.

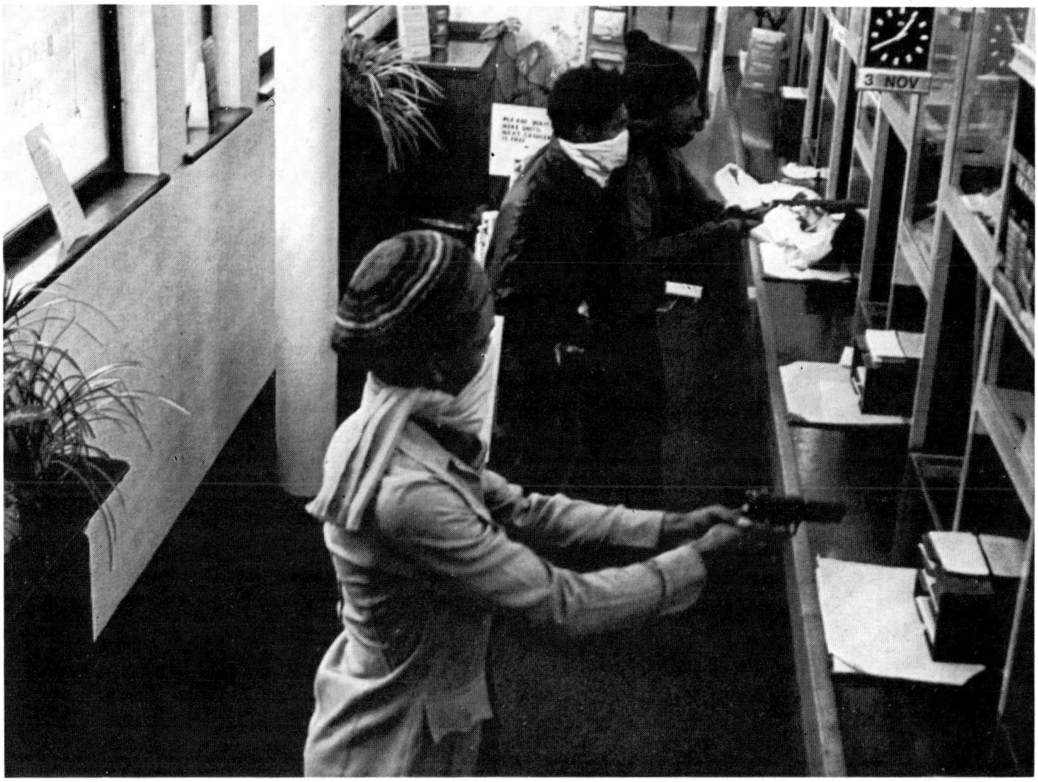

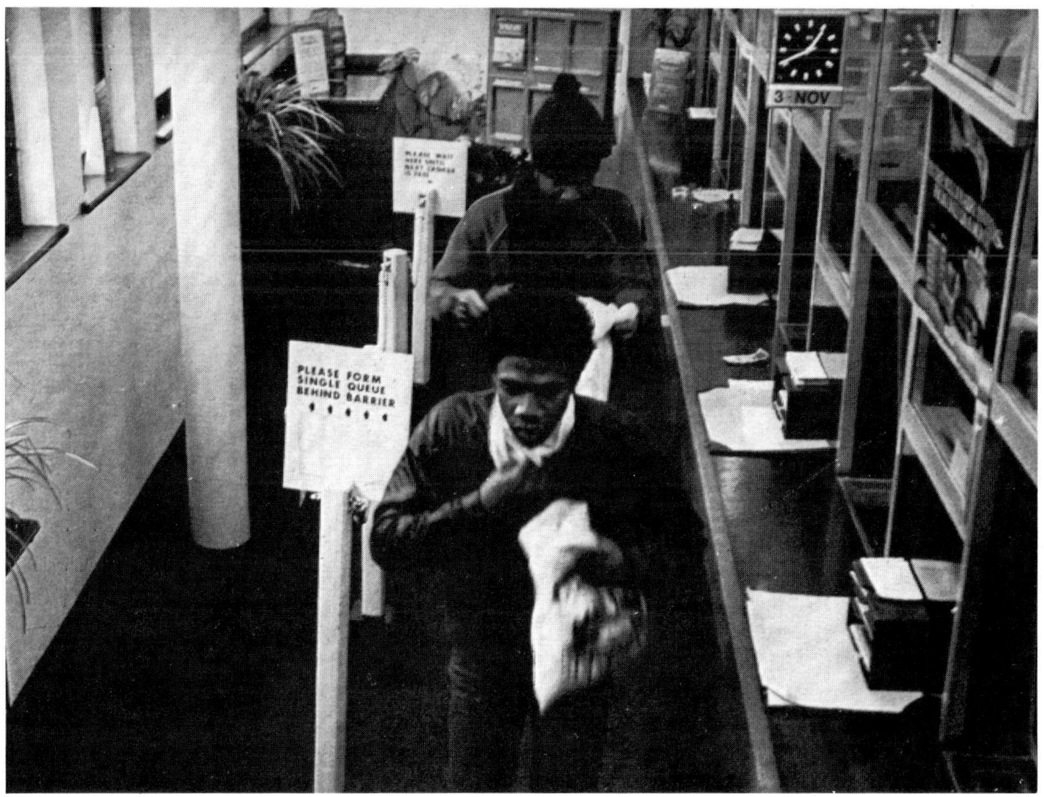
3 NOV
PLEASE FORM
SINGLE QUEUE
BEHIND BARRIER

Screen games

Video games are the new addiction of the eighties. Not only have they revolutionized entertainment, with machines blasting out intergalactic warfare everywhere from the arcade to supermarket, they have also spearheaded the march of computer technology into the home.

In 1972 Pong, a screen table tennis, became the first home video game. Designed for use in conjuction with a television, it consisted of a console containing a collection of integrated circuits to produce the picture, plus rotary hand controls to manipulate the bats on the screen. This, and many similar games that followed, all had the game selection built in, finite and permanent; a system described as 'dedicated' and still used, albeit with refinements, in arcade video games.

For home video games the next step was the semi-programmable game, introduced in 1975 and made possible by the use of individual silicon chips, custom-designed to generate several games. The largest number of games these systems offered was ten.

More far-reaching was the following move to separate the components, making the expensive bulk of the system constant and thus a once-only purchase, and leaving the individual game programs as a series of cartridges (usually called carts), to be slotted in. The constants are a master console, containing the microcomputer for interpreting the carts; the power supply; the television game switch box for attaching to the VHF antenna connection of the television; and the assorted hand controls. The only variables are the chip-containing carts, but the information they contain is permanent, like an audio disk, not re-recordable like a tape. Further refinements have permitted one cart to contain several games with many permutations in the level of difficulty and number of players; most sophisticated are the programmable consoles.

Video games of the future will allow opponents thousands of miles apart to compete, with the aid of a telephone link-up, and have displays that react to the spoken word or the point of a finger. Aside from recreation, or even education, video games promise to prove themselves increasingly valuable medical aids for relieving depression, and helping in the rehabilitation of stroke patients through the relearning of manipulative skills and hand-eye coordination.

Caring for video games
Video game carts are tough enough to withstand rough handling, even by small children, but following these rules will prolong their lives:–

1 Keep game carts clean and dry. Dirt, damp and grease can all impair action.
2 When not in use, keep game carts in their storage cases or in a covered box.
3 Do not let carts become baked by the sun or by the heat from a radiator as this may distort them.
4 When changing games, always switch off the master console to protect both the carts and the console.
5 Do not use force when inserting or removing game carts.
6 For the best game image, turn down the regular TV contrast and manually tune the picture on the display.
7 If games are to be played on more than one TV set in the house, fit each TV with a TV/game switch box to save moving this from set to set.
8 Make sure that players do not tug on the wires connecting the hand controls to the console.

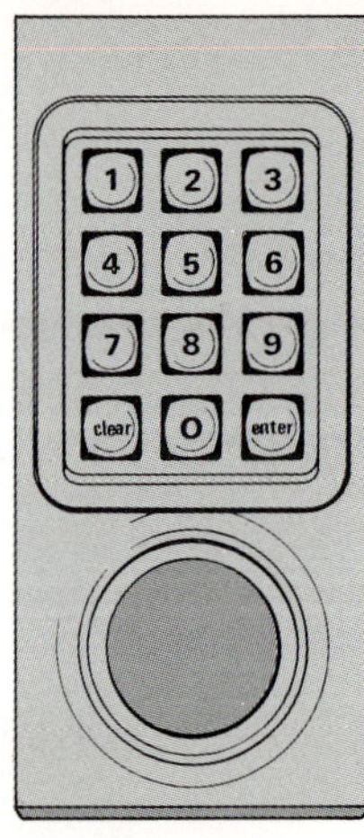

The Intellivision games system has hand controls, *above*, which reflect some of the newest trends in home video games. Each control consists of a key pad with 12 keys numbered 0 to 9, plus keys labelled 'Enter' and 'Clear'. Below is a direction disk; when touched it can sense 16 different directions.

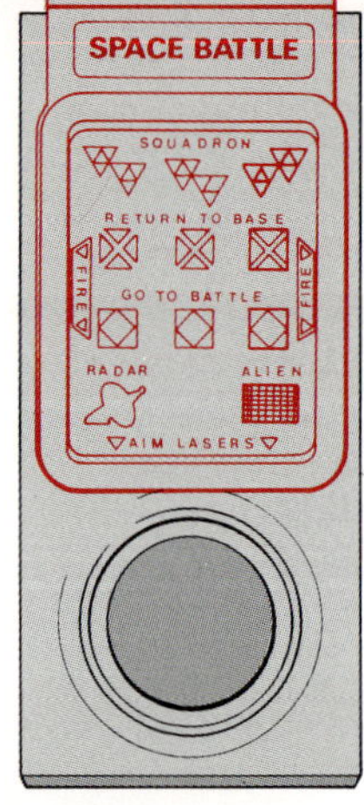

To play a specific game such as 'Space Battle' the player of Intellivision slots a special overlay on top of the keypad. Used in conjunction with a detailed instruction booklet, this overlay gives precise guidance, permitting the player to determine many different aspects of one game.

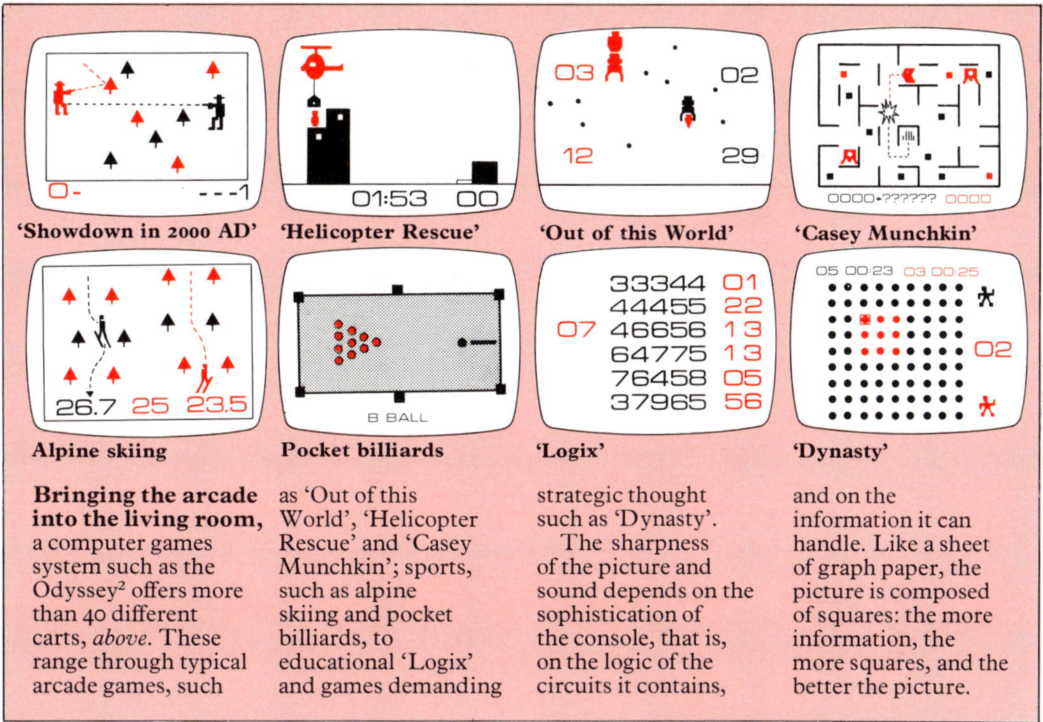

'Showdown in 2000 AD' **'Helicopter Rescue'** **'Out of this World'** **'Casey Munchkin'**

Alpine skiing **Pocket billiards** **'Logix'** **'Dynasty'**

Bringing the arcade into the living room, a computer games system such as the Odyssey² offers more than 40 different carts, *above*. These range through typical arcade games, such as 'Out of this World', 'Helicopter Rescue' and 'Casey Munchkin'; sports, such as alpine skiing and pocket billiards, to educational 'Logix' and games demanding strategic thought such as 'Dynasty'.

The sharpness of the picture and sound depends on the sophistication of the console, that is, on the logic of the circuits it contains, and on the information it can handle. Like a sheet of graph paper, the picture is composed of squares: the more information, the more squares, and the better the picture.

The heart of a computerized games system is its microcomputer or master console, *right*, plugged into the TV via a games switch box. The keys on the console allow games to be programmed to select different speeds and degrees of difficulty: A 'Basic' (computer language) cart converts the console into a small home computer.

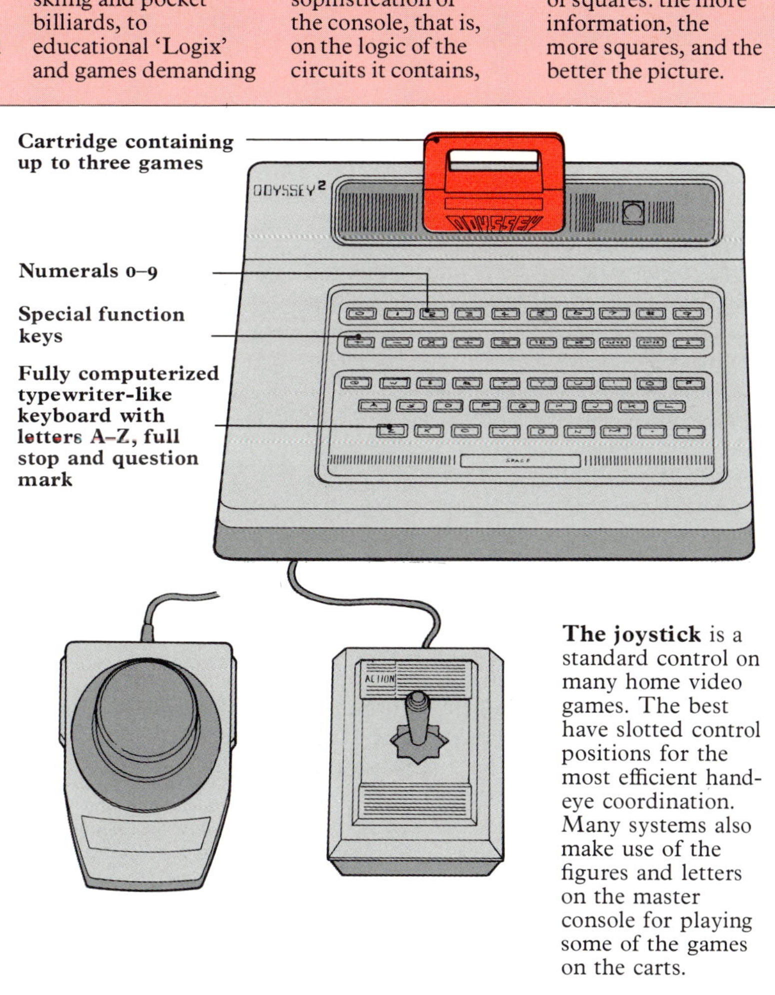

The hand control, which rotates freely through 360°, is used with many bat and ball games. The cart, which is controlled by the master console, supplies the ball to the screen, while the player turns the control to alter the position of the bat in an attempt to hit the ball.

The joystick is a standard control on many home video games. The best have slotted control positions for the most efficient hand-eye coordination. Many systems also make use of the figures and letters on the master console for playing some of the games on the carts.

Intelligent video

The computer is the superchild of electronics, the technology fundamental to both computers and video. It is a high-speed calculator with an electronic memory. Very detailed instructions are typed in special languages which bridge the gap between English and the binary machine code in which the computer 'thinks'.

There are already some computers which can receive and reply in spoken English, but the program which enables them to do this must still be written in a symbolic code, where words, letters and numbers stand for specific instructions. These are translated, via binary, into thousands of electrical currents. The signals thus built up allow the computer to calculate, and to operate, through electrical switches, all kinds of machinery.

A computer is no longer just the fortress-like mainframe (a massive machine using far more complex processors and needing a team of expert operators) beloved of science fiction movies. This is the age of the microcomputer, looking like a typewriter and using a single large chip which is called a microprocessor. It is slower than a mainframe, but quite fast enough to be useful in the home or to the small business user.

What could a personal computer do for you and your video? For a start, it could index all your recordings and locate each one instantly; keep a record of unused hours of tape; turn your television on and sound a bleep to tell you a film has started.

In the future there could be a choice of over 100 channels. If a key signal were broadcast, the computer could easily identify wanted material: all kinds of sport or, for instance, if the key signal provided sufficient information, just highlights of the ladies' long jump. It could even edit out commercials.

Computers have already proved effective as teaching aids. They do not suffer from boredom or impatience, and slow pupils get on as well as quick ones, each learning at his or her own speed. Whether used with a library of video tapes or as part of a fully interactive system, which can test the student, rerun the sequence and answer queries, computers, teamed up with video, will give a more personal education. It is well known that children have none of the techno-fear that besets adults.

Basic is the name of one of the most popular languages used to program microcomputers. A program is a set of instructions which are entered via the keyboard into the computer's memory circuits. Once entered, the program can be activated by keying in another instruction, already known to the computer, which will then automatically go through a series of events it has memorized.

This is a program for a simple guessing game you can play with a computer:

```
1ϕ PRINT "HIGH-LOW NUMBER GAME"
2ϕ A = INT (RND(1)*1ϕϕ): B = ϕ
3ϕ INPUT "YOUR GUESS, BETWEEN 1–1ϕϕ"; C:
   1ϕϕ LET B = B + 1
4ϕ IF C > A, THEN PRINT "TOO HIGH": GO TO
   3ϕ
5ϕ IF C < A, THEN PRINT "TOO LOW": GO TO
   3ϕ
6ϕ IF C = A, THEN PRINT "CORRECT, IN"; B;
   "TRIES": END
```

1ϕ is a number identifying a line of the program. The symbol ϕ distinguishes zero from the letter O. Quotation marks tell the computer that when the program is activated it must print everything within the marks.

In line 2ϕ, A is an unspecified figure which the computer must define. It must be an integer, or whole number (INT), and

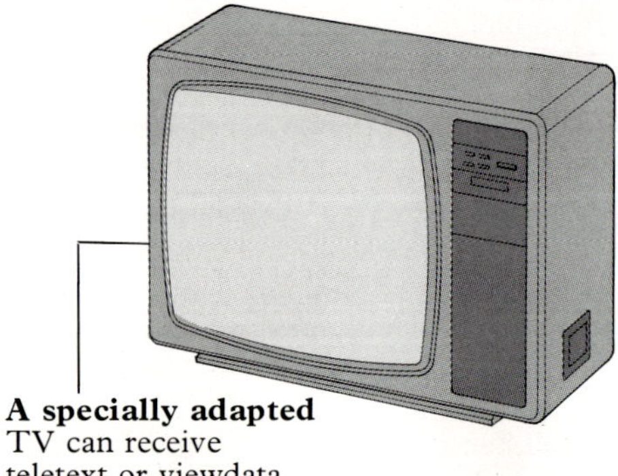

A specially adapted TV can receive teletext or viewdata. Modified further, it can receive computer data through either system, This can be fed directly into a computer memory.

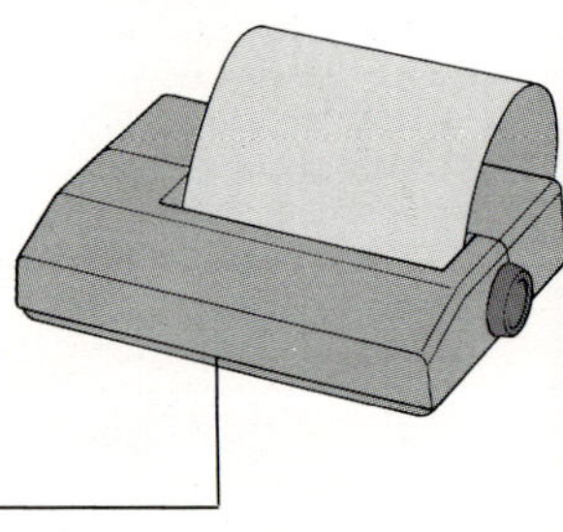

A hard copy printer transfers data from the computer directly on to paper, either by printing like a typewriter, or by burning characters on to heat-sensitive paper with a tiny electric current.

must be chosen at random (RND) between 1 and 1ϕϕ. The computer already knows that it must deal with the command in the inner brackets first, and work outward. When generating the random number, it first takes a fraction of 1 (that is, between ϕ and 1). It then multiplies this (* means multiply) by 1ϕϕ to give a number between 1 and 1ϕϕ, which it rounds to an integer. The colon marks the start of a new instruction in the same line. Another unspecified figure, B, must be equal to zero at this point.

In line 3ϕ, the computer is instructed to print out an invitation to guess a number. C must be defined by the player, and so stands for his guess. Any unspecified figure preceded by "if" in this program must be specified from outside the program, that is, by the operator. At this point, the value of B is increased by 1, thus keeping count of the player's guesses.

In lines 4ϕ and 5ϕ, the computer is told to return the player to line 3ϕ if the guess is too high or too low.

In line 6ϕ, if the guess equals the computer's random number A, the number of guesses made by the player is displayed on the screen at B. The player with the lowest score is the winner.

As satellite and cable TV become widespread, computer data will be sent by the same channels.

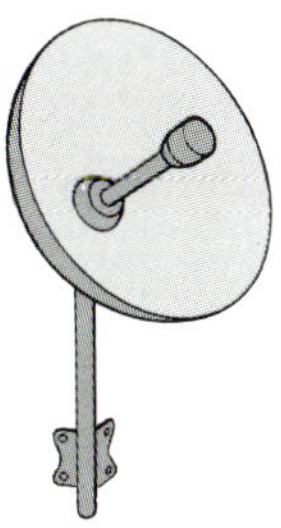

Home owners could use a video door phone to speak a code to their computer to gain entrance.

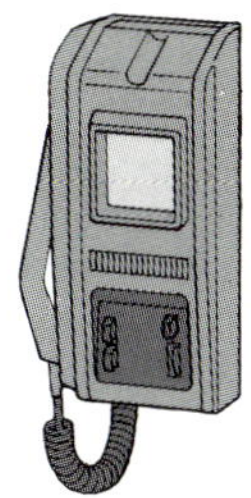

This Apple II, *below*, has a VDU (a monitor designed to be used with computers) and two disk drive units. Flexible (floppy) disks are the usual form of information storage for microcomputers. Illustrated here are some devices which could be linked to a home computer.

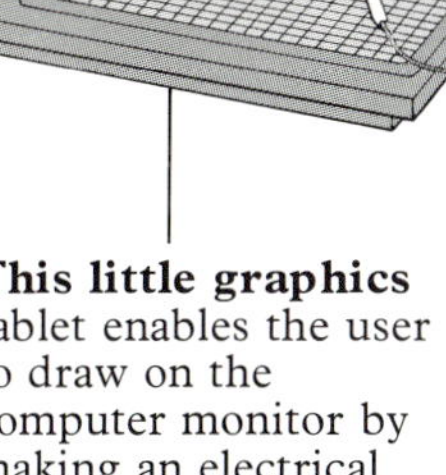

This little graphics tablet enables the user to draw on the computer monitor by making an electrical contact with the tablet and a stylus.

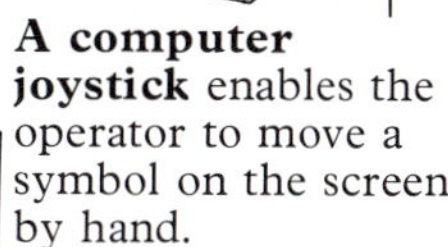

A computer joystick enables the operator to move a symbol on the screen by hand.

Computers will soon recognize visitors (via a video camera) by their appearance. A time-lapse VTR could then record strangers.

Once it becomes possible to use a VCR for data storage and access, a home VCR may be the first step towards a home computer system.

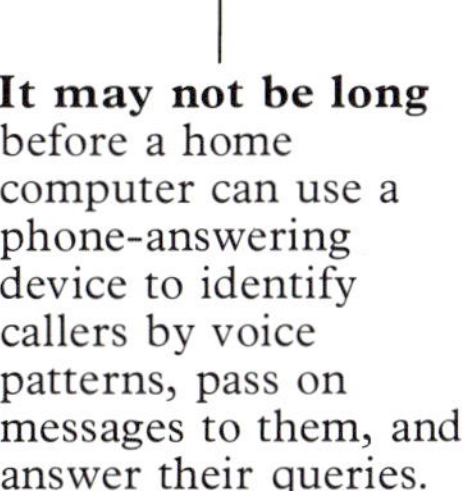

It may not be long before a home computer can use a phone-answering device to identify callers by voice patterns, pass on messages to them, and answer their queries.

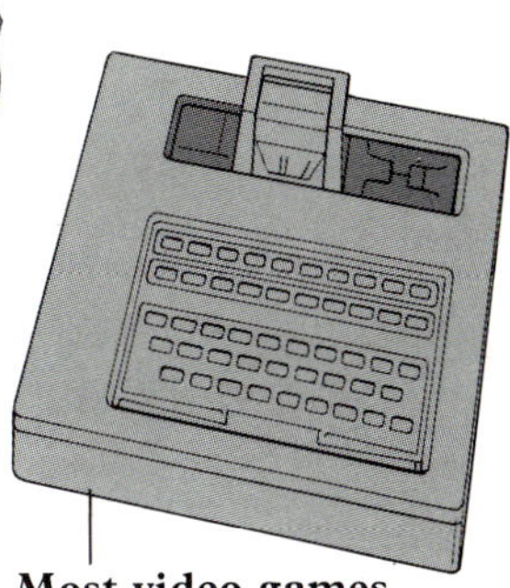

Most video games have an internal microprocessor, but some are so sophisticated that they are almost computers.

Visual display

Whether you start with a VCR and move on to a microcomputer or vice versa, the visual centre of the system will be a television screen. The video signals feeding a computer monitor are the same as those used in broadcast television. Only differences in the way the signals pass from one to the other limit the interaction of VCR, computer and monitor and these differences may vanish before long. Most microcomputers are designed to use a television screen. Some will only display in monochrome, but some can use colour as well. The future may see one large screen, capable of being split into many small screens, handling all the home's video.

A VCR is not fully compatible with a microcomputer. It can be used as an archive memory, but needs a special interface (a device which translates signals from one type of machine to another). However, new interfaces are under development which may allow a home VCR to be used as a data store for a microcomputer.

The advent of viewdata is bringing television and computers closer together. It will soon be routine for microcomputers to order and receive ready-made computer programs by viewdata – perhaps the most direct form of direct mail yet invented.

Some home microcomputers can generate coloured graphics, like those used for teletext, on a colour television or VDU. Coloured captions and pseudo-animations can be created, and even transferred to video with a camera. Some owners may find that their VCRs can record their graphics directly from a television in the normal way.

A professional art computer, by contrast, provides hundreds of colours, can imitate the strokes of a paint brush and could be used to broadcast pictures directly on a television network.

Computer-controlled video can be seen daily on television in the form of multiple or rotating pictures. These need a frame store, a computer which scans a frame of video point by point, gives each point digital coordinates for location, brightness and colour, and immediately displays these on a screen in any order and position. In this way it can move the image around the screen. The speed needed to do so much, so quickly, calls for an expensive multiprocessor computer, so this electronic trickery will remain beyond the microcomputer user.

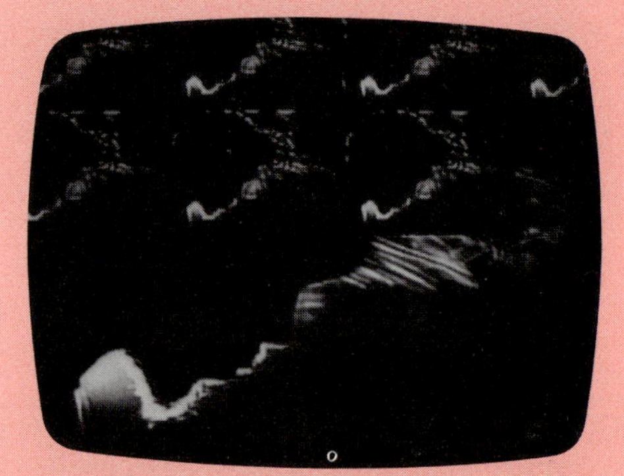

This picture shows an effect devised by using a frame storer. Each point of a single frame is encoded digitally. Enough points are then scanned to reproduce the picture in miniature.

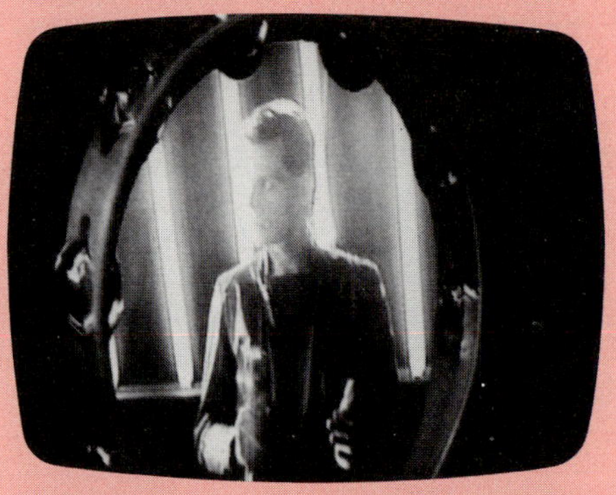

A single frame is here scanned into an area defined by a tambourine. As the tambourine turns, fewer of the horizontal points of the inside frame are scanned, and it also appears to turn.

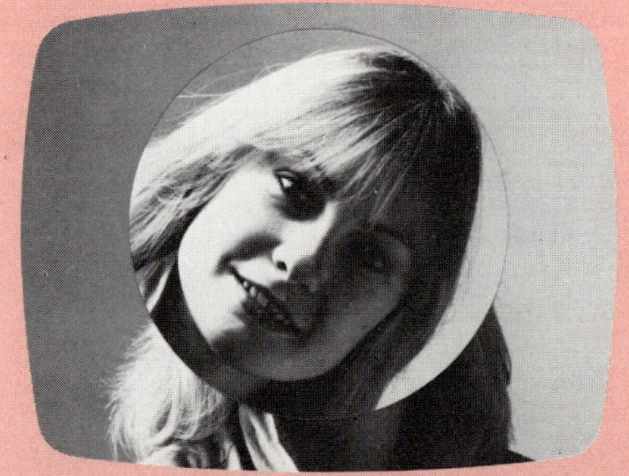

The rotation of a frame in the same plane as the screen is a newly developed effect. It is also possible to rotate only part of the picture, and increase and decrease the size of the turning area.

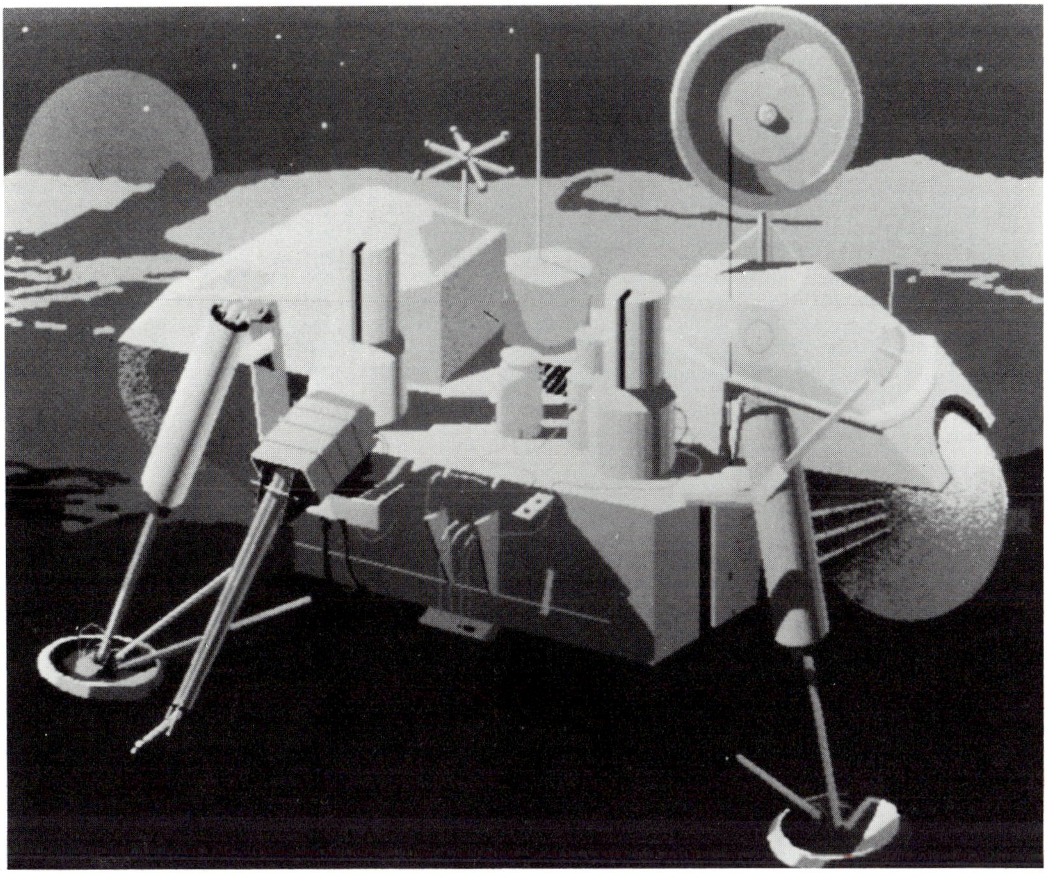

Artists creating computer graphics today are no longer restricted to defining each point on their drawing through a keyboard. Instead, the work area might be an electronically sensitive tablet upon which lines can be drawn with a special stylus, or a light-pen. The signals generated are interpreted by a computer and the picture displayed on a colour monitor.

An area of the tablet will be designated as the paintbox: by touching the right spot, the artist can call up a specific colour on to the screen. Once there, the artist can use it freely until another colour is called up by touching the paintbox on the screen. There may also be a mix mode, so that several colours can be blended, either on a special area of the tablet designated as a palette, or via the drawing surface on the screen.

The range of effects available from the most advanced systems is startling; paint textures and even the handling characteristics of different kinds of paint, pen and crayon, can be imitated.

The picture, *above*, was generated on Flair, a system in current commercial use. It offers a paintbox of 256 colours, and the blending is made more natural by a process called anti-aliasing: each dot on the monitor screen can be varied from its own colour to that of the next, as required. This kind of process is not simple. It is said that computer art enjoys the most advanced electronics outside space science laboratories.

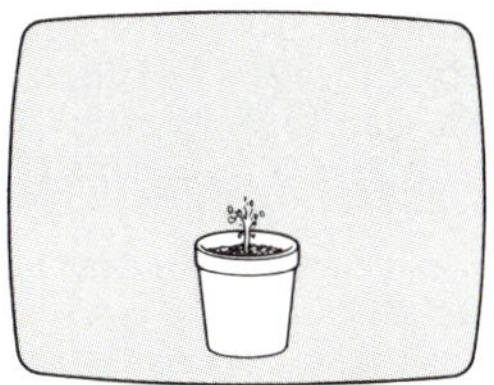
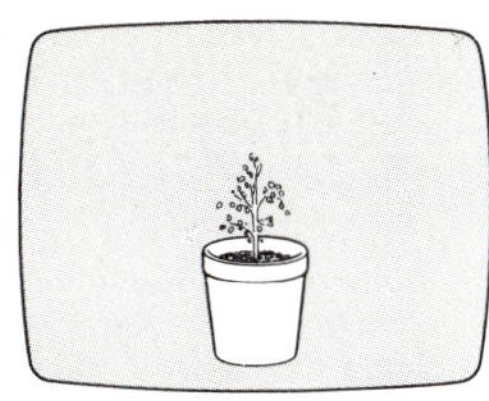
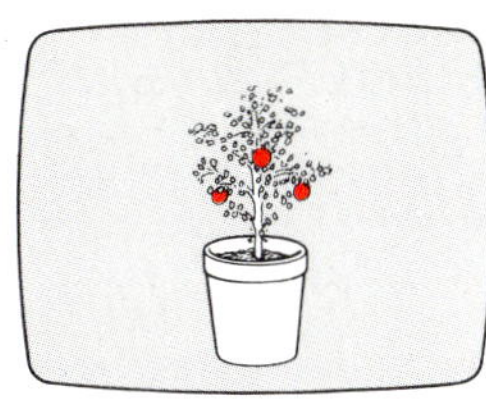
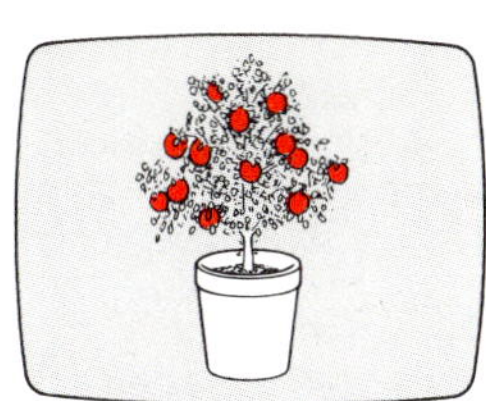

The sequence, *above*, shows how a computer graphic might be drawn in order to rerun the sequence to make a simple animation. This is known as a pseudo-animation. A true animation is a series of still frames, but this is actually a real-time recording of a piece of art-work developing gradually, line by line.

Understanding a specification

All video hardware, be it VCR, camera or power adaptor, has its technical specification. This is usually printed in the owner's manual, and in the manufacturer's brochures.

The specification, or 'spec', is a technical description of a machine's capabilities. Some of the details will be common to all machines working on one set of television standards, American NTSC for instance, and others will outline the capabilities of the particular model being described.

This mass of figures can be daunting, but with a little effort the video buyer can put the data to good use when comparing one machine with another, or when choosing items to make up a complete video system. Here, some sample features taken from hypothetical VCR and camera specifications are explained.

Video recorders and adaptors

Power source: 120v AC ± 10%

The VCR runs on mains (alternating current electricity), at the voltage given, with plus or minus 10% allowance for AC voltage fluctuations. Some VCRs have alternative settings, or wide tolerances, between 220 and 240 volts for instance. Portapacks and cameras use DC (direct current) at 12 volts.

Frequency: 50 Hz (UK); 60 Hz (USA)

Electrical frequency is measured in cycles per second (Hertz or Hz). The VCR must be designed to work on the local supply frequency. Some have a general tolerance between 48 and 62 Hz.

Operating position: Horizontal ± 10%

Home decks need to operate horizontally or very nearly so. 10% is a variation of up to 6 degrees of tilt in either direction.

Relative humidity: 30% to 80%

This gives the percentage of moisture in the air, below condensation levels, in which it is safe to operate the VCR. If condensation is forming in the room, it may also form inside the VCR, causing damage. If the air is dry, as for instance in a closed, centrally heated room, this may cause warping inside the VCR, which can also cause malfunction.

Video recording system: Rotary two-head helical scan, slant azimuth.

This describes the recording system used by all home video systems. Helical scan describes the slanted path the tape takes around the circular head drum; slant azimuth means that the video heads record at an angle greater or smaller than 90 degrees to the tape. Some machines have two extra heads for still frame.

Video signal to noise: More than 42 dBs

Signal to noise (S/N) is the ratio of video signal (picture information) to noise (interference) which the system produces. 'dB' stands for decibel, a measurement of ratio such that with every additional 3 dB the signal strength doubles. A 42-dB S/N ratio is a good reading for a VCR; below 40 would be poor. The higher the figure, the better the picture will be. This ratio is given for the luminance (brightness) part of the video signal, which is more crucial to the sharpness of focus and fineness of contrast than the chrominance (colour) signal.

Resolution: 190 lines centre screen
***or*: 3 MHz (−15 dB)**

If a television camera is trained on a test pattern of fine black lines on a white background, the lines will be distinguishable at the centre of the screen. The more lines visible, the higher the resolution and the sharper the picture.

The resolution may be given in MHz (megahertz). At 3 MHz, the VCR is capable of receiving signals over a bandwidth of three million Hz. The higher the bandwidth of the VCR, the more finely distinguished the signals it can receive and, therefore, the higher the resolution. A bandwidth of 3 MHz is normal for a VCR, and is equivalent to about 190 lines.

'−15 dB' means that the measurement was taken when the VCR was producing a video signal 15 dB below the maximum signal output. A maximum signal of 0.7 V (volts) gives true white on the screen, and 0 volts a true black. A signal 15 dB down on the maximum represents variations between light and dark grey, so that although lines will be visible, they will not be sharply distinguished. Resolution readings are normally given for the luminance signal. Chrominance (colour) signals do not need such a high resolution, as the human eye is less sensitive to the focus of colour.

Video input and output: 1 V p–p, 75 ohms, unbalanced

This is standard for all video signals. When transmitted along a cable, the strength of a video signal is one volt from its highest to its lowest points, from −0.3 V to +0.7 V. '−0.3' describes not an absence of voltage, but a voltage in the reverse direction. Zero volts is the point at which the television screen is perfectly black, and is called the black level. Below 0 volts is the synchronization signal, called the 'sync pulse', which ensures that the electron beam which scans the screen begins its path at the same two points every time it scans. (As a television picture, or frame, is broadcast in two fields, each covering an alternate set of lines on the screen, the sweep pattern has two alternate starting points, one at the top left, one at top centre). This gives a perfectly aligned frame, or a '2:1 fully interlaced picture'. Some very cheap cameras give a random interlaced picture, which is not so good. The video signal is also sometimes described as 'sync negative'.

75 ohms (Ω) is the standard measure of the electrical resistance used in television systems. A cable has little resistance, but its input and output connections must have the appropriate resistance. If this is too low, the signal will rush through, stretching out and losing definition; if too high, the signal will bounce back, sending reflections down the cable and causing ghost images on the screen.

Since the signal is varying, the resistance is different from that for a constant current and is called impedance. 75 ohms impedance is standard for television cable, as the impedance of the cable must match the input and output resistance of the video equipment.

Frequency response: 60 to 10,000 Hz ± 6 dB

Audio, as well as video, signals are measured in

Hz. The higher the frequency, the higher the pitch of the sound transmitted and vice versa. Therefore, the wider the frequency response, the better the sound quality.

'± 6 dB' means that the relative loudness of one part of the sound signal over another can vary by as much as 12 dB.

Distortion: less than 7%
The purity of the waveform from the audio track will be disrupted by electrical noise picked up inside the machine by less than 7%. (Video sound is low fidelity: high fidelity (hi-fi) would allow only around 0.05% distortion.)

Audio input: 100 k ohms, −10 dB
or: **−20 dB (low impedance)**
(0 dB = 0.775 rms)
For professional and hi-fi audio, a reference level for signal strength is set at 0.775 volts, and taken to be 0 dB. 'Rms' is a method of measuring the voltage of an alternating signal.

Video sound is usually transmitted at a signal strength below that used for hi-fi. This is not a measure of sound quality, but means that if sound is recorded from a hi-fi source the signal may be strong and likely to distort.

Audio sources, such as microphones, are described in terms of their impedance. 100 kilohms (100,000 ohms) is a high impedance; 600 ohms is a common low impedance. A low impedance source may work with a high impedance input; a high impedance source will not work with a low impedance input.

Audio signal to noise: More than 50 dB
This is a good audio S/N ration for a VCR. Less than 44 dB would be poor; hi-fi, however, requires around 70 dB.

External microphone input: −66 dB, 2 k ohms
'−66 dB' is, in this instance, the minimum input level the microphone must give for good results. 2 k ohms indicates that the input is low impedance, but, being borderline, will probably tolerate some high impedance microphones.

In audio specifications, a distinction is made between 'mike' (a microphone connection) and 'line' (a cable connection between two machines).

Antenna: 75-ohm external terminal (VHF)
300-ohm external terminals (UHF)
or: **75-ohm asymmetrical. Max input voltage 25V (sync level)**
These are standard impedence values needed for VHF and UHF broadcasts. Asymmetrical refers to the unbalanced coaxial cable. The maximum voltage the antenna input can receive is 25 millivolts. For technical reasons, the video signal is broadcast inverted, with the picture information at the lowest power and the sync pulse and black level at the highest power. Hence the highest voltage the antenna can receive is at sync level.

Input sensitivity: 120 μV rms
or: **TV channels 21 to 68 UHF**
In the UK, broadcast transmissions go out on channels designated numbers 21 to 68. 120 μV is the signal power the antenna must supply to the VCR from the signal it receives. A good antenna can develop more power from a weak signal. The lower this figure, the more likely the VCR is to give an acceptable picture from a weak signal. Check the S/N ratio to see how good a picture is being claimed by the manufacturer.

Modulator output voltage: 3mV rms ± 3 dB
The RF output from the VCR to the television gives a signal of 3 millivolts with a variation over a 6 dB range. If the VCR gives out much more than 5 millivolts the television may be damaged.

Modulator frequency: Channel 30 to 43 (factory set to channel 36)
In order to transmit recorded video from tape to a television receiver, the RF modulator transfers the signal to a channel to which the receiver is tuned. This will be one of the television frequencies in use locally, tuned in at the factory, but may need re-tuning if that frequency is actually being occupied by a local television station.

Cameras
These are some points from the specification which apply solely to cameras:

Pickup system: single-frequency carrier
This merely describes the standard system used with the local colour system. This example is used with NTSC.

Viewfinder: 1.5-in electronic
1.5 in is the normal screen size for television-type viewfinders. If the viewfinder is optical or TTL (through the lens) check to see how accurately it shows the picture you are shooting, especially when zooming.

Colour temperature: indoor/outdoor
This describes the range of settings over which the camera can adjust the picture colour in different lights. Some cameras have four or five settings for colour temperature, but in general 'indoor'/'outdoor' or 'daylight'/'tungsten' are the minimum variations which a good video camera should have.

Synchronization: internal
All home video cameras generate their own sync pulse internally. Professional cameras can be synchronized externally from another piece of video equipment.

Resolution: 270 lines
The resolution figure for a camera is higher than that for a VCR but, in practice, this makes little difference to the picture finally obtained on video tape.

Minimum illumination: 100 lux, F1.4
Lux is a measurement of the amount of light needed; below this amount, the camera will not record a reasonable picture.

F1.4 describes the aperture setting needed for shooting at 100 lux. It is the widest setting; a telephoto setting would give a poorer picture.

Lens: f1.4 × 6 zoom lens; f = 11 to 70 mm
This means that the image is magnified six times between the wide angle and telephoto setting. 'f' designates the focal length of the camera, here in millimetres.

Most home video cameras have fixed lenses. A few have removable lenses, which normally use a mount called a C-mount.

TV standards

The technology by which television signals are broadcast has been developed over a long period. Consequently there are many different methods for transmitting the colour and black and white signals, and for transmitting these signals through the electronics inside the television receiver to its screen.

When video equipment is bought to be used within an area which uses television standards compatible with it, all is well. Problems of compatibility arise when equipment is bought to be taken overseas, or when recorded tapes are sent for replay in another country. In the former instance, the answer is to establish what systems are used in the place of purchase and the eventual destination and, if they are different, to consult a specialist video dealer about buying the most suitable equipment and making any necessary adjustments and modifications. This is often a straightforward process, but requires advice.

In the latter instance, a solution may be harder to find. An ordinary VCR cannot be set up to replay tapes on more than one system. A partial answer is a triple-standard machine, which, however, does not give equally good results with all television systems. Sometimes, an extra gadget can be added to the VCR to allow it to replay on one system tapes recorded on a variant of that system.

The most crucial differences between one set of television systems and another are in the colour standards, as the colour signal is the hardest to encode and decode without loss of picture resolution. There are three basic colour systems: NTSC (National Television Standards Committee), used primarily in the USA and Japan, PAL (Phase Alteration Line) used in the UK, much of Europe and South America, and Australia, and SECAM (Sequential Colour with Memory) used in France, other parts of Europe and Africa, the Middle East and the USSR. There are variations: PAL-M uses a different colour subcarrier, and SECAM is divided into SECAM V (vertical) and SECAM H (horizontal), along with other minor modifications. There are also six monochrome systems: British, Belgian, CCIR (International Radio Consultative Committee), French, OIRT (International Radio and Television Organization) and USA. The differences between these are less crucial than the differences between the colour systems. In addition, there are fourteen separate systems for broadcasting the whole television signal, including sound, and these are designated A to N.

Table 1: World television standards

A specialist can modify video equipment for use with a set of television systems for which it was not designed, once the colour standard, television system and the numbers of the local television channels are known. Table 1, *below*, gives this information for a sample selection of nations where video is becoming popular. Also listed are the electricity supply voltages, on which the equiment must be able to operate, and the frequency in Hertz (cycles per second) of the AC supply. This is, with few exceptions, the same as the number of fields per second broadcast by the local television system, as the number of fields was originally derived from the supply frequency.

Country	System	Channels	Voltage	Hz	B/W
NTSC					
Bahamas	M	A13	110/240/415	60	USA
Canada	M	A2–13, A14–40	120/240	50	USA
Hawaii	M	A2–13	120/240	60	USA
Japan	M	J1–12; 27–62	100/200	50; 60	USA
Mexico	M	A2–23; A14–45	110/220	60	USA
Philippines	M	A2–13	110/220/240	60	USA
(US Forces)	(M)	(A8, 28–74)			
USA	M	A2–13, A14–83	120/230/240	60	USA
PAL					
Algeria	B	E5–11	127/220/380	50	CCIR
Argentina	N	A2–13	220/380	50	USA
Australia	B	Aus 0–11	240/415	50	CCIR
Austria	B;G	E2–12; 21–68	220/380	50	CCIR
Azores	B	E7–9	110/220/380	50	USA
Bahrain	B	E4	220/240	50	CCIR
Belgium	B;H	E2–11; 21–69	220/380	50	Belgian
Denmark	B	E3–10	220/380	50	CCIR
Germany, Fed. Rep.	B;G	E2–11; 21–60	220/380	50	CCIR
Gibraltar	B	E6–11	240/415	50	CCIR
Hong Kong	I	E21–60	200/220/346	50	British
Iceland	B	E3–11	220/380	50	CCIR
Indonesia	B	E4–10	127/220/380	50	CCIR
Ireland, Rep.	I	A–J, 29–43	220/380	50	Belgian
Israel	B;G	E5–11; 25–56	230/400	50	CCIR

Italy	B;G	It A–H; 21–34	127/220/380	50	CCIR
Jordan	B	E3–9	220/380	50	CCIR
Luxembourg	C;G	E7, 27	110/220	50	Belgian
Malaysia	B	E2–10	230/240/415	50	CCIR
Netherlands	B;G	E2–12; 21–60	127/220/380	50	CCIR
New Zealand	B	NZ1–10	230/400/415	50	CCIR
Nigeria	B	E2–12	230/400	50	CCIR
Norway	B;G	E2–12; 44	230	50	CCIR
Pakistan	B	E4–10	230/400	50	CCIR
Portugal	B;G	E2–11; 25–46	220/380	50	CCIR
Singapore	B;G	E5–12; 21–34	230/400	50	CCIR
South Africa	I	14–13, 21–68	220/230/380	50	British
Spain	B;G	E2–11; 21–68	127/220/380	50	CCIR
Sudan	B	E5–7	240/415	50	CCIR
Switzerland	B;G	E2–12; 21–63	220/380	50	CCIR
Syria	B	E4–9	115/220/380	50	CCIR
United Kingdom	I	21–68	240/415	60	British

SECAM

Cyprus	B(H)	E5–11	115/400	50	CCIR
France	L(V)	21–65	127/220/380	50	French
Greece	B;G(H)	E5–12; 21–69	220/380	50	CCIR
Iraq	B(H)	E5–12	110/220/380	50	CCIR
Lebanon	B(V)	E2–11	110–190	50	CCIR
Luxembourg	L(V)	21	380	50	Belgian
Saudi Arabia	B(H)	E5–10	120/208	50+60	CCIR
United Arab Republic	B(V)	E3–11	110/220/380	50	OIRT
USSR	D;K(V)	R1–12; 21–69	127/220/380	50	OIRT
Zimbabwe	B	E2–11	230/240	50	CCIR

Table 2: World television systems

In the first column of Table 1, *above*, is listed the designation letter, A to N, of the world's television broadcast systems. In Table 2, *below*, a selection of the most important characteristics of these systems is listed. Systems A, C, E, F and H are now rarely used.

The number of lines describes the number of times the electron beam in a television receiver sweeps across the screen in a single frame, and the number of fields is given for one second in time (a field is half a frame). The colour subcarrier frequency is chosen to separate the colour from the brightness in the television signal. Systems M and N have a lower frequency because their vision signal bandwidth is narrower than normal. The vision and sound separation is very important, since, once correctly set up, it is possible to tune the set to different television channels keeping the sound automatically tuned into the vision.

The vision modulation tells an engineer whether the brightest (white) part of the signal or the darkest (sync pulse) is broadcast at the highest signal power, and sound modulation tells whether the sound is amplitude modulated (by varying the strength of the signal) or frequency modulated (by varying the frequency of the signal). With these basic facts, a television engineer can assess a piece of video equipment for modification to deal with local television conditions.

System	Number of lines	Number of fields	Colour subcarrier	Vision/sound separation	Vision modulation	Sound modulation
A	405	50	none	−3.5 MHz	Positive	AM
B	625	50	4.43 MHz	+5.5 MHz	Negative	FM
C	625	50	4.43 MHz	+5.5 MHz	Positive	AM
D	625	50	4.43 MHz	+6.5 MHz	Negative	FM
E	819	50	none	+11.15 MHz	Positive	AM
F	819	50	none	+5.5 MHz	Positive	AM
G	625	50	4.43 MHz	+5.5 MHz	Negative	FM
H	625	50	4.43 MHz	+5.5 MHz	Negative	FM
I	625	50	4.43 MHz	+6 MHz	Negative	FM
K	625	50	4.43 MHz	+6.5 MHz	Negative	FM
K1	625	50	4.43 MHz	+6.5 MHz	Negative	FM
L	625	50	4.43 MHz	+6.5 MHz	Positive	AM
M	525	60	3.38 MHz	+4.5 MHz	Negative	FM
N	625	50	3.38 MHz	+4.5 MHz	Negative	FM

Names and addresses

Listed below is a selection of addresses in the UK which the video owner might find useful in the case of query or complaint. Each address is followed by a brief note of the range of the company's services, or the goods manufactured. Normally, the owner's first point of enquiry would be his or her own video dealer, but if contacting a manufacturer, ask for the customer liaison department or the product manager. Manufacturers find it easier to be helpful if the customer is as clear as possible about the facts surrounding the enquiry or complaint when writing or calling.

Agfa-Gevaert Ltd., Great West Road, Brentford, Middlesex
(01) 560 2131
Video tapes (VHS, Betamax, LVC, SVC, U-matic).

Akai (UK) Ltd., 12 Silver Jubilee Way, Haslemere Heathrow Estate, Hounslow, Middlesex TW4 6NF
(01) 897 7171
VCRs (VHS), portable systems, cameras, video tapes (VHS), televisions.

Association of Video Dealers, Greystones, Brighton Road, Godalming, Surrey GU7 1PL
Godalming (04868) 23429
Queries and complaints concerning member dealers.

AV Distributors (London) Ltd., 26 Park Road, Baker Street, London NW1 4SH
(01) 935 8161
Agents for Bilora tripods and lighting equipment.

BASF (UK) Ltd., Haddon House, 2–4 Fitzroy Street, London W1P 5AD
(01) 388 4200
Video tapes (VHS, Betamax, VCC, VCR).

R.R. Beard, 10 Trafalgar Avenue, London SE15 6NR
(01) 703 3136/9638
Lighting equipment.

Bell & Howell A-V Ltd., Alperton House, Bridgwater Road, Wembley, Middlesex HA0 1EG
(01) 903 5411
Agents for JVC professional video equipment.

BIB Audio/Video Products Ltd., Kelsey House, Wood Lane End, Hemel Hempstead, Herts HP2 4RQ
Hemel Hempstead (0442) 61291
Video accessories.

British Video Association, 10 Maddox Street, London W1R 9PN
(01) 499 3131
Investigates video piracy and breach of copyright in prerecorded tapes.

Burton Manor, Burton, South Wirral, Cheshire
(051) 336 5172
Short and weekend video courses.

Colortran UK, P.O. Box 5, Burrell Way, Thetford, Norfolk IP24 3RB
Thetford (0842) 2484
Lighting equipment.

Comprehensive Video Supply (UK), 565 Kingston Road, London SW20 8SA
(01) 543 3131
Video accessories from lighting equipment to labels.

The Consumers' Association, 14 Buckingham Street, London WC2N 6DS
(01) 839 1222
Publishes *Which?* magazine; general advice on purchasing and consumers' rights to members.

Council for Educational Technology (CET), 3 Devonshire Street, London W14 2BA
(01) 580 7553
Provides information on media courses and specifications for audiovisual equipment.

Educational Television Association, 86 Micklegate, York YO1 1J2
York (0904) 29701
Professional association to bring together institutions and individuals using television in education and training.

EMI Tapes Ltd., Alma Road, Windsor, Berks SL4 3JA
Windsor (95) 59171
Videotapes (VHS, Betamax)

Grundig International Ltd., Newlands Park, London SE26 5NQ
(01) 659 2468
VCRs (Video 2000), cameras, video tapes (VCC), video disk (LaserVision).

Hitachi Denshi (UK) Ltd., Lodge House, Lodge Road, London NW4 4DQ
(01) 202 4311
Professional video cameras and control systems.

Hitachi Sales (UK) Ltd., Hitachi House, Station Road, Hayes, Middlesex UB3 4DR
(01) 848 8787
VCRs (VHS), portable systems, cameras, video tape (VHS), televisions, accessories.

Introphoto Ltd., Prior's Way, Maidenhead, Berks SL6 2HP
Maidenhead (0628) 7441
Velbon tripods, Hoya photographic filters.

ITT Consumer Products (UK) Ltd., Chester Hall Lane, Basildon, Essex
Basildon (0268) 3040
VCRs (Video 2000, VHS), televisions.

JVC UK Ltd., Eldonwall Trading Estate, Staples Corner, London NW2
(01) 450 2621
VCRs (VHS), portable systems, cameras, video tapes (VHS), televisions, accessories, video disk (VHD).

Kennett Engineering Co. Ltd., The Lodge Works, Drayton Parslow, nr. Milton Keynes, Bucks
Mursley (029672) 605
Tripods.

Maxell (UK) Ltd., 1 Tyburn Lane, Harrow, Middlesex HA1 3AF
(01) 423 0688
Video tapes (VHS, Betamax, U-matic).

Memorex (UK) Ltd., Memorex House, 94–104 Church Street, Staines, Middlesex TW18 4XU
Staines (0784) 51488
Video tapes (VHS, Betamax).

Michael Cox Electronics Ltd., Hanworth Trading Estate, Hampton Road West, Feltham, Middlesex TW13 6DH
(01) 898 6091
Includes mixing and switching equipment suitable for the smaller professional unit.

Mitsubishi Electric (UK) Ltd., Otterspool Way, Watford WD2 8LD
Watford (0923) 40566
VCRs (VHS), televisions.

The National Audio-Visual Aids Centre (NAVAC), Paxton Place, London SE27
(01) 670 4247
Short courses in video production.

National Panasonic Ltd., 308–318 Bath Road, Slough
Slough (0753) 34522
VCRs (VHS), portable systems, cameras, video tape (VHS), televisions, accessories.

Nordmende (UK) Ltd., Units 8–9 Faraday Road, Rabans Lane, Aylesbury, Bucks
Aylesbury (0296) 20501
VCRs (VHS), portable systems, cameras, video tape (VHS), televisions, accessories.

North East London Polytechnic, Short Courses Unit, Longbridge Road, Dagenham, Essex RU8 2AS
(01) 597 7591
Short courses in video production and technique for amateurs and professionals.

Pelling & Cross Ltd., 104 Baker Street, London W1M 2AR
(01) 487 5411
Industrial video hire, including VHS, Betamax; agents for Gitzo, Manfrotto tripods.

Philips Video/Audio, City House, 420–430 London Road, Croydon, Surrey CR9 3QR
(01) 689 2166
Philips Video: VCRs (Video 2000), cameras, video tape (VCC, VCR, LVC), televisions, video games
Philips Audio: video disk (LaserVision).

Polar Video Ltd., 12–18 Brook Mews North, London W2
(01) 724 3736
Video mixer.

Pye Ltd., 137 Ditton Walk, Cambridge, CB5 8QD
Teversham (02205) 2781
VHS (Video 2000), cameras, video tape (VCR, VCC), televisions.

Rank Strand Electric, P.O. Box 70, Great West Road, Brentford, Middlesex
(01) 568 9222
Lighting equipment.

Sanyo Marubeni UK Ltd., Sanyo House, 8 Greycaine Road, Watford, WD2 4UQ
Watford (0923) 46363
VCRs (Betamax), portable systems, video tape (Betamax), televisions.

Sharp Electronics UK Ltd., Sharp House, Thorp Road, Manchester M10 9BE
(061) 205 2333
VCRs (VHS), cameras, televisions.

Sony UK Ltd., Communications Systems Division, Pyrene House, Sunbury-on-Thames, Middlesex
Sunbury-on-Thames (09327) 81211
VCRs and video tape (U-matic; Sony industrial and educational video equipment.

Sony UK Ltd., Consumer Products Division, 134 Regent Street, London W1
(01) 439 3874
VCRs (Betamax, U-matic), portable systems, cameras, video tape (Betamax, U-matic), televisions, telecine, special effects generator.

Survey & General Instrument Co. Ltd., Firecroft Way, Edenbridge, Kent
Edenbridge (0732) 864111
Fujinon, Miller, Quickset tripods.

TDK Tape Distriburors (UK) Ltd., 8th Floor, Pembroke House, Wellesley Road, Croydon CR0 9XW
(01) 680 0023
Video tape (VHS).

Thorn Consumer Electronics Ltd., Great Cambridge Road, Enfield, Middlesex
(01) 363 5353
VCRs (VHS), portable systems, cameras, televisions, video tape (VHS).

3M UK PLC, Recording Materials and Consumer Products Division, P.O. Box 1, Bracknell, Berks
Bracknell (0344) 36726
Video tapes (VHS, Betamax, VCC, U-matic, professional).

Toshiba (UK) Ltd., Toshiba House, Frimely, Camberley, Surrey GU16 5JJ
Camberley (0276) 6222
VCRs (Betamax), video tapes (Betamax), televisions.

Video Electronics Ltd., (VEL), Wigan Road, Atherton, Manchester
Atherton (0942) 882332
Video mixers.

The Video School, 36 Lorne Park Road, Bournemouth, Dorset BH1 1JL
Bournemouth (0202) 28786
Short- and long-term courses in video production.

W. Vinten Ltd., Western Way, Bury St. Edmunds, Suffolk IP33 3TB
Bury St. Eds. (0284) 2121
Tripods.

Glossary of terms

A

Adaptor
A device for converting the format of a plug or socket, for example, from a ¼-in (6-mm) jack to a mini-jack.

Amplifier
An electronic device that increases the strength of an electronic signal.

Analog See **Digital**

Aperture
The size of the opening, or iris, in a lens, known as the 'f-stop number'.

Aspect ratio
The horizontal and vertical proportions of a screen. The standard TV format is 4 units of width to 3 units of height.

B

Back projection
The projection of a film or slide on to a translucent screen so that the image can be viewed from the opposite side of the screen to the projector.

Back-space editing
A facility on some VCRs which rolls the tape back slightly each time the camera button is pressed, to give a clean transition from one shot to the next.

Bandwidth
The range of signal frequencies which a piece of equipment can encode or decode.

Boom
An extension arm allowing a microphone to be brought closer to a performer.

C

Camera tube
A glass cathode ray tube using an electron beam to translate the perceived image into electronic signals.

Capacitor
A component used in electronic circuits to store and release voltages.

Cathode
The electrode within a camera or TV tube which emits a beam of electrons to scan or recreate an image.

CCD (Charge coupled device)
A device that replaces the camera tube with a light-sensitive matrix which translates the light falling on each minute portion into an electric charge.

Chrominance signal
The part of the video signal containing the colour signal.

Colour temperature
The colour tonality of a light source, analogous to the colour of a material burning at a specified heat, measured in Kelvins (K).

Commercial cutter
A device for automatically pausing the VCR during commercials.

Control track
A series of signals recorded on the video tape for the VCR to lock onto in playback, to ensure a stable picture.

Crimping tool
A tool for changing a cable end into a connector.

Cycle
In AC (alternating current) a cycle is a complete change from zero voltage to maximum positive voltage, through zero to maximum negative voltage, and back to zero. UK voltage operates on 50 cycles per second; USA voltage operates on 60 cycles per second.

D

Depth of Field
The area (in depth) of a picture which is in focus at any given aperture setting. The smaller the aperture, and the bigger the f-number, the deeper the area in focus.

Dichroic Mirror
A mirror which reflects certain wavelengths of light and allows others to pass through.

Digital
A method of representing a signal by a set of precise numerical values as opposed to a fluctuating current or voltage (analog signal).

Dipole
A VHF antenna.

Director
A reflector plate on a UHF antenna which focuses the signal on the elements.

Dolby System
The trade name of an audio noise-reduction system.

Domestic standard
A standard of technical quality referring to ½-in (12.5-mm) format VCRs designed for the home market. These technical standards are lower than those used in professional equipment.

E

Electrode
A plate at which an electric current is changed into a stream of electrons, or vice versa.

Electron
A negatively charged subatomic particle. (A stream of electrons constitutes an electric current.)

F

Fibre optic
A glass fibre strand capable of transmitting light.

Filter
A transparent material capable of regulating light which passes through it.

Frame
A complete TV picture, composed of 2 fields, which takes 1/25 of a second (PAL) or 1/30 of a second (NTSC) to scan.

Frequency
The number of times a signal vibrates in a second, expressed as Hertz (Hz) or cycles per second.

Fresnel lens
A lens used to focus light into an even concentrated beam.

H

Head
An electromagnetic device that records on, or receives signals from magnetic tape.

Hertz See **Frequency**

Hi-fi
High fidelity audio reproduction as produced by the better quality home audio systems.

Hologram
A 3-dimensional image created by the interaction of two laser light sources.

I

Image enhancer
An electronic device for smoothing out irregularities in the video signal to improve picture definition.

Impedence
A measure of the total electrical resistance (*q.v.*) of a circuit, measured in ohms. Used in describing some electrical equipment, particularly microphones, to ensure connection with compatible equipment.

Infra-red
Electromagnetic waves beyond the red end of the visible spectrum.
Interactive video
Equipment capable of eliciting a response from the user and selecting the information to be displayed according to the user's response.

L

Lacing
The mechanical action of a VCR threading video tape round the record heads.
Lag
An image on the pickup tube of the camera which lingers. This is caused by over-bright subjects.
Laser
A device in which excited atoms produce an intense beam of radiation.
LED (Light emitting diode)
A semiconductor which lights up when a current of electricity is passed through it.
Limiter
An electronic circuit which automatically adjusts audio or video signal levels to a predetermined setting.
Luminance
The brightness of a TV picture.

M

Magnetic field
The area around an object influenced by its magnetism.
Matrix
A rectangular array of qualities or items in rows and columns to make up a single whole.
Microchip
A microprocessor or other electronic component made up of a chip of silica with microelectronic circuitry photographically printed on its surface; a 'silcon chip'.
Microprocessor
A microchip capable of doing complex calculations, used as the 'brain' of a microcomputer or other logic-controlled device, such as a VCR timer; also (loosely) a microcomputer.
Modulate
To alter a signal by adding the waveform (*q.v.*) of another signal to it. To broadcast a TV signal, a plain carrier signal is modulated with video, audio and control signals.

N

Noise
Unwanted audio or video signals which interfere with the normal signal.

P

Parallax
The apparent movement of an object caused by the shifting of the observer's viewpoint.
Photo-conductivity
Electrical conduction caused by the action of light on a surface.
Photodiode
A light-sensitive diode which conducts a current when light falls on it.

R

Real time
A recording or playback which takes the same time as the original action recorded, rather than in slow or fast motion.
Resistance
The extent to which an electrical conductor resists the flow of electricity; measured in ohms.
RF converter
A device to convert audio and video signals into a combined RF signal suitable for reception by a standard TV set.

S

SEG See **Special effects generator**
Signal splitter
A device for dividing an RF signal to feed more than one VCR or TV set.
Silica gel
Quartz crystals with the ability to absorb moisture from the surrounding air.
Special effects generator (SEG)
A device used to mix, switch or process video signals from different sources.
Solenoid
An electro-magnetic device for switching a current on and off.
Standby
The mode in which the VCR has the necessary circuits activated in readiness for instructions.
Switcher
A simplified SEG which selects video signals from two or more sources.
Sync pulse
Synchronization pulse; a signal generated inside a camera or separate generator which ensures that each picture frame is scanned consistently.

T

Telecine converter
A device for transferring filmed material to video tape.
Time-shifting
Recording a broadcast from the TV for later viewing.
Tolerance
The amount by which any technical requirement is permitted to vary without impairing its function.
Transformer
A device for reducing or increasing the voltage of an electric current.
Tuner
A unit incorporated into most domestic VCRs to receive and decode RF signals, from an antenna or other RF sources, into separate audio and video signals.

U

Ultrasonic sound
Sound pitched above the higher limit of human hearing.

V

VCR (video cassette recorder)
A video recorder with tape stored in cassettes.
Vectorscope
An oscilloscope with a round screen, producing information from three colour signals, to allow precise adjustment of professional three-tube cameras.
Video
An electronic signal carrying picture information.
VTR (video tape recorder)
A video recorder with video tape stored on open reels.

W

Waveform
A practical representation of the changing values of a signal.
White light
An equal combination of light of all the colours of the visible spectrum.
Writing speed
The speed at which the video heads revolve, relative to the speed of the tape passing them.

Bibliography

Anderson, C. *The Electric Journalist* Praeger Publishers, New York, 1973.
Baddeley, W.H. *Documentary Film Production* (Fourth edition) Focal Press, London and New York, 1975.
Bensinger, C. *The Home Video Handbook* Video-info Publications, Santa Barbara, California, 1979.
Bensinger, C. *The Video Guide* (Second edition) Video-info Publications, Santa Barbara, California, 1979.
Buckwalter, Len *Video Games* Today Press, Gosset and Dunlap, New York, 1977.
Chorafas, D.N. *Interactive Videotex: The Domesticated Computer* Petrocelli Books, Princeton, N.J. 1981.
Consterdine, G. and Nicholson, R. *The Prestel Business* Northwood Books, London, 1980.
CTL Electronics Inc. *Video Tools* CTL Electronics, New York (Annual).
Davis, D. *The Grammar of Television Production* Barrie and Jenkins, London, 1978.
Fedida, S. and Malik, R. *The Viewdata Revolution* Associated Business Press, London 1979; John Wiley and Sons, New York, 1980.
Foss, H. *How to Make Your Own Video Programmes* Elm Tree Books/Hamish Hamilton, London, 1982.
Foss, H. (Ed.) *Video Production Techniques* Kluwer Publishing, Brentford, Middlesex (Loose-leaf, biannual).
Frost, J.M. (Ed.) *World Radio TV Handbook* Billboard, London, Watson Guptill, New York, 1982 (Annual).
Greenfield, A. and Maltin, L. *The Complete Guide to Home Video* Crown Publishers/Harmony Books, New York, 1981.
Hunt, A. *The Language of Television* Eyre Methuen, London, 1981.
Jones, P. *The Technique of the Television Cameraman* (Third edition) Focal Press, London and New York, 1972.
JVC *Video the Better Way* Victor Company of Japan Ltd, Tokyo, 1980.
Kehoe, V.J.R. *The Technique of Film and Television Make Up for Colour and Black and White* (Second edition) Focal Press, London, 1969; Hastings House, New York, 1969.
Kybett, H. *The Complete Handbook of Video Cassette Recorders* TAB Books, Blue Ridge Summit, PA, 1971.
Kybett, H. *Video Tape Recorders* (Second edition) Howard W. Sams, Indianapolis, IN, 1978.
Murray, M. *The Videotape Book: A Basic Guide to Portable TV Production* Bantam Books/Taplinger Publishing, New York, 1975.
MacRae, D.L., Monty, M.R. and Worling, D.G. *Television Production: An Introduction* Methuen, Ontario, 1979.
Millerson, G. *The Technique of Television Production* (Tenth edition) Focal Press, London and New York, 1979.
Money, S.A. *Teletext and Viewdata* Newnes Technical Books/Butterworth, London, 1981.
National Cable Television Association *Cable Television and Education* NCTA, Washington DC, 1973.
Overman, M. *Understanding Sound and Video Recording* Lutterworth Press, Guildford, Surrey, 1977; TAB Books, Blue Ridge Summit, PA, 1978.
Petzold, P. *The Photoguide to Moviemaking* Focal Press, London and New York, 1975.
Quick, J. and Wolff, H. *Small-Studio Video Tape Production* (Second edition) Addison-Wesley Publishing, Reading, MA, 1976.
Robertson, A. *From Television to Home Computer: the Future of Consumer Electronics* Blandford Press, Poole, Dorset, 1979; Sterling Publishing, New York, 1979.
Robertson, A. (Ed.) *The Home Video Yearbook* Link House Magazines, Croydon, Surrey (New edition annually).
Robertson, A. *International Video Yearbook* Blandford Press, Poole; Sterling Publishing, New York (Annual).
Robinson, J.F. *Videotape Recording: Theory and Practice* (Third edition) Focal Press, London and New York, 1981.
Robinson, R. *The Video Primer; Equipment, Production and Concepts* Quick Fox, New York, London and Tokyo, 1978.
Sexton, B. *First Steps in Television* Fountain Press, Kings Langley, Herts, 1975.
Sigel, E. (Ed.) *Video Discs: The Technology, the Applications and the*

Future Knowledge Industry Publications, New York, 1981.
Sigel, E. *Videotext* Crown Publishers/Harmony Books, New York, 1981.
Sigel, E. (Ed.) *Videotext: Coming Revolution in Home/Office Information Retrieval* Knowledge Industry Publications, New York, 1980.
Shamberg, M. *Guerilla Television* Holt, Rinehart and Winston, New York, 1971.
Sloan Commision on Cable Communication *On the Cable: The Television of Abundance* McGraw-Hill Publishing, New York, 1971.
Smallman, K. *Creative Film-making* Bantam Books/MacMillan, New York, 1972.
Smith, R.L. *The Wired Nation* Harper and Row, New York, 1972.
Utz, P. *Video User's Handbook* Prentice-Hall, Englewood Cliffs, NJ, 1980; Hemel Hempstead, Herts, 1981.
Wilkie, B. *Creating Special Effects for Film-TV* Hastings House, New York, 1977.
Wilkie, B. *The Technique of Special Effects in Television* Focal Press, London and New York, 1971.
Winsburg, R. *Viewdata in Action A Comparative Study of Prestel* McGraw-Hill (UK), Maidenhead, Berks, 1981.
Videography Magazine Editors *The Video Handbook* (Third edition) United Business Publications, New York, 1977.
Zmijewsky, B. *The Consumer's Guide to Video Tape Recording* Stein and Day, New York, 1979.

Periodicals

The Big Reel Drawer B, Summerfield, NC 27358, USA (Monthly; film journal).
Bulletin for Film & Video Information 80 Wooster Street, New York, NY 10012, USA (Monthly).
Cable Information Newsletter Room 852, 475 Riverside Drive, New York, NY 10027, USA (Monthly; cable TV).
Educational and Industrial Television 607 Main Street, Ridgefield, CT 06877 (Monthly)
Filmakers' Newsletter PO Box 46, New York, NY 10012, USA (Monthly).
Hampton's Guide to Pre-Recorded Video PO Box 684, Southampton, New York, NY 11968, USA (Annual; software).
Hampton's Official Video Buyers' Guide PO Box 684, Southampton, New York, NY 11968, USA (Annual; home video).
Home Video 475 Park Avenue South, New York, NY 10016, USA (Bimonthly).
Media and Methods 134 North 13th Street, Philadelphia, PA 19107, USA (Monthly during school year; education).
Movie Maker PO Box 35, Bridge Street, Hemel Hempstead, Herts, UK (Monthly).
Panorama 850 Third Avenue, New York, NY 10022, USA (Monthly; TV).
Popular Video 30 Wellington Street, Covent Garden, London WC2, UK (Monthly).
Television King's Reach Tower, Stamford Street, London SE1 9LS, UK (Monthly).
Television and Home Video Link House, Dingwall Avenue, Croydon, Surrey CR9 2TA, UK (Monthly).
Video 235 Park Avenue South, New York, NY 10003, USA (Monthly).
Video Link House, Dingwall Avenue, Croydon CR9 2TA, UK (Monthly; professional video).
Video A–Z 22 Albany Road, London W13 8PG, UK (Quarterly).
Video Buyer's Review PO Box 684, Southampton, New York, NY 11968, USA (Quarterly).
The Video Exchange Directory 261 Powell Street, Vancouver, British Columbia, Canada V6A 1G3 (Annual).
Hi-Fi for Pleasure 40 Long Acre, London WC2E 9JJ, UK (Monthly; home video).
Videography 475 Park Avenue South, New York, NY 10016, USA (Monthly).
The Videophile 2003 Apalachee Parkway, Tallahassee, FL 32301, USA (Bimonthly; home video hardware).
Video Review 325 East 75th Street, New York, NY 10021, USA (Monthly; software).
Video Review Surrey House, 1 Throwley Way, Sutton, Surrey, SM1 4QQ, UK (Monthly).
Video Today 145 Charing Cross Road, London WC2H 0EE, UK (Monthly).
Video World Galaxy Publications Ltd, Hermit Place, 252 Belsize Road, London NW6 4BT, UK (Monthly).
What Video? 11 St Bride Street, London EC4, UK (Monthly).
Which Video? 145 Charing Cross Road, London WC2H 0EE , UK (Monthly).

Index

The most important entries are indicated in bold type; illustrations are indicated by page numbers in italics.

D

E

F

G

H

I

Q

R

S

Acknowledgements

Writers: Bernard Nyman (Video and the law pp 48–49); Ruth Binney (The TV connection pp 80–81; Optional extras pp 178–179); Shelley Turner (Viewing data pp 86–87; Screen games pp 202–203); Andrew Emmerson (Giant screens pp 88–89; Skybirds pp 90–91; TV tomorrow pp 92–93); Helen Armstrong (Video security pp 200–201; Intelligent video pp 204–205; Visual display pp 206–207)
Editors: Ruth Binney; Ian Graham (Assistant Editor, Which Video?); Cynthia Fraser.

The Publishers received invaluable help from the following people and organizations:– Mike Aarons, Stanmore Video Services Ltd (Sonic Sound Audio Group); Akai America Ltd; Akai (UK) Ltd; George Allinson, PTS Electronics Ltd; Ampex Great Britain Ltd; Andy Armstrong, Monode Ltd; Bell and Howell A-V Ltd; John Cammish, Visiting Lecturer, The School of Film and Television, The Royal College of Art; Dr J O Clarke, Head of Resources, North E Wales Institute; English Electric Valve Company Ltd; Fantasy Factory Video Ltd; J O Grant and Taylor (London) Ltd (Agents for Grundig Electronic, West Germany); Grundig International Ltd; Hitachi Denshi (UK) Ltd; Hitachi Sales (UK) Ltd; International Television Association (UK); Introphoto Ltd; Introvideo Ltd; JVC (UK) Ltd; David Klein, Sales Director, Oracle (UK); Maurice Langton, HTV Ltd; Mac MaGowan, MVP Studios (UK); Ken Marsom, DATS Video Ltd; Richard Maybury, *Video Today*; A J Mitchell, Special Effects Director, The Moving Picture Company; National Panasonic (UK) Ltd; Panasonic Co (USA); Eric Parsloe, Managing Director, Epic (Eric Parsloe Industrial Communications) Ltd; Philips Electronics, Video Division; Philips LaserVision; RCA SelectaVision Videodiscs; Cris Rhodes, Comprehensive Video Supply Europe; Sanyo Marubeni (UK) Ltd; School of Communication, Polytechnic of Central London; Sharp Electronics (UK) Ltd; Jim Slater, Senior Engineering Information Officer, Independent Broadcasting Association (UK); Sony Consumer Products Company, Sony Corporation of America; Sony (UK) Ltd, Communication Systems Division; Sony (UK) Ltd, Consumer Products Division; Dave Swan, Head of the Camera Department, Trilion Video Ltd; Technicolor Audio-Visual (USA); Technicolor Viditronics Ltd; Thames Video-Thames TV International Ltd; Thorn EMI Ferguson Ltd; Toshiba (UK) Ltd; Ray West and Frank Wright, PTS Electronics Ltd.

Agfa-Gevaert Ltd; AICO International (UK); Ambico Inc; Ampex Corporation (USA); Apple Computer Inc; The Association of Video Dealers Ltd; Audio Visual (UK); AV Distributors (London) Ltd; Barco Electronic nv (UK); BASF UK Ltd; R R Beard Ltd; BIB Audio/Visual Products Ltd; Braun Electric (UK) Ltd; CEL Electronics (Harlow) Ltd; Cine 60 Inc; Colortran Inc; Colortran UK; Comart Ltd; Computer Aided Design Centre (UK); The Computer Corner (USA); Concord Lighting International Ltd; Creative Film Makers Ltd; Dicoll Electronics Ltd; EDC (Elkom Design Ltd); Educational Television Association (UK); EMI Tape Ltd; ERCO Lighting Ltd; Ercotron AB; Evershed Power-Optics Ltd; EVF Manufacturers Ltd; Leo Fabbri, Sales Director, Video City Productions Ltd; For-A Company Ltd; GEC Computers Ltd; Alan Groves, Chief Engineer, Transcan Video Ltd; Infopress Ltd, PR Consultants to Mitsubishi Electric (UK) Ltd; ISSCO (UK) Ltd; ITT Consumer Products (UK) Ltd; John Garbett (Audio-Visual) Ltd; Keen Computers Ltd; Kennett Engineering Company Ltd; Labgear Ltd; Roland Lewis, Lecturer in Film-making, Polytechnic of Central London; Lightolier (USA); Link Electronics Ltd; Lowel-Light Manufacturing Inc; Lucas (UK) Ltd; Magnavox Consumer Electronics Company (USA); Malham Photographic Equipment Ltd; Marconi Communication Systems Ltd; The Metropolitan Police Crime Prevention Service; Michael Cox Electronics Ltd; Michael Turner and Associates (Public Relations) Ltd; Molinare Ltd; Mothercare Ltd; MPW Ltd; Nicolet Zeta Corporation (USA); Nordmende (UK) Ltd; Online Publications Ltd; Optical & Textile Ltd; Packaged Lighting Systems Inc; PAG Power Ltd; Ripley Pedro, Pedro Computer Services (UK); Pelling and Cross Ltd; Personal Software Inc; Photopia Ltd; Polysales Photographic Ltd; Pye Ltd; Rank Strand (UK); R K Industries (UK); Robert Bosch Ltd; Quantel Ltd; Shinecrest Ltd; Shure Electronics Ltd; Sony Broadcast (UK); Ian Stokes; Survey and General Instrument Company Ltd; Tanenbaum Services (USA); TDK Tape Distributors (UK) Ltd; Techex Ltd; Tele-Jector Ltd; 3M United Kingdom PLC; Times Square Theatrical and Studio Supply Corp (USA); Thomson-CSF Components and Materials Ltd; Thomson-CSF/DTE/France; Total Video Supply (USA); Trilog Inc (USA); Michael Turner and Associates (Public Relations) Ltd; Velbon Tripod Company Ltd (USA); Versatec Electronics Ltd; Vicon Industries (UK) Ltd; Video Electronics Ltd; The Video School (UK); Vidicraft Inc; W Vinten Ltd; Welt/Safe-Lock Inc.

Acknowledgements

Picture credits

b = bottom c = centre l = left r = right t = top

Pages 12–13: Toshiba (UK) Ltd; 16: Schöner Wohnen/Camera Press; 24: National Panasonic (UK) Ltd; 27: Peter Smith Studio; 34t: Steve Back; bl: Image Bank; br: Sony (UK) Ltd, Consumer Products Division; 36tl: Schöner Wohnen; tr: Peter Smith Studio; 2nd l: Schöner Wohnen; 2nd r: Julian Calder; 3rd l: Peter Smith Studio; 3rd r: Spectrum Colour Library; bl: Peter Smith Studio; br: Steve Back; 37: JVC (UK) Ltd; 43: Sony (UK) Ltd; 45: John Sanders; 52: Sony (UK) Ltd, Consumer Products Division; 53: Courtesy of Sony Corporation of America; 54t: Sanyo Marubeni (UK) Ltd; b: Courtesy of Sony Corporation of America; 55tl: Sanyo Marubeni (UK) Ltd; Toshiba (UK) Ltd; tr, br: Sony (UK) Ltd, Consumer Products Division; 56: JVC (UK) Ltd; 58t: Akai (UK) Ltd; b: JVC (UK) Ltd; 59tl; Akai America Ltd; bl: National Panasonic (UK) Ltd; tr, br: JVC (UK) Ltd; 60: Grundig Video (UK) Ltd; 62: Philips Electronics (Video Division); 63tl, tr: Technicolor Audio-Visual (USA); c: ITT Consumer Products (UK) Ltd; b: Pye Ltd; 64: Bell and Howell A-V Ltd; 66t, b: National Panasonic (UK) Ltd; 67tl: Bell and Howell A-V Ltd; bl: Sony (UK) Ltd, Consumer Products Division; br: Courtesy of Sony Corporation of America; tr: JVC (UK) Ltd; 70: Barry Fox; 71: Philips LaserVision Division; 72: Ampex (UK) Ltd; 74: Peter Smith Studio; 76: BASF Aktiengesellschaft; 89: Schöner Wohnen; 90: Science Photo Library; 93tl, tr: Lucas (UK) Ltd; 95: Peter Smith Studio; 96–97: Hitachi Sales (UK) Ltd; 98t, c, b: Steve Back; 99tl, tr: Paul Brierley; c, b: Steve Back; 100t: Hitachi Sales (UK) Ltd; b: JVC (UK) Ltd; 101tl: Hitachi Denshi (UK) Ltd; bl, tr: National Panasonic (UK) Ltd; br: JVC (UK) Ltd; 102–103 Sony (UK) Ltd, Consumer Products Division; 103: Peter Smith Studios; 104t: Philips Electronics Video Division; b: JVC (UK) Ltd; 105tl: JVC (UK) Ltd; bl: Hitachi Sales (UK) Ltd; tr: Sony (UK) Ltd, Communications Systems Division; br: Sony (UK) Ltd, Consumer Products Division; 106: artwork by permission Sony (UK) Ltd; 111: Peter Smith Studio; 112–113: Schöner Wohnen/Camera Press; 114–115: Steve Back; 116–117: Steve Back; 119: Steve Back; 120: Schöner Wohnen/Camera Press; 121: Steve Back; 122–123: Steve Back; 126: Peter Smith Studio; 127: Schöner Wohnen/Camera Press; 128: Peter Smith Studio; 129: Schöner Wohnen/Camera Press; 132: Spectrum Colour Library; 133t: Julian Calder; b: Spectrum Colour Library; 135t: Steve Back; b: Peter Smith Studio; 136t, c: Spectrum Colour Library; b: Paul Brierley; 137tl: Steve Back; tr: Spectrum Colour Library; b: John Garrett; 140: Spectrum Colour Library; 141tl: G.P. Eisen, *The Daily Telegraph* Colour Library; tr, bl, br: Spectrum Colour Library; 142–143: Spectrum Colour Library; 144: Spectrum Colour Library; 145t, bl, br: Spectrum Colour Library; 146–147 Leo Mason; 148t: Leo Mason; b: Jerry Young; 149: Jerry Young; 151t: Spectrum Colour Library; c, b: Mark Dunton; 152: artwork by Alexander O'Donnell; 156–157: Tony Duffy/All Sport; 158t: Steve Back; b: Paul Wilkinson; 160: Peter Smith Studio; 171: Max Hole/Camel; 176: Ampex Great Britain Ltd; 177tr: Ampex Corporation (USA); br: The Moving Picture Company; 179: Steve Back; 180: Molinare Ltd; 181: McCann Ericksen (Nederlands) bv; 183: Peter Smith Studio; 184–185: Sally & Richard Greenhill; 186l: *The Daily Telegraph* Colour Library; r: The Image Bank; 188–189: *The Daily Telegraph* Colour Library; 190–191: Octopus Books; 192–193: Nils Christie/Craig Dodd; 196–197: Josiah Wedgwood and Sons Ltd; 199: Peter Smith Studio; 201t, b: Popperfoto; 206t, c: Molinare Ltd; b: Peter Smith Studio; 207: Logica Ltd.